international®
AIR POWER
REVIEW

AIRtime Publishing
United States of America • United Kingdom

international® AIR POWER REVIEW

Published quarterly by AIRtime Publishing Inc.
US office: 120 East Avenue, Norwalk, CT 06851
UK office: CAB International Centre, Nosworthy Way,
Wallingford, Oxfordshire, OX10 8DE

© 2003 AIRtime Publishing Inc.
Mirage 2000 and Tu-16 cutaways © Mike Badrocke
Photos and other illustrations are the copyright
of their respective owners

Softbound Edition ISSN 1473-9917 / ISBN 1-880588-56-0
Hardcover Deluxe Casebound Edition ISBN 1-880588-57-9

Publisher
Mel Williams

Editor
David Donald   e-mail: airpower@btinternet.com

Assistant Editor
Daniel J. March

Sub Editor
Karen Leverington

US Desk
Tom Kaminski

Russia/CIS Desk
Piotr Butowski, Zaur Eylanbekov   e-mail: zaur@airtimepublishing.com

Europe and Rest of World Desk
John Fricker, Jon Lake

Correspondents
Australia: Nigel Pittaway
Belgium: Dirk Lamarque
Brazil: Claudio Lucchesi
Bulgaria: Alexander Mladenov
Canada: Jeff Rankin-Lowe
France: Henri-Pierre Grolleau
Greece: Konstantinos Dimitropoulos
India: Pushpindar Singh
Israel: Shlomo Aloni
Italy: Luigino Caliaro
Japan: Yoshitomo Aoki
Netherlands: Tieme Festner
Romania: Danut Vlad
Spain: Salvador Mafé Huertas
USA: Rick Burgess, Brad Elward, Mark Farmer (North Pacific region),
   Peter Mersky, Bill Sweetman

Artists
Chris Davey, Zaur Eylanbekov, Mark Rolfe, John Weal,
Keith Woodcock, Vasiliy Zolotov

Designer
Zaur Eylanbekov

Controller
Linda DeAngelis

Origination by Universal Graphics, Singapore
Printed in Singapore by KHL Printing

International Air Power Review is published quarterly in two editions (Softbound and Deluxe Casebound) and is available by subscription or as single volumes. Please see details opposite.

Acknowledgments

We wish to thank the following for their kind help with the preparation of this issue:

Paul E. Eden
Ulf Hugo/FMV
Vadim Yegorov

The author of the Mirage 2000 feature would like to thank the following for their kind help: Gérard David, Philippe Deleume, Thierry Duvette, Patrick Masse, Paul Mustacci, and Yves Robins of Dassault; capitaine Solano and capitaine Dumas, of the SIRPA Air; Patrick Wang, François Brugier and Simon Chiu of the Taiwanese Institute in Paris; Eric Hsu of the Government Information Office; and Colonel Chih-Chung Chen of the Taiwanese Ministry of Defence. Special thanks to Major-General Yu-Chi Lei and to all ROCAF personnel who provided help during the author's visit to Hsinchu. Thanks also to Jean-Marc Gaspéri, Alain Martel, Philippe Rebourg, Nicolas Renard, Yves Robins and Danièle Viaud for their kind help in organising the demonstration flight

The editors welcome photographs for possible publication but can accept no responsibility for loss or damage to unsolicited material.

Subscriptions & Back Volumes

Readers in the USA, Canada, Central/South America and the rest of the world (except UK and Europe) please write to:
AIRtime Publishing, P.O. Box 5074, Westport, CT 06881, USA
Tel (203) 838-7979 • Fax (203) 838-7344
Toll free 1 800 359-3003
e-mail: airpower@airtimepublishing.com

Readers in the UK & Europe please write to:
AIRtime Publishing, RAFBFE, P.O. Box 1940,
RAF Fairford, Gloucestershire GL7 4NA, England
Tel +44 (0)1285 713456 • Fax +44 (0)1285 713999

**One-year subscription rates (4 quarterly volumes),
inclusive of shipping & handling/postage and packing:**
Softbound Edition
USA $59.95, UK £48, Europe EUR 88, Canada Cdn $99,
Rest of World US $79 (surface) or US $99 (air)

Deluxe Casebound Edition
USA $79.95, UK £68, Europe EUR 120, Canada Cdn $132,
Rest of World US $99 (surface) or US $119 (air)

**Two-year subscription rates (8 quarterly volumes),
inclusive of shipping & handling/postage and packing:**
Softbound Edition
USA $112, UK £92, Europe EUR 169, Canada Cdn $187,
Rest of World US $148 (surface) or US $188 (air)

Deluxe Casebound Edition
USA $149, UK £130, Europe EUR 232, Canada Cdn $246,
Rest of World US $187 (surface) or US $227 (air)

Single-volume/Back Volume Rates by Mail:
Softbound Edition
US $16, UK £10.95, Europe EUR 18.50, Cdn $25.50 (plus s&h/p&p)
Deluxe Casebound Edition
US $20, UK £13.50, Europe EUR 22, Cdn $31 (plus s&h/p&p)

All prices are subject to change without notice.
Canadian residents please add GST. Connecticut residents please add sales tax.

**Shipping and handling (postage and packing) rates
for back volume/non-subscription orders are as follows:**

	USA	UK	Europe	Canada	ROW (surface)	ROW (air)
1 item	$4.50	£4	EUR 8	Cdn $7.50	US $8	US $16
2 items	$6.50	£6	EUR 11.50	Cdn $11	US $12	US $27
3 items	$8.50	£8	EUR 14.50	Cdn $14	US $16	US $36
4 items	$10	£10	EUR 17.50	Cdn $16.50	US $19	US $46
5 items	$11.50	£12	EUR 20.50	Cdn $19	US $23	US $52
6 or more	$13	£13	EUR 23.50	Cdn $21.50	US $25	US $59

Volume Nine
Summer 2003

CONTENTS

MAJOR FEATURES PLANNED FOR VOLUME TEN
Focus Aircraft: Boeing C-135 family, **Warplane Classic:** Vought A-7 Corsair II – US Navy, **Variant File:** Short Sunderland,
Pioneers & Prototypes: Yakovlev Yak-36, -38 and -41, **Air Combat:** 67th TRW in Korea,
Type Analysis: de Havilland Hornet, **Special Report:** Spanish F-5s

PROGRAMME UPDATE

RQ-4A Global Hawk

On 14 February 2003 the seventh Northrop Grumman RQ-4A Global Hawk unmanned aerial vehicle (UAV) touched down at Edwards AFB, California, ending a delivery flight that began at Air Force Plant 42 in nearby Palmdale. The air vehicle is the final advanced concept technology (ACTD) platform and will be used as a test vehicle in support of developmental and upgrade efforts. The air vehicle incorporates a number of new design features that will be incorporated on production Global Hawks. Included is a new integrated mission management computer system, which controls all of the flying and navigation operations of the aircraft. Several of the improvements have resulted from Global Hawk's early operational debut in support of Operation Enduring Freedom. The first two production Global Hawks, built under a low-rate initial production

(LRIP) contract are slated for delivery to the USAF by September and December 2003.

The USAF recently conducted a series of test flights of an RQ-4A Global Hawk unmanned air vehicle equipped with an electronics intelligence (ELINT) payload developed by EADS. Conducted at Edwards AFB, California, the test flights demonstrated the UAV's ability to transmit data, collected by the sensors, to a ground station. Northrop Grumman is proposing a development of the RQ-4A known as the Euro Hawk as a replacement for Germany's signals intelligence system, beginning in 2008. Currently, Germany employs sepcially modified Peace Peek Atlantics for the provision of Sigint.

The Global Vigilance Combined Test Force, which is based at Edwards AFB, California, includes the 452nd Flight Test Squadron (FLTS) and 31st Test and Evaluation Squadron (TES), recently began training the initial cadre of RQ-4A Global Hawk operators from the 12th Reconnaissance Squadron (RS). The squadron, which is based at Beale AFB, expects to have 17 personnel trained by mid-summer, prior to delivery of the first operational unmanned aerial vehicle. Three personnel from Beale were already qualified to operate the Global Hawk development systems, which had primarily been operated by a cadre from the Air Force Materiel Command (AFMC). As part of the training students receive academics, 18 simulator sessions and complete three actual flights.

In early 2003 the USAF's only mission-capable RQ-4A was deployed to Al Dhafra in the UAE for use during Operation Iraqi Freedom, where it operated in the SCAR (Strike Co-ordination and Reconnaissance) role. A mission-capable rate of 74.1 percent was recorded.

Lockheed Martin F-16 updates

Lockheed Martin and the USAF have concluded development flight-testing of new conformal fuel tanks (CFTs) for the F-16C/D fighter. Attached to the upper surface of the fighter's fuselage, the CFTs are being installed on advanced Block 50/52 and Block 60 aircraft, which are currently in development and initial production. The development flight-tests were conducted using an F-16C operated by the Air Armament Center's 40th Flight Test Squadron (FLTS) at Eglin AFB, Florida, which conducted 54 test flights and compiled 135 flight test hours. The first production CFTs were recently flown for the first time on newly built F-16C/Ds destined for the Hellenic Air Force. The CFTs are carried on the left- and right-side of the fighter's upper fuselage and each shipset carries 450 US gal (1703 litres) or 3,060 lb (1388 kg) of fuel. The first production shipset of CFTs was delivered to the contractor's Fort Worth, Texas, facility in January 2003. Although the Hellenic Air Force is the lead customer for the CFTs, all advanced Block 50/52 and 60 F-16 are compatible with the tanks.

The first F-16C updated by the Ogden Air Logistics Center (OG-ALC) under the Common Configuration Implementation Program (CCIP) was delivered to the 57th Wing at Nellis AFB, Nevada, where the first Phase IA Block 50/52 aircraft is now undergoing operational test and evaluation. The 389th Fighter Squadron at Mountain Home AFB, Idaho, will be the first squadron to be equipped with the modified aircraft. The configuration incorporates the AN/APX-113 air-to-air interrogator and the ability to autonomously identify targets, along with

PROJECT DEVELOPMENT

India

HJT-36 starts flight development
A successful first flight was achieved at Bangalore on 7 March by HAL's HJT-36 intermediate jet-trainer, launched as a Rs1.8 billion ($38 million) project only in 1997. Powered by a single Snecma Larzac 04H-20 turbofan, the tandem-seat straight-wing HJT-36 was designed to replace the IAF's 200 or so HAL HJT-16 Kiran basic jet trainers flown since the late 1960s.

As a joint project with India's Aeronautical Development Agency and the Defence Research and Development Organisation, three more HJT-36 prototypes are currently taking shape. A second prototype, due to fly

in August, will incorporate full digital avionics. Many US and European companies are partnering HAL in supplying sub-systems and components.

The first of an initial 16 production HJT-36s, from total requirements expressed in January 2001 for 211, including 24 for Indian Naval Aviation, is due for delivery in 2005. Output should then increase to about 20 per year, with the first 50 planned for delivery by 2007, at a reported fly-away unit cost of only $5.2 million.

AJT proposal cleared
Some progress with India's long-delayed procurement plans for new advanced jet trainers (AJT), for which the BAE Hawk was selected last year, may have been achieved in April.

Defence Minister Fernandez then announced that his department had cleared the $2 billion AJT proposal, and sent it for approval to the Finance Ministry. In turn, Finance Minister Jaswant Singh maintained that his ministry would not delay taking a decision, although he claimed that the Cabinet Committee on Security was responsible for final AJT procurement approval. And no preference had then been announced between the $18 million Hawk 100 or cheaper ($12 million) Aero Vodochody L-159B trainer for the AJT requirement.

This has been in prospect since the mid-1980s, since when more than 250 IAF aircraft have crashed, following about 200 in the previous decade. About 70 percent of these have been MiG-21s, largely through pilot error and mechanical failure, compounded by the lack of an IAF advanced trainer, since mid-retirement of two-seat Hunter T.Mk 7s in the early 1990s. Other IAF accident losses since the late 1980s to date have also included four Mirage 2000Hs, 31 Jaguars, 56 MiG-23s, and at least 29 MiG-27s.

In May 2003 Lockheed Martin delivered the 14th F/A-22 to the USAF. The aircraft will join Raptors 12 and 13 with the 422nd Test and Evaluation Squadron at Nellis AFB, which will have six Raptors for tactical development work by the end of the year.

Parliamentary figures quoted in February indicated that the IAF lost 68 aircraft, or an average of nearly two per month, and 31 pilots killed in crashes in the past three years. Itemised totals comprised 27 aircraft and 15 pilots lost in 2000/01; 20 aircraft and nine pilots in 2001/02; and 21 aircraft and seven pilots in 2002/03.

International

European turboprop for A400M
The long-awaited powerplant selection for its multi-national A400M transport aircraft, announced on 6 May by Airbus Military Company (AMC), favoured the 7908-kW (10,600-shp) TP400-D6 turboprop project proposed by Europrop International (EPI) in Germany, rather than Pratt & Whitney Canada's cheaper and more powerful (8952-kW/12,000-shp) PW180 bid. European engine preference was perhaps politically inevitable, although AMC earlier quoted a price differential of as much as 20 percent between the TP400 and the PW180. Snecma chairman/CEO Jean-Paul Bechat said that selection of the revised European powerplant bid would increase the Euro 18.2 billion ($20.64 billion) A400M programme costs by only one to two percent.

Initially, more than 750 three-shaft TP400-D6 powerplants, claimed as the largest western turboprops yet, and costing Euro 2 billion ($2.27 billion),

Conformal fuel tanks are carried by these two production Peace Xenia III F-16 Block 52+ aircraft for the Elliniki Polemiki Aeroporia, photographed during a test flight from the Fort Worth plant.

the capability of operating either the HARM missile targeting system (HTS) pod or a FLIR targeting pod. Alternatively, the new Sniper XR Advanced Targeting Pod can also be carried. The fighters were previously equipped with new colour displays under Phase I. Phase II modifications, which will be fielded in July 2003, will incorporate the NATO-standard Link 16 data link, the joint helmet-mounted cueing system (JHMCS) and an electronic horizontal situation indicator. Beginning in 2005, Block 40/42 aircraft will receive these modifications under Phase III. The CCIP will eventually be installed in 650 Block 40/42/50/52 F-16C/Ds.

The Ogden Air Logistics Center (ALC) at Hill AFB, Utah, has begun regeneration work on the first of 30 former USAF F-16As and four F-16Bs destined for the Aeronautica Militare Italiana (Italian Air Force), under a leasing arrangement. Five airframes were initially trucked to Hill after being removed from storage at the Aerospace Maintenance and Regeneration Center (AMARC) at Davis Monthan AFB, Arizona, and the ALC was scheduled to receive one additional airframe every two weeks until all are delivered. After the overhaul and test flights have been completed, the fighters will be delivered to Italy in groups of five and the last example is scheduled for delivery on 4 October 2004.

Wearing Italian roundels and serial MM7238 (with original USAF serial 80-0615), this refurbished F-16A for the Aeronautica Militare Italiana is seen at Hill AFB during the official roll-out on 8 May 2003.

---

are scheduled for the A400M production programme. Under the management of OCCAR, the European military procurement agency, this now involves commitments from the seven participating countries for 180 aircraft. Parliamentary funding ratification was still awaited in May from Germany, however, following Portugal's cancellation of three planned A400Ms on economic grounds.

Engine design, development and production are being shared between the EPI partners in proportion to their shareholdings. These comprise 28 percent each by MTU Aero Engines, Rolls-Royce and Snecma Moteurs; and the remaining 16 percent by Spain's Industria de Turbo Propulsores (ITP). Driving a 5.3-m (17-ft 4.5-in) diameter propeller, the TP400-D6 benefits from the proven technologies developed in many civil and military engine programmes, and aims for wider applications in both fields at a later stage. The prototype A400M is due to fly in September 2007, for initial deliveries from 2009.

Italy

Batch 2 MB-339CD progress
Flight-development and service trials are now well advanced with the first of the Aeronautica Militare Italiana's (AMI) 15 follow-on upgraded Aermacchi MB-339CD lead-in fighter trainers (CS X606), which first flew

On 28 April 2003 the KAI /Lockheed Martin T-50 Golden Eagle notched up its 100th test flight, and on the same day reached Mach 1.2, the highest speed yet attained. Two prototypes are flying. Meanwhile, on 25 April, the airframe durability vehicle completed one 'lifetime' of testing, equivalent to 8,334 flight hours. This trial began on 22 July 2002 at the Agency for Defence Development at Taejon. The second 'lifetime' of testing is due to be completed in April 2004, with production deliveries due from 2005.

from the Venegono factory on 24 May 2002. Compared with the AMI's first 15 MB-339CDs, operated by the 61st Wing at Galatina, Lecce, since 1998, the second batch (CD2) features enhanced avionics for improved training and operational capabilities.

In addition to new GATM-standard UHF/VHF radio and IFF systems, new equipment in X606 includes a digital map generator linked with the existing multi-function cockpit displays; modified internal and external lighting for use with night-vision goggles; and new safety systems.

New features also include embedded simulation of electronic warfare and tactical co-operation scenarios, and provision for an air combat manoeuvring instrumentation (ACMI) pod. Full operational capability clearance for Batch 2 MB-339CDs is expected later in 2003.

Japan

T-4 production completed
Delivery was completed in February to the JASDF of the last of 212 T-4 twin-turbofan intermediate jet trainers, including four prototypes, designed and produced by Kawasaki at its Gifu facility since the programme was launched in 1981. Powered by two 16.37-kN (3,680-lb) Ishikawajima Harima F3-IHI-30 engines, the T-4s replaced the Fuji T-1A/Bs and Lockheed T-33As formerly operated by the JASDF's 31st and 32nd Training Wings.

South Korea

AWACS development
The Republic of Korea's Defense Ministry has announced plans to develop an airborne warning and

control system (AWACS) beginning in 2005. The $1.5 billion program, which is known as the E-X project, includes four AWACS aircraft that will enter operational service in 2011.

United States

CC-130J austere field tests
Lockheed Martin and the USAF recently completed austere field take-off and landing operations trials with the stretched CC-130J in Yuma, Arizona, as part of the airlifter's developmental military utility tests. The trials, which included the first dirt landing by a USAF flight crew, were conducted at Tyson Airfield on the US Army's Yuma Proving Grounds. Austere field trials with the standard length C-130J were completed during 1999. The CC-130J will undergo operational testing and evaluation in 2003.

In May 2003 the USAF awarded contracts to several suppliers to begin development of the equipment for the MC²A (Multi-sensor Command and Control Aircraft), now designated E-10A. Boeing will build one Boeing 767-400ER demonstrator (left). Currently the only military 767s in service are the E-767 AWACS aircraft of the JASDF's Dai 601 Hiko-tai (above).

MH-60R deployed for testing

Two MH-60R Seahawks and their crews from test and evaluation squadron HX-21 recently deployed from NAS Patuxent River, Maryland, to the US Navy's Atlantic Undersea Test and Evaluation Center (AUTEC) on Andros Island, Bahamas. Over a three-week period the crews conducted a series of tests in support of the helicopter's developmental test (DT) programme. These included the 'Romeo's' first shipboard landing aboard the guided missile cruiser USS *Gettysburg* (CG-64), and its first inflight launch of a sonobuoy. In another test the helo's airborne low frequency sonar (ALFS) was used for the first time to locate a submerged submarine, while its multi-mode radar (MMR) and electronic surveillance

measures (ESM) simultaneously performed surface sweeps. The crews logged 126 flight hours in the two helicopters while deployed.

Assessment for upgraded H-1s

Pilots and enlisted personnel assigned to the H-1 Operational Test Team recently conducted the H-1 Upgrade Program's first operational assessment. Both the UH-1Y and AH-1Z aircraft flew missions from NAS Patuxent River, Maryland, to MCAF Quantico, Virginia, MCAS New River, North Carolina, and the US Army Aberdeen Proving Ground in Maryland. During the missions the No. 1 UH-1Y accumulated 12.7 flight hours and flew simulated reconnaissance team insertion and extraction, aerial reconnaissance, airborne forward air control

and escort missions. The No. 3 AH-1Z tallied up 11.8 hours of flight time while conducting simulated armed reconnaissance, air interdiction, close air support, forward air control and escort missions. Prior to making the flights, the pilots underwent simulator training at Bell Helicopter's facility in Texas.

Development testing also continues and the No. 1 UH-1Y and AH-1Z were subsequently ferried to Alamosa, Colorado, where high-altitude testing was conducted, demonstrating the aircraft's ability to operate at higher density-altitudes than can be accomplished at Pax River.

Osprey completes detachment

The V-22 Integrated Test Team (ITT) recently completed a series of tests

that validated the Osprey's ability to serve as an aerial delivery platform. As part of the tests the V-22 ITT deployed Osprey No. 21 to the ranges around Fort Bragg, North Carolina, where it verified the tiltrotor's ability to deploy cargo and personnel via parachute. During the tests dummies weighing 68-163 kg (150-360 lb) and 227- and 454-kg (500- and 1,000-lb) containers were deployed. In other developments, Osprey No. 8 is nearing the completion of Phase I of the high rate of descent testing at NAS Patuxent River, and Osprey No. 10 has tested the V-22's handling characteristics in the low airspeed regime. Osprey No. 22, which is the second LRIP example, was recently delivered and received modifications that will allow it to join the active test force.

UPGRADES AND MODIFICATIONS

Australia

Hawk air refuelling trials

A two-week series of flight trials were completed in December 2002 by one of the RAAF's 33 BAE Hawk Mk 127 lead-in fighter trainers (LIFT), equipped with a fixed refuelling probe forward and to starboard of the windscreen. Five sorties were completed by BAE Systems and Australian service test pilots from Williamtown air base between 3-12 December, to assess the Mk 127's handling qualities and fuel system in contacts with one of the RAAF's four Boeing 707-338C tanker/transports. The trials marked completion of the RAAF's Project Air 5367 requirement, to add a new capability to Australia's fleet of advanced LIFT aircraft, for converting pilots to fly its Boeing/MDC F/A-18A/B combat aircraft.

Czech Republic

'Hind' upgrades

The CLPO is planning in-country avionics and weapons upgrades by its military airframe and engine depots (Letecke Opravarny Kbely/Malesice – LOK/LOM), of its new and 24 or so existing Mi-24D/V 'Hind-D/Es'. This programme will be undertaken within a co-operative agreement reached by four Visegrad countries, also including Poland, Hungary and Slovakia, for limited upgrades of about 100 Mi-24s operated between them.

These include about 40 in Poland, 30 or more in Hungary, and some 10 in Slovakia, to achieve enhanced combat capabilities, and meet NATO

interoperability requirements. Upgrade offers from Russia, which is insisting on Mil OKB involvement in the Visegrad project, are based on national Mi-24PK-1 and VK-1 programmes, involving new avionics, sensors and weapons packages. Several western contractors are also competing for this potential $400-540 million Visegrad contract, including BAE Systems Avionics, SAGEM in France, Elbit Systems, IAI's Tamam division, and South Africa's Advanced Technologies & Engineering (ATE).

Malaysia

Modified Hercules into service

The Royal Malaysian Air Force has placed its first 'stretched' C-130H-30 in service following a fuselage modification programme. This incorporated a stretch kit that added individual sections forward and aft of the wings, extending the aircraft by 4.57 m (15 ft). The mods provide the aircraft with a 32 percent increase in cargo volume capacity. A second kit is currently being installed in another C-130H, while other C-130s are receiving tanker capabilities, large observation windows and other modifications.

In May 2003 the Bulgarian air force launched a limited navigation equipment upgrade for some of its aircraft. The enhanced suite includes TACAN, ILS/DME and Trimble ASN-173 GPS, and provides NATO/ICAO compatibility, in turn allowing aircraft to be used outside Bulgaria. This Su-25UBK from the 22nd ShtAB at Bezmer was one of the first aircraft to be upgraded.

Slovakia

MiG-29 and Mi-17 upgrades

Negotiations were being finalised in February 2003 by the Slovakian government for a $56 million maintenance contract with MiG RSK in Russia, to achieve economies of up to 40 percent in operating Slovakia's force of 16 MiG-29 fighters and two MiG-29UB two-seat combat trainers. LPSOS chief Maj. Gen. Jozef Dunaj said that contract signature was then imminent, and its cost offset against some of Russia's trade debts to Slovakia.

Against current MiG-29 requirements for a complete airframe overhaul every 800 flight hours, from an overall life of 2,400 hours, he said new procedures and components, including revised periodic checks at 100-200 hours, would extend overall

life to 4,000 hours. Similar maintenance plans were already in operation for Hungarian and Polish MiG-29s, he added. Maintenance and overhaul work on 15 of the LPSOS MiG-29s' twin Klimov RD-33 turbofans and 11 gearboxes had also been completed from an earlier RSK MiG contract.

Two of 15 Mil Mi-17 'Hip-H' transport helicopters operated by Slovakian forces since 1992 were being upgraded earlier this year by Letetske Opravovne in Trenchin (LOT), for UN service in Bosnia. Installation of NATO-compatible navigation and communications equipment was supplemented by cockpit armour and possible defensive EW systems, from an 8 million koruna ($263,426) contract. LOT is also planning to work with MiG RSK on LPSOS MiG-29 upgrades.

Left: This B-52H is testing the Avionics Mid-life Improvement (AMI) upgrade with the 419th FLTS at Edwards AFB.

Below: This USCG HH-65 Dolphin carries a 0.50-in (12.7-mm) M240D machine-gun in the door. The Coast Guard has recently armed its helicopters with this weapon to enhance their law enforcement capabilities.

United Kingdom

Hawk upgrades completed

Completion was expected ahead of schedule in June of the RAF's long-term BAE Hawk life-extension programme from upgrades of 80 of its 132-strong long-serving fleet. This began in the early 1990s with replacement wings produced by the-then British Aerospace, fitted to the Hawks by the MoD's Defence Aviation Repair Agency (DARA) facility at RAF St Athan, in South Wales. BAe was then contracted to manufacture new centre- and rear-fuselage sections for the 80 high-time RAF Hawks, for similar St Athan installation.

Finally, BAE Systems received a £10.5 million five-year UK MoD Defense Logistics Organisation contract, for production of 130 new and structurally-improved export-standard tailplane and ballast modification kit-sets to complete RAF Hawk life-extension upgrades. An initial contract covered 117 new-build kits to replace original RAF Hawk T.Mk 1 components, then nearing their design fatigue life limits. The order was increased in March 2000 by 13 more tailplanes for the Red Arrows display team, to extend planned RAF Hawk retirement to early 2011.

Some tailplanes in BAE's contract were also incorporated into new-build Hawks for overseas customers, resulting in transfer of their production to the countries concerned. An initial 53 kit-sets built by BAE at Brough, in Yorkshire, were followed by another 37 by Hawker de Havilland in Australia. Of 33 RAAF advanced Hawk 127 lead-in fighter trainers (LIFT) ordered from a $A850 million (now $507 million) 1997 contract, and recently all delivered, 21 underwent final assembly at BAE's LIFT support facility at the Williamtown fighter base in New South Wales, following the first dozen from Brough.

Production of the last 40 Hawk tailplanes was then transferred to South Africa's Denel Aerospace, in which BAE has a 30 percent share-holding, and has been completed about five months ahead of schedule.

United States

Coast Guard arms helicopters

The US Coast Guard has begun arming its fleet of HH-65A/B Dolphin and HH-60J Jayhawk helicopters. In support of this programme the Naval Air Warfare Center Aircraft Division at NAS Patuxent River, Maryland, certified both airframes to operate the M240D machine-gun. Although the HH-60J is similar to the US Navy's HH-60H, which was already certified to carry machine-guns, the HH-65 was

not. The gun mounts installed in the Dolphins were developed by the Naval Surface Warfare Center in Crane, Indiana. NAVAIR ground and flight testing was completed in mid-February 2003. In addition to the gun mounts, NAVAIR is equipping the aircraft with a flashing blue law enforcement light and white night-lighting that will illuminate the Coast Guard insignia. The service intends to arm as many as 170 helicopters in support of law enforcement and national defence efforts.

Tomcats armed with JDAM

Naval Air Systems Command (NAVAIR) recently equipped the fleet's F-14D fighters with an upgrade that allowed the aircraft to deliver GBU-31 Joint Direct Attack Munitions (JDAM). The initial software and hardware modifications were conducted aboard the USS *Theodore Roosevelt* (CVN 71) in early February 2003 when the fighters assigned to VF-213 were modified. By mid-February the team had modified the forward deployed F-14Ds assigned to VF-2 and VF-31and trained more than 90 aircrew and maintenance personnel on JDAM employment. On 1 March 2003 the first operational sortie was conducted. The Tomcat is capable of carrying four GBU-31s.

Gunship modification

Boeing has been awarded a $17.4 million contract to begin conversion of C-130H transport to AC-130U gunship configuration. The contract initially covers the induction and preparation of the aircraft and the actual modification will be funded separately. A second $18.9 million contract covers the design and development effort associated with the conversion of C-130H serial 89-1056 to the latest gunship configuration.

B-52H avionics update

The 419th Flight Test Squadron, a component of the 412th Test Wing at Edwards AFB, California, has begun flight tests of a B-52H that has been modified under the B-52 Avionics

Midlife Improvement (AMI). Begun in mid-December 2002, the flight test programme will continue through March 2004. It includes 80 sorties that will average eight hours in length, along with several global missions that will last more than 24 hours. The AMI primarily updates the bomber's offensive avionics system and comprises modifications to the inertial navigation system (INS), avionics control unit, data transfer system and associated hardware and software. The $8.5 million AMI was developed by Boeing at its Wichita, Kansas, facility.

B-2 test aircraft flying again

Spirit of New York, the only B-2A dedicated to testing duties, returned to the air at Edwards AFB, California, during December 2002. The aircraft rejoined the 419th Flight Test Squadron (FLTS) at the conclusion of a modification period that began in January 2002. As part of the modifications the aircraft was equipped with new fuel cells that match the Block 30 production configuration of the 20 operational bombers assigned to the 509th Bomb Wing. The aircraft will continue to support B-2 flight-testing at Edwards through 2009 and upcoming tests will provide the aircraft with enhanced beyond line-of-sight communications capability. The USAF and Boeing have begun flight-testing the aircraft's compatibility with the 227-kg (500-lb) GBU-38 JDAM. As part of the initial separation tests, conducted at Edwards, 16 of the

lightweight JDAMs were dropped from four 'smart' bomb racks. Developed by Boeing, the new racks allow the Spirit to carry up to 80 GBU-38s.

First AMP Galaxy flies

The first C-5B Galaxy modified under the Avionics Modernization Program (AMP) recently made its initial flight from Lockheed Martin's facility at Dobbins Air Reserve Base in Marietta, Georgia. During the five-hour mission the joint USAF/contractor flight crew demonstrated the aircraft's basic flying qualities, and verified the airworthiness of the new avionics suite. The initial aircraft, serial 85-0004, entered AMP upgrade on 13 June 2002.

Last command post modified

The US Navy has awarded Raytheon E-Systems an $8.4 million modification to an existing contract covering the conversion of one E-6A to E-6B configuration. It is the final Mercury to aircraft to be equipped with the airborne command post kit. The aircraft will be delivered in late 2003.

Serving with Dai 8 Hiko-tai (part of Dai 3 Koku-dan at Misawa), this F-4EJ Kai wears an experimental camouflage, similar to that worn by the Mitsubishi F-2. Since 1996 the unit has been the sole operator of the Phantom in the Fighter Support role. Anti-ship missions form an important part of this tasking, for which the radar-guided ASM-1C and IR-guided ASM-2 (illustrated) are available.

PROCUREMENT AND DELIVERIES

Algeria

Mi-17 reinforcements

Latest recipient for new-build Mil Mi-17-1V transport and assault helicopters from Russia is the Algerian air force (QJJ). Deliveries of more than 40 Mi-17s are involved in the new $180 million contract, which will almost double the QJJ's procurement of this type, operated alongside earlier Mi-8 'Hip-C/Fs' since 1994.

Australia

New trainers for RAAF

The Royal Australian Air Force has leased seven Raytheon Super King Air 350 twin-turboprops as replacements for six HS.748s and four older leased King Air B200s used as navigator trainers. The King Airs will be leased from Hawker Pacific and operated by 32 Squadron in support of the School of Air Navigation at RAAF Base East Sale, Victoria. Although the first Super King Airs were delivered to Hawker in March 2003, training will not begin until navigator-training stations have been installed in the cabin. Under the terms of the 10-year, $95 million agreement Hawker Pacific will also provide all maintenance and support.

Austria

Eurofighter interest confirmed

Austria's planned Eurofighter procurement programme, suspended last August following heavy summer floods in Central Europe, was reinstated on 16 May, after a further review of the contending bids. As expected, however, Austrian Defence Minister Guenther Platter said that programme economies would be achieved by reducing the full air force (OeLk) requirement for 24 air defence Eurofighters to 18 single-seat versions, costing Euro 1.96 billion ($2.22 billion).

Similar numbers of new or remanufactured Gripens have also been offered to Austria in a new bid from SAAB/BAE and the Swedish government. Some review is further expected of original Eurofighter contract negotiations with the EADS-led consortium,

The Hellenic Air Force's 356 Mira operates the service's two EMBRAER ERJ-135s. One is an EMB-135LR (without winglets), delivered in early 2000, and the other is this later Legacy Executive (EMB-135BJ), delivered in 2002.

Seen in the US prior to delivery, this Bell 412EP is one of 16 ordered by the Royal Saudi Air Force for SAR duties with No. 33 Squadron. The aircraft is fitted with a FLIR turret and Spectrolab Nitesun.

involving agreed industrial offsets for Austria totalling 200 percent of the contract value, for parliamentary approval in June.

First C-130K transferred

Formal hand-over of the first of three ex-RAF C130K Hercules transports bought by the OeLk through a government-to-government sales agreement between the MoD's Disposal Services Agency and the Austrian Government signed on 22 May 2002, took place at Marshall Aerospace's Cambridge headquarters, on 19 March. The three C-130Ks involved were withdrawn from RAF service in 2000/01, after replacement by new Lockheed Martin C-130Js, and are undergoing extensive structural and avionics upgrades by Marshall before flight-tests and delivery.

In addition to new communications and navigation systems, engine instrumentation displays, fuel gauges and weather radar, Marshall is also installing a flight management system to replace a navigator. Provision is also made for extensive flight and ground-crew training and familiarisation by Marshall, as well as on-the-job training with the RAF at Lyneham.

Brazil

Air Force accepts final EMB-145

Embraer has delivered the last of five EMB-145SA (Surveillance Aircraft) aircraft to the Força Aérea Brasileira (FAB) or Brazilian Air Force. The aircraft are a component of the Sistema de Vigilância da Amazônia (SIVAM) or Amazon Surveillance System that also includes three EMB-145RS (Remote Sensing) aircraft. The EMB-145SA airborne early warning and control aircraft (AEW&C), which has been designated R-99 by the FAB, is equipped with an Ericsson Erieye airborne radar. It is also equipped with systems that intercept radar and communications signals.

Canada

C-17 plans abandoned

One of the main casualties of revised Canadian military budget plans, now giving priority to a $C700 million intel-

ligence, surveillance, targeting, acquisition and reconnaissance (ISTAR) programme approved in February, was the long-discussed CAF requirement for up to six Boeing C-17 Globemaster III outsize-load transports. Instead, Canada has elected to join France, Germany, the Czech Republic, Denmark, Hungary, Luxembourg, Norway, Poland and Portugal, in a late 2002 agreement in Prague, to explore proposals for a pooled NATO strategic heavy-lift transport force. If it materialises, this could be based on C-17s, although such alternatives as the Antonov An-124 and possibly the Airbus A400M may also be evaluated for the required roles.

Challengers delivered

Two new CC-144 Challenger 604s were recently delivered to the Canadian Air Force's 412 Transport Squadron (TS) at MacDonald Cartier IAP, Ottawa, which is a component of 8 Wing at CFB Trenton, Ontario. The Challengers comprise serial 144617 and 144618.

Chile

US offers Bell UH-1s

Recent US military aid to Chile has included offers of six surplus Bell UH-1H utility helicopters. These are expected to be cannibalised for spares to return to service some of the FACh's 20 or so UH-1Hs, of which only about a half-dozen have been recently operational. Funding and technical problems had earlier resulted in abandonment of FACh plans to acquire 12 Sikorsky S-70 Black Hawks, and then a similar number of used Bell 412 utility and SAR helicopters from US civil sources in May 2002.

Colombia

Spanish Mirage F1s vetoed

An offer from Spain to the Colombian government in February for the transfer at nominal cost of eight surplus Dassault Mirage F1EDA/DDA air defence fighters, together with two CASA C-212 light turboprop transports equipped for medical evacuation, was apparently rejected by FAC C-in-C Gen. Hector Fabio Velasco. The Mirages were among the 13 transferred from Qatar to Spain in 1994, and withdrawn from EdA service in August 2002. They were due for overhaul before transfer to Colombia, but were consid-

ered to be too costly to operate by Gen. Velasco. He also favours less sophisticated aircraft for the FAC's narcotic interdiction requirements.

More US aid

Four Basler Turbo AC-47 turboprop conversions of the venerable Douglas Dakota operated by the FAC since the late 1980s on narcotic interdiction operations are to be supplemented by an additional armed BT-67 version, costing $3.2 million, in a new $532 million US military aid programme confirmed by Congress in February. This also includes overhauls of the earlier AC-47s, as well as the supply of four refurbished C-130Hs, together with 10 Bell UH-1H utility helicopters, probably upgraded to Huey II standards, from surplus US stocks.

Czech Republic

UK Tornado F.3 lease discussed

Recent discussions in Prague with the UK's Defence Export Service Organisation concerning possible collaboration on Czech air defence improvements have included a possible lease of surplus RAF Tornado F.3 interceptors. Up to 14 F.3s could be made available for Czech operation by 2006, following the return in the coming year of 24 from a similar interim lease by the Italian air force.

Replacements are required for some 30 or more Czech Air & Air Defence Forces (CLPO) MiG-21MF 'Fishbed-Js', due for retirement in early 2005. Funding for next-generation combat aircraft is not currently available, however, following last year's widespread floods. These resulted in deferment of Kcs50 billion ($1.7 billion) CLPO acquisition plans for 24 JAS 39C/D Gripens, and consideration of leased NATO F-16s or F/A-18A/Bs, or locally-deployed alliance fighters, as alternatives to the Tornado F.3s.

Egypt

G400s ordered

The USAF formally ordered two Gulfstream G400 (formerly Gulfstream IV-SP) aircraft on behalf of the Egyptian government. Purchased from Gulfstream Aerospace under the Peace Lotus III foreign military sales (FMS) programme, at a cost of $59.2 million, the aircraft will be assigned to Egyptian Air Force's presidential airlift fleet at Cairo International Airport.

The RAF's latest recruit is the Bell Griffin HAR.Mk 2 (Bell 412EP), four of which have been procured under a COMR (civilian-owned, military registered) contract to provide SAR coverage for Cyprus with No. 84 Squadron. Initial crew training was undertaken at RAF Valley with temporary civil registrations and the aircraft were scheduled to arrive in Cyprus in June. Since the retirement of No. 84 Squadron's Wessexes on 31 January 2003, the SAR tasking has been undertaken by a detachment of No. 203 (Reserve) Squadron Sea Kings.

Hawkeye 2000 arrives

The Egyptian air force has taken delivery of the first of six E-2C Hawkeye 2000 aircraft from Northrop Grumman. The aircraft was built to an international configuration and had previously seen service with the US Navy. It was remanufactured and upgraded by the contractor at its St Augustine, Florida, facility where it flew for the first time on 8 October 2002. The Hawkeye subsequently left on 11 February enroute to Cairo. Egypt purchased five E-2Cs in the early 1990s and the first of these is currently undergoing the same modifications and will be delivered in early 2004.

Ethiopia

More 'Flankers' delivered

Deliveries of seven Sukhoi Su-27s last year to the Ethiopian air force (YIAH) reported by Rosoboronexport included at least four attrition replacements from original procurement of eight single- and two-seat 'Flankers' operated since 1998. The YIAH Su-27s were among 60 Sukhoi combat aircraft exported in 2002, compared with 48 in the preceding year, and represented a major proportion of the $4.8 billion from 2002 Russian arms sales.

India

IAF gains air refuelling

Formal acceptance took place on 28 February in Tashkent, Uzbekistan, by Indian Defence Minister George Fernandez, of the first of six IAF Ilyushin Il-78 'Midas' air refuelling tankers bought from a $166 million February 2001 agreement. Developed from the Il-76 'Candid' heavy-lift transport, the Il-78 has a single UPAZ-1A hose-drum unit on the port side of the rear fuselage, plus two underwing pods in the Il-78M, to transfer up to 64000 litres (14,080 Imp gal) of fuel from fuselage tanks.

Over 30 Il-78s are believed to have been converted from Il-76s built at the Tashkent Chkalov Aviation Production Association (TAPO) factory. Nearly 950 'Candid' transport versions had been built there since 1974, including over 100 for export, when Uzbekistan gained independence from the USSR in 1992. Long-term Russian plans to restart Il-76 production at the Voronezh Aircraft Building Stock Company plant, delayed for lack of funding, are now being implemented, to produce the stretched Il-76MF. Russo/Uzbek agreement was also recently reached to market future Il-76/78 sales through Russia's Rosoboronexport state arms agency.

Ukraine reportedly acquired at least a dozen Il-78s, and about 20 are believed to be operated in Russia. Operating initially from Agra, the IAF Il-78s, delivered by Indian crews from 3 March onwards about every six weeks, will support Jaguars, Mirage 2000s (including four new single-seat and six two-seat 2000Hs due this year) and Sukhoi Su-30MKIs. The latter will then have a sortie capability of up to 8000 km (4,320 nm). Mirage 2000-5s are also favoured for a new $8 billion IAF requirement for 125 more advanced combat aircraft.

India's Il-78s will supplement the IAF's 16 Il-76MDs, which transferred from Allahabad, in northern India, to Sonegaon, in centrally located Nagpur, as their main operating base late last year. Flown by Nos 25 and 44 Squadrons, the Il-76s were replaced at Allahabad by a new UAV unit.

Funded IAF procurement costing nearly $1 billion of three Il-76s modified by Israel with the installation of Phalcon phased-array radar and associated equipment for airborne early-warning roles has been delayed, pending Russian government/industry clearance and programme participation agreement for the aircraft modifications. Settlement of this situation, involving royalty payments to new Russian industrial programme participants, was reportedly reached in March, allowing a go-ahead for the Israeli and Indian partners.

Indonesia

Su-27 procurement revived

A revival in Russian military equipment procurement by Indonesia was signalled in April, following a visit to Moscow of Indonesian president Sukarnoputri. Agreement was then reached for the eventual purchase of up to 48 Sukhoi Su-27/30-series advanced combat aircraft for the TNI-AU. In August 1997, the Indonesian government had announced the planned procurement of 12 KnAAPO-built Su-30KIs, which incorporated the main features of the basic two-seat Su-30K, including a retractable port-side air refuelling probe, but were single-seat versions. Serialled 'Blue 27', the first of these flew in prototype form, but the programme was then suspended, following the late 1990s collapse in Far Eastern currencies and economies.

Indonesia's new initial Russian contract is for two Su-27SKs and two two-seat Su-30MKs from KnAAPO, with options for another eight, plus two Mil Mi-35V attack helicopters, costing some $197 million, for delivery from September. Orders for 44 more Su-27/30s to equip four TNI-AU squadrons are expected by 2007, similarly funded by only 12.5 percent in hard currency. The remaining payments will be in traditional Indonesian products, including bauxite, tea, coffee, rubber, and palm oil.

Deliveries are also expected from October by Indonesian naval aviation (TNI-AL) of two Mil Mi-17s and eight smaller Mi-2s ordered in 2000-01.

Korean KT-1 trainer deliveries

Formal TNI-AU acceptance took place on 25 April at Sacheon Air Base of the first of seven Korea Aerospace Industries KT-1B turboprop basic trainers ordered from a $60 million Indonesian contract in 2001. After further flight development, the TNI-AU's first KT-1Bs (LO-0101) were due to be shipped to Indonesia from July, to begin replacing the Beech T-34Cs of No. 102 Squadron at Adisucipto. Follow-on orders are expected from 13 additional options in the original contract.

Japan

Tanker transport programme

Contract signature, worth about $200 million, was announced in St Louis on 4 April between Boeing and the Japanese Defence Agency, together with the Itochu trading company, for the first of four B767 tanker-transport aircraft and associated support required by the Japanese Air Self Defence Force (JASDF). Its Boeing B767TTs will be military derivatives of the proven 767-200ER commercial aircraft, configured with an advanced Boeing air refuelling boom and remote operator system. Their non-combi convertible freighter configuration will offer operational flexibility to carry either cargo or passengers, for planned Japanese support of UN-sponsored peace-keeping and humanitarian activities.

The JASDF became the second B767TT customer, when it competitively selected the aircraft in December 2001. The Italian Air Force will receive the first of its four more advanced B767TTs in late 2005, followed by the JASDF's first aircraft in spring 2007. Delivery of all four JASDF B767TTs is planned by 2010. Boeing

NATO Channel Flights

Since November 2001 the BVVS (Bulgarian air force) has been operating a transport aircraft on behalf of NATO on its Channel Flights, which provide a personnel and cargo transport service throughout the Balkans region, primarily between Macedonia, Bosnia and Kosovo, with occasional trips to NATO's operational headquarters in Naples. The Bulgarian government has allocated one An-26 ('Red 087') and its seven-man crew to the mission, the aircraft parented by the 16th TrAB at Sofia-Vrazhdebna. By late April 2003 the aircraft had notched up 1,100 hours on this mission. It is seen (right) at Vrazhdebna prior to another typical Channel Flight service to Skopje, Naples, Sarajevo and Banja Luka, before returning to Sofia.

Alexander Mladenov

Anafa retirement

On 31 December 2002 the Bell 212 Anafa (Heron) was retired from Israel Defence Force/Air Force (IDF/AF) service. Having entered IDF/AF service in 1975 with the 'Rotor and the Sword' Squadron, the Bell 212 succeeded the Bell 205 as the standard IDF/AF utility helicopter. The transition was completed in 1978 when the 'Desert Birds' Squadron converted to the type. Additionally, a small number of Anafa helicopters served with the 'Light Helicopter' Squadron at Sde Dov between 1975 and 1989, principally as VIP transports.

Three of the helicopters were lost in action during fighting in Lebanon (two in 1982 and a third in 1985) and at least three others were lost in accidents. The IDF/AF fleet flew a total of 263,000 flight hours between 1975 and 2002, for an average of almost 10,000 flight hours per year. However, it must be pointed out that the annual number of hours flown during the initial introduction process and the final retirement process were significantly less than the overall average. The service entry of the Sikorsky UH-60A Yanshuf (Owl) in 1994 marked the beginning of the end for the venerable twin-rotor Bell. By 1998 the 'Rotor and the Sword' Squadron completed conversion to the Sikorsky assault helicopter, and during 2002 the 'Desert Birds' Squadron followed suit, the unit becoming operational with the Yanshuf on 1 January 2003.

Shlomo Aloni

Bell 212s are seen in storage at Hatzerim after retirement (left), while the 'Desert Birds' Squadron is now operational on the UH-60A (above).

claims air-tanker pre-eminence, from production of almost 2,000 tanker-transports, as well as 99 percent of all air refuelling booms made to date.

Jordan

Fighters delivered
On 29 January 2003 six F-16A fighters were delivered to the Royal Jordanian Air Force at Shahid Muwafak al-Salti Air base. The fighters had seen previous service with USAF and Air National Guard units and had been declared surplus. They were turned over to Jordan as part of a military assistance agreement that includes additional aircraft along with Patriot air defence missile batteries. Jordan had previously taken delivery of 16 surplus F-16A/B(ADF) fighters from the USAF and had recently announced the need for 10 additional F-16A/Bs, comprising four two-seat F-16Bs and six single-seat F-16As. The air force would eventually like to obtain as many as 70 F-16s, which would allow it to replace older F-5 and Mirage F-1 aircraft.

Transports ordered
Two C-295 military transports were ordered for RJAF from CASA during February 2003 at a cost of $45 million.

More EC635s
Orders announced last year by the RJAF for an initial batch of nine Eurocopter EC635 eight-seat twin-turboshaft light utility helicopters are to be increased by follow-on contracts for a further seven. Deliveries from the initial order were due to start in the summer of 2003.

Malaysia

New fighter plans trimmed
While apparently going ahead with its ambitious dual-type procurement plans of Boeing F/A-18EF Hornets and Sukhoi Su-30MKM multi-role fighters, the RMAF is scaling down its initial orders from 18 to 12 aircraft each, because of budget limitations. The original Hornet package, including weapons, was officially costed at $1.48 billion, with possible reductions if the RMAF had traded-in, as discussed, its current eight F-18Ds.

Unofficial estimates for scaled-down Su-30MKM 'Flanker' procurement, apparently regarded as a priority RMAF requirement, are around $720 million instead of $960 million, paid mostly from commodity sales to Russia.

Some of the planned fighter savings will help finance the RMAF's associated requirement for three or four AWACS aircraft. Current RMAF evaluations favour such lower-cost types as the EMBRAER/Ericsson EMB-145SA or surplus USN Northrop Grumman E-2Cs upgraded to Hawkeye 2000 standard, rather than the Boeing 737AEW&C. Similarly, the RMAF is considering acquiring the RNZAF's 17 recently-retired Aermacchi MB-339CB lead-in fighter trainers, instead of new-build BAE Hawks, to supplement its eight earlier Hawk MB-339As.

Mexico

IDF/AF E-2Cs acquired
Three of four Northrop Grumman E-2C Hawkeye airborne early-warning aircraft operated by the Israeli Defence Force/Air Force between 1978 and 1996, and then placed in storage at Hatzerim, were acquired in March by Mexican Naval Aviation (AAM). The reported $30 million contract includes refurbishment and installation of new and additional sensors by Israel Aircraft Industries, prior to delivery in the coming year, for AAM operation in Mexico's maritime surveillance and anti-narcotic campaign.

This is being further expanded by the recent AAM receipt of three Boeing/MD Helicopters MD 902 Combat Explorer NOTAR (no tail-rotor) gunships, armed with a GD 12.7-mm (0.5-in) GAU-19/A rotary multi-barrel cannon, and provision for seven 70-mm (2.75-in) rockets.

Morocco

Panthers equip naval air arm
Initial deliveries of the first of a small batch of Eurocopter AS 565MB Panther patrol and surveillance helicopters late last year marked the formation of an air arm of the Moroccan navy (Marine Royale). At least three Panthers, equipped with nose radar, a side-mounted multi-sensor FLIR/TV turret, and advanced navigation systems, are being acquired by the Marine Royale, for operation from its new 'Mohammed V'-class frigates.

Nepal

Skytruck delivered
Poland's PZL (Polskie Zaklady Lotnicze) aircraft manufacturing plant in Mielec recently delivered a single example of the M28.05 Skytruck tactical transport aircraft to the Royal Nepalese Army.

New Zealand

New RNZAF helicopters sought
Replacements are planned of five remaining Bell 47G-3B-2 Sioux basic training helicopters from 13 delivered to the RNZAF from 1965, and of 14 Bell UH-1H utility Iroquois operated since 1966. Priority is being given by the NZ Defence Forces and MoD to Project Kea, for an initially interim helicopter training capability, to replace the unrepresentative piston-engined Bell 47Gs.

Longer-term solutions to both Bell 47G and Bell UH-1H replacements are also being considered by the Project Warrior programme, for the NZDF's Rotary-Wing Transport Force. Limited funding to replace failing UH-1H structural components, including fin spar and main rotor blades, will extend RNZAF Iroquois operating lives by several years, for various civil and military tasks, such as SAR and disaster relief.

Operational analysis is being conducted to ascertain the ideal fleet size and fleet mix of generic helicopter types to meet NZDF and other government department requirements. Meanwhile, the RNZ Navy started acceptance trials on 2 February of its fifth and final Kaman SH-2G(NZ) Seasprite ship-based ASW helicopter, although one has been returned to its US manufacturer for repairs after a heavy landing. When ashore, the Seasprites also operate from Whenupai.

In the aftermath of Operation Iraqi Freedom US President George W. Bush visited USS Abraham Lincoln on its return to US waters. He arrived on board in this VS-35 S-3B, with the appropriate callsign NAVY 1 and his name on the starboard side. The aircraft is earmarked for preservation at Pensacola.

Updates for transport fleet

The Royal New Zealand Air Force purchased two Boeing 757-200 airliners that will replace a pair of 24-year-old Boeing 727-100s in service. The aircraft comprise B757-2K2 PH-TKA (c/n 26633/519) and PH-TKB (c/n 26634/545), which were previously operated by Dutch airline Transavia before being purchased by GE Capital Aviation Services for the Royal New Zealand Air Force. The first 757 was delivered to the RNZAF on 7 May 2003 and the second was due to follow in June. Although the 727s will continue operations they will be placed in short-term storage and eventually sold. The 757s will initially operate in an all-passenger configuration but will later be equipped with cargo doors for operation in all-cargo or 'combi' configurations.

Under the $340 million plan the RNZAF's fleet of five C-130H Hercules airlifters will undergo a structural life extension programme that will allow the transports to remain in service until 2017. Mechanical and avionics systems will also be upgraded as part of the programme. The C-130 upgrade will begin in late 2004 and will be completed by 2007.

Peru

Deliveries and withdrawals

Recent FAP deliveries from the US have included two refurbished Cessna A-37B Dragonfly light attack aircraft for use in counter-narcotic operations. These replace two A-37Bs lost earlier, and supplement 10 others being similarly refurbished from a $20 million US grant aid contract. A single light twin-turbofan Cessna 550 Citation II, equipped with sensors, signals intelligence and communications systems, has also been supplied by the US for airborne control of FAP drug interdiction and interception aircraft.

Other FAP deliveries in the past year or so, from eastern Europe, have included two more Antonov An-32 high-altitude twin-turboprop transports, increasing overall Batallon de Aviones 811 totals to six, and a twin-turboprop LET-410UVP. From an earlier strength of about 85 aircraft in all, including 12 transports, 62 helicopters and 11 trainers, the FAP had reportedly withdrawn from use in late 2002 eight Aermacchi MB.339As, one DHC-5, one DHC-6 and two Lockheed C-130s.

Poland

F-16 contract agreement

As expected, agreement was reached on 18 April between the US and Polish governments for the planned $3.5 billion Peace Sky procurement of 36 F100-PW-229 engined Lockheed Martin Advanced Block 50/52M+ F-16Cs and 12 two-seat F-16Ds, plus associated equipment and weapons. F-16s were selected on 27 December to meet long-standing Polish air force (PWLOP) requirements for 48 new combat aircraft, in preference to rival bids of Gripen International JAS 39C/Ds, and Dassault Mirage 2000-5EPL/DPL Mk 2s.

A key factor in PWLOP selection was unprecedented US government support, comprising a $3.8 billion 15-year Foreign Military Financing (FMF) low-interest loan to Poland, backed by industrial offset commitments totalling at least $6 billion. PWLOP F-16 deliveries, initially to Krzesiny air base, are scheduled between June 2006 and late 2008.

Saudi Arabia

Bell 412EPs delivered

Deliveries will be completed this year to the Royal Saudi air force (RSAF) of 16 upgraded Bell 412EP helicopters, ordered in February 2001, from a 562 million riyal (then $150 million) contract. IFR-equipped for search and rescue (SAR) missions, with FLIR and radar sensors, the RSAF Bell 412EPs are being built jointly by Bell Helicopter Textron in Canada and AgustaWestland in Italy, with deliveries starting in 2002.

RSAF Bell 412EP enhancements include new Pratt and Whitney Canada PT6T-9 turboshafts with electronic fuel-control for improved tropical and single-engine performance. A new glass cockpit incorporates four 6 x 8-in (15.2 x 20.32-cm) Rogerson-Kratos liquid-crystal display units, backed up by stand-by airspeed, altitude and attitude instruments.

Super Mushshak ordered

The RSAF was expected to be the recipient of four PAC Super Mushshak primary trainers, powered by 194-kW (260-hp) Textron Lycoming IO-540-V4A5 engines, ordered by Saudi Arabia from Pakistan in April. These could be initial replacements for the dozen or so Cessna 172 four-seat light aircraft operated by the RSAF for grading training since 1967, and further Saudi orders may follow. Saudi Arabia thus becomes the fourth Middle Eastern customer, including 25 to Iran, three to Oman, and six to Syria.

South Africa

Hawk procurement doubled

Confirmation emerged in early 2003 from BAE Systems of SAAF intentions to exercise its options on another dozen Hawk Mk 120 lead-in fighter trainers (LIFTs), increasing total South African procurement to 24. Financial provision for the options was included in the SAAF's December 1999 R4.7 billion (now $584.7 million) orders for 12 Hawk 120 LIFTs, for delivery from 2005, and accompanied by offset contracts initially totalling R8.6 billion ($1.07 billion) for local industry.

Earlier this year, BAE Systems transferred ZJ951, its upgraded Hawk new demonstrator aircraft (HNDA), to the SAAF's Test Flight Development centre, near Cape Town, for intensive flight trials. Apart from the new digital mission systems avionics, including a wide-angle HUD, three multi-function cockpit displays, hand-on-throttle-and-stick (HOTAS), on-board oxygen generation, and air refuelling capability already installed, the HNDA will have a Rolls-Royce Turboméca Adour Mk 951-standard turbofan installed in South Africa.

With full-authority digital engine control (FADEC) and a new combustor, the Adour 951 combines increased power from 25.8 kN (5,800 lb) to 28.9 kN (6,500 lb) with a doubled 4,000-hour overhaul life. After completing its South African flight development programme this autumn, the HNDA will become BAE's main Hawk development and demonstrator aircraft. Structural modifications are further planned for improved operational performance and reliability.

Despite acute defence funding problems, the South African government is also going ahead with long-deferred plans to acquire four Agusta

This colourful F/A-18A is the squadron commander's aircraft of VMFA-321, a Marine Corps Reserve unit based at NAF Washington (Andrews AFB), Maryland.

Westland Super Lynx ASW helicopters, for operation from its new Meko A200 naval patrol corvettes. A R1.1 billion ($150 million) contract was imminent earlier this year for the Super Lynx, to supplement Denel Oryx and Alouette III helicopters of the SAAF's No. 22 Squadron.

BBJ delivered

The South African Air Force has taken delivery of a new intercontinental Boeing Business Jet. The 737-7ED (BBJ) (c/n 32627/826) is registered ZS-RSA and named *Inkwazi*, a Zulu word for an African Fish Eagle. It is based at Waterkloof Air Base and assigned to No. 21 Squadron as a presidential transport. The 737 replaced a Dassault Falcon 900 that had previously operated in this role. Equipped to carry 18 passengers and a crew of six, its interior is divided into four main sections comprising a lounge with a convertible bed, an adjoining master office, a senior executive office and executive space. A crew rest area, storage area, full galley and lavatory are also provided. The jet, which was modified by Jet Aviation Business Jets in Basel, Switzerland, has a range of 11500 km (6,210 nm) and can fly from South Africa to London without refuelling.

Thailand

UH-60s sought

Recent Thai government equipment requests made to the US in March have included two Sikorsky UH-60L Black Hawk transport helicopters, from a potential $35 million FMS contract. Endorsing the request, the US Defense Security Co-operation Agency said that, in addition to the two UH-60Ls, this would include two spare GE T700-701C engines, an M-130 chaff dispenser, and spares and support equipment.

Nose art and special markings have made something of a comeback in ANG units, as demonstrated by these F-15As. Above is an aircraft from Oregon's 114th FS/173rd FW with a shark's mouth, while at right is a 110th FS/131st FW aircraft from the Missouri ANG with a shark holding a missile and the name Top-o-the Food Chain.

Korea's Northrop trainers

The Northrop F-5 was first delivered to South Korea in 1965, and has been a major element in the country's defence ever since. Two-seat F-5s support front-line squadrons (F-5F right, 8 Wing at Wonju) and equip advanced training units (F-5B below, at Kwangju. Note instrument flying hood in rear cockpit). Advanced training is also the role of the Northrop T-38As leased from the US (below right) and based at Yecheon.

United Arab Emirates

AEW choice imminent
Selection of a new airborne early warning aircraft for the UAE's defence forces is expected in 2003, according to Air Force and Air Defence chief Maj. Gen. Khaled Al-bu Ainain. Up to five AEW aircraft are required by the UAE, which was then finalising evaluations of the leading contenders. Preliminary reports indicate that the upgraded Northrop Grumman E-2C Hawkeye 2000 was favourably placed to meet UAE performance, cost and delivery requirements.

CASA 295s ordered
The United Arab Emirates has announced plans to purchase four maritime patrol variants of the EADS/CASA C-295 airlift aircraft at an estimated cost of $215 million.

United States

Multi-year C-130J contract
The USAF has awarded Lockheed Martin a $4 billion multi-year contract covering the purchase of 60 C-130J airlifters for the USAF and USMC. The six-year programme will provide the USAF with 40 stretched CC-130Js and the USMC will obtain 20 KC-130Js. To date only $271.6 million has been obligated toward the purchase of four KC-130Js for the Marines.

In related news, Lockheed Martin rolled out the 100th C-130J Hercules at its Marietta, Georgia facility on 17 February 2003. The event involved HC-130J serial 2003 (c/n 5534), which is slated for delivery to the US Coast Guard later in 2003. The aircraft will be the third HC-130J for the USCG.

Kiowa Warriors ordered
The US Army has awarded Bell Helicopter a contract to convert 22 additional OH-58A/C Kiowa helicopters to the OH-58D Kiowa Warrior configuration. Although the full contract is valued at $50.3 million, the Army has thus far only committed just $16.9 million.

Citation Encore ordered
The US Army, acting on behalf of the US Navy, has awarded Cessna Aircraft a $7.4 million contract covering the purchase of a single UC-35D Citation Encore. The aircraft, which was authorised as part of the 2003 defence budget, will be delivered to the Marine Corps in early 2004.

Advanced Apache Longbow
Boeing has delivered the first of 217 Block II AH-64D Apache Longbows to the US Army. This helicopter is the first production Apache Longbow to be equipped with advanced avionics, digital enhancements and communications upgrades intended to improve situational awareness. The upgrades include the modernised tactical acquisition designation sight (M-TADS), TADS electronic display and control (TEDAC) and a modernised pilot night vision system (M-PNVS).

Air ambulances ordered
Sikorsky Aircraft has been awarded a $69.2 million modification to an existing contract covering the conversion of nine UH-60L aircraft to HH-60L air ambulance configuration. The aircraft had been purchased under a separate contract.

Clipper ordered
Boeing has received a $61.0 million contract covering the purchase of one C-40A Clipper airlifter for the US Naval Reserve. Delivery of the aircraft will take place by November 2004.

Blackhawks and Knighthawks
The US Army Aviation and Missile Command has awarded Sikorsky Aircraft a $270.5 contract on behalf of both the Army and the US Navy. The contract covers the construction of 12 UH-60L for the former and 15 MH-60S for the latter service.

First JSTARS delivered to ANG
Northrop Grumman recently delivered the 15th E-8C Joint surveillance target attack radar system (JSTARS) aircraft to the Georgia Air National Guard's 116th

Air Control Wing (ACW) at Robins AFB, Georgia. This example is the first to be delivered since the mission transitioned to the Air National Guard late in 2002. The E-8Cs are now assigned to the 116th ACW, which is a 'blended wing' that includes both active duty Air Force and Air National Guard personnel. The E-8C is the fifth example built to Block 20 configuration and the contractor is currently updating the earlier aircraft as part of the Block 20 upgrade programme. Two updated examples have already been delivered and two are currently in work.

Global Hawks ordered
Northrop Grumman has received a $307 million modification to an existing contract associated with Lot 2 low-rate initial production (LRIP) of the RQ-4A Global Hawk unmanned aerial vehicle. The modification covers the purchase of four RQ-4A air vehicles, three integrated sensor suites, two electro-optical infrared sensors, one launch recovery element and one mission control element for the USAF, and two RQ-4A air vehicles, two integrated sensor suites, one mission control element and two launch recovery elements for the US Navy.

Modernised Chinooks ordered
The US Army has issued a $140 million contract to Boeing covering the remanufacture of seven CH-47D Chinooks to the new CH-47F and MH-47G special operations configurations. Under the low-rate initial production (LRIP) contract, Boeing will deliver one CH-47F and six MH-47Gs, and the first delivery will take place in September 2004. As part of the remanufacturing process the airframes will be equipped with new cockpit sections, updated avionics and more powerful Honeywell T55-GA-714A engines with full authority digital engine control (FADEC). The CH-47F avionics include advanced cockpit management systems, a digital mission management suite with a moving map display and digital modem. The MH-47G will feature the Army's

advanced common avionics architecture that equips all special operations helicopters.

Boeing tanker lease still on hold
The US Department of Defense recently announced that it is still reviewing the USAF's plan for leasing 100 so-called Boeing KC-767A tanker aircraft. The leasing plan, announced early in 2002, has received considerable resistance from US lawmakers who would prefer the USAF purchase the aircraft.

USCG agrees to helicopter lease
The US Coast Guard and AgustaWestland have agreed to a follow-on lease for the eight MH-68A helicopters operated by Helicopter Interdiction Tactical Squadron Ten from Cecil Field Airport in Jacksonville, Florida. The renewed five-year lease includes an option for two more helicopters and modifications to the existing fleet. The previous two-year lease expired in January 2003.

Creeks ordered
The US Army Aviation and Missile Command has awarded Bell Helicopter Textron a $9.6 contract covering six new TH-67A+ training helicopters for the US Army Aviation Center (USAAVNC) at Fort Rucker, Alabama. The helicopters will support the school's transformation to Flight School XXI and allow the retirement of older aircraft.

C-40B enters service
The Boeing C-40B, Air Mobility Command's newest aircraft, formally entered service with the 89th Airlift Wing (AW) on 24 January 2003 following a ceremony at Andrews AFB, Maryland. Serial 01-0040, which is operated by the 1st Airlift Squadron (AS), was actually delivered on 6 December 2002 but completed a 45-day initial operating capability assessment before being declared ready for operations. Based upon the Boeing 737-700 BBJ, the C-40B will

China has converted at least two Tu-154 airliners into Sigint (signals intelligence) gatherers, with a series of radomes covering antennas along the centreline.

ANG combat units undertake their OREs (Operational Readiness Exercises) at Volk Field Combat Readiness Training Center in Wisconsin. The facility offers an ACMI range and the Hardwood Bombing Range within a large airspace over sparsely populated country. Shown here during an ORE is the squadron commander's aircraft from the 170th FS/183rd FW, Illinois ANG, complete with the unit's new fin stripe.

operate with up to 11 crew and 26 passengers in support of global operations and the Unified Combatant Commanders. During the IOC assessment flight crews logged over 200 flight hours and more than 20,000 miles. A second C-40B was delivered to the 15th Air Base Wing (ABW) at Hickam AFB, Hawaii on 9 January 2003 for the 65th AS.

T-38Cs delivered to Moody
In early November 2002 personnel from the 415th Flight Test Flight (FLTF) (an Air Force Reserve Command unit) at Randolph AFB, Texas, delivered a modified T-38C to the 479th Flying Training Group at

Moody Air Force Base, Georgia. The Talon was the first modified under the Propulsion Modernization Program (PMP), which provides the Talon with redesigned intakes and modified engines. Externally, the most obvious change is the larger inlets – however, improved components include a redesigned compressor. The PMP modifications improve the T-38's take-off performance by increasing the thrust and will extend the service life of the T-38C through the year 2020. More than 500 aircraft and 1,200 engines will be modified under the programme. Lear Siegler Services is carrying out the modernisation programme at Randolph under a 10-year contract.

100th Globemaster III delivered
Boeing delivered the 100th C-17A to the USAF on 8 November 2002 at its

facility in Long Beach, California. The aircraft, which is the 93rd production example, departed the same day for Charleston AFB, South Carolina, where it joined the 437th Airlift Wing. The contractor is currently building C-17As at a rate of 15 per year, and has already received contracts covering the production of 180 examples from the USAF. The Globemaster III is also in service with the UK's Royal Air Force, which operates four examples.

Venezuela

National Guard helicopters
Enstrom Helicopter has delivered four piston-powered 280FX training helicopters to the Venezuelan National Guard's (Guardia Nacional) Air Support Command. Equipped with dual controls and capable of carrying up to three persons, the helicopter is powered by a single Textron Lycoming HIO-360-F1AD rated at 168 kW (225 hp).

Air Force buys AMX-T
Embraer announced that it has signed a contract with the Venezuelan Air Force covering the purchase of 12 AMX-T aircraft. The two-seat multi-role aircraft will replace the FAV's aging fleet of Rockwell T-2D Buckeye trainers and deliveries are expected to begin in 2005. This event marks the first export sale for the AMX.

AIR ARM REVIEW

Albania

Combat aircraft retirement
Plans for extensive reorganisation of the severely cash-strapped Albanian defence forces announced earlier this year are expected to involve early withdrawal of the army air wing's few remaining MiG-21s, Shenyang F-6s and Chengdu F-7s, and its expansion as a mainly rotary-wing force. Four recently-delivered Agusta-Bell AB 206s from Italian army stocks are already being supplemented by a further three, and have been followed by seven larger Agusta-Bell AB 205s. Air force personnel strength will be reduced to about 1,400, operating from only two air bases at Rinas, near Tirana, and Kucove in the south.

Czech Republic

L-159 lost
With deliveries of 58 of 72 Aero L-159As on order to several operating and training units, the Czech air force (CLPO) lost the first of these single-seat light-attack aircraft in a fatal crash on 24 February, in central Bohemia. The aircraft was being flown by a CLPO test-pilot on air-to-ground firing trials with a 900-kg (1,984-lb) under-fuselage twin-barrelled 20-mm (0.787-in) ZVI Vsetin Plamen cannon pod.
Preliminary accident investigations attributed the loss to pilot error, from an attempted low-altitude loop. ZVI Vsetin was awarded a $65 million government contract in 2001 for 144 Plamen cannon and ammunition, for

the CLPO's L-159 programme. Some development delays are anticipated, however, since the cannon destroyed in the accident was apparently the sole working prototype.

Pakistan

C-in-C killed in F27 crash
Air Chief Marshal Mushaf Ali Mir, C-in-C of the Pakistan air force, his wife and 15 other people, reportedly including two PAF Air Vice-Marshals, were killed in the crash of a Fokker F27 twin-turboprop transport on 20 February. PAF spokesman Air Commodore Sarfraz Ahmed Khan said that the debris of the F27 turboprop, described as a military aircraft, was located near high ground some 27 km (14.5 nm) east of Kohat in northwest Pakistan. The F27 was en route to the PAF's Kohat air base, in apparently poor weather conditions. Deputy C–in–C Air Marshal Syed Qaser Hussain was appointed in succession, initially on an acting basis.
Two other PAF aircraft were lost in subsequent accidents, both on 12 April. One of the PAF's 40 or so ageing Nanchang A-5 ground-attack fighters crashed near the town of Pindigheb in central Punjab province,

BO 105P 86+66 was given this special scheme to commemorate the 30th anniversary of the formation of Heeresfliegerversuchsstaffel 910. The German army's trials unit was formed at Celle in 1973, and moved to its current base at Bückeburg in 1993.

after the pilot ejected safely. An instructor was killed and his student injured in the crash of a PAC MFI-17 Mushshak primary trainer from the PAF Air Academy, near its Risalpur air base in northern Pakistan.

Philippines

Budget problems force cuts
Economic problems in the Philippines have resulted in defence cuts which have effectively grounded about 150 of some 220 PhilAF aircraft, with little immediate prospects of new replacements or availability improvements. Military equipment appropriations in 2002 totalled only 5 billion pesos ($91.32 million). Coupled with growing insurgency from communist and militant Islamic groups, the Philippine forces are becoming increasingly dependent on additional US aid, totalling $135 million in the current year, or other foreign assistance.
Of the PhilAF's 13 Lockheed C-130B/H and L-100 Hercules, only

three are currently in service, alongside two of nine Fokker F27s, and about 15 of 55 Bell UH-1D/Hs. Planned PhilAF procurement of four ex-RAF C-130Ks taken back by Lockheed Martin in part exchange for new C-130Js, has also been abandoned for financial reasons. Ten UH-1Hs are currently being refurbished, however, from a 600 million peso contract with US aid, which in 2002 totalled about $70 million. This will contribute towards PhilAF acquisition of up to 12 more MD-520G attack/scout helicopters, to supplement about 20 previously delivered.

Qatar

Combat element withdrawn?
Reported offers by the Qatar Emiri air force of its nine multi-role Dassault Mirage 2000-5EDA fighters and three Mirage 2000-5DDA two-seat operational trainers to India, to supplement the IAF's expanding Mirage 2000H fleet, would appear to indicate aban-

Spring Flag 2003

Exercise Spring Flag is Italy's largest and most important annual air exercise. It is a Red Flag-type exercise, albeit on a much smaller scale, which aims to train and integrate all the Italian air defence and attack forces. Organised by the COFA (Comando Operativo Forza Aerea, or Operational Command of the Air Force), it was first held in 2000, at Piacenza air base, with the main purpose to train Tornado ECR and ADV crews for their ensuing participation in the USAF's Green Flag 2000 exercise. Since 2002 the main base hosting the exercise has been Decimomannu in Sardinia, which offers a large site, easily able to accommodate dozens of aircraft and hundreds of personnel. In addition, the surrounding area has air-to-air and air-to-ground instrumented ranges, suitable also for the release of both inert and live ammunition, a large area of unrestricted airspace over the Tyrrhenian Sea, and good weather throughout the year.

Between 4 and 17 April, Spring Flag 2003 involved 40 aircraft of various types, and included participation by other forces: the Italian Navy, which sent AV-8B Plus Harriers; and the French Air Force, present with two Mirage 2000-5s and one E-3F AWACS. Other 'firsts' were the carriage by many aircraft of the new autonomous BGT ACMI pod, the delivery of precision-guided munitions (PGMs) with the aid of ground special forces teams for lasing, and the use of satellite systems (SICRAL and HELIOS) for communication, intelligence and targeting purposes. It also was the first Spring Flag exercise to involve slow-movers and AMI Special Forces. Observers from the German and Swiss Air Forces were present to evaluate the possible participation of their assets in the 2004 exercise.

Dott. Riccardo Niccoli

Spring Flag 2003, aircraft participants

Tornado ADV	4	36° Stormo	AMI
Tornado ECR	4	50° Stormo	AMI
Tornado IDS	4	36° Stormo	AMI
AMX	10	32° Gr, 51° Stormo	AMI
F-104S.ASA-M	2	37° Stormo	AMI
MB.339CD	2	61° Stormo	AMI
G.222TCM	1	46ª Brigata Aerea	AMI
G.222VS	1	14° Stormo	AMI
B.707T/T	1	14° Stormo	AMI
C-130J	1	46ª Brigata Aerea	AMI
HH-3F	2	15° Stormo	AMI
AV-8B Plus	4	Grupaer	MMI
SH-3D	1	1° Grupelicot	MMI
Mirage 2000-5	2	EC 2/2	AdA
E-3A	1	NAEWF	NATO
E-3F	1	EDCA 36	ADA

Easily distinguished by its fin-top radome and farm of underfuselage antennas, the G222VS is flown by 14° Stormo's 71° Gruppo Guerra Elettronica on Elint missions.

Top: AMXs from 32° Gruppo prepare to launch for a COMAO (combined air operations) exercise with Grupaer AV-8Bs. The aircraft are carrying the new BGT ACMI pod.

Below: The Tornado ADV is in its last full year of AMI operations, the leased fleet being returned to the UK on the arrival of the Peace Caesar F-16s, the first of which was rolled out after upgrade on 8 May.

---

donment of full-scale combat roles allocated to the QEAF. Delivered only in 1997 to the QEAF, the Mirage 2000-5s replaced its 11 earlier Mirage F1EDAs and two two-seat F1DDAs, purchased by Spain in late 1994. Apart from utility helicopters and transports, the QEAF now operates only six Dassault/Dornier Alpha Jets, with limited light armament, as its fixed-wing combat aircraft inventory.

Russia

VVS takes over Army Aviation

Further consolidation of the Russian armed forces took place on 1 January this year, with transfer of the helicopter-equipped Army Aviation (Armeiskaya Aviatsiya) to air forces (Voyenno Vozdushnye Sily – VVS) command, control and administration. Russian Army Aviation is currently reported to have an operating strength of around 1,000 Mil Mi-24 series attack helicopters, and similar Mi-8 totals for transport, utility, command and support roles.

Other types in AA service include about 100 smaller Mi-2s for training and communications, plus about 30 heavy-lift Mi-26s and declining numbers of slightly smaller Mi-6s. Funding limitations and spares shortages have resulted in very low availability and serviceability rates down to 20 percent for all types in recent years, however, together with prolonged delays in planned combat helicopter re-equipment by such types as the Kamov Ka-50 and Mil Mi-28.

Army Aviation procurement was taken over several years ago by the VVS, which has also been supervising long-overdue upgrade plans for the AA's mainstay Mi-8s and Mi-24s. Funding has now finally been approved for new night/all-weather mission systems avionics in 120 Mi-8s and Mi-24s, for which prototypes are reaching completion of service trials at the VVS Akhtubinsk flight-test centre.

A common Mi-8MTKO/Mi-24V/PN avionics installation developed with the Mil Design Bureau (OKB) and Rostvertol integrates Krasnogorsk Zarevo/Nokturn electro-optical sighting and targeting systems, liquid-crystal cockpit displays, new-generation Geophizika-NV night-vision goggles and KBM 9M120 Ataka (AT-9 'Spiral 2') tube-launched ATMs in the Mi-24PK-1. VVS tests are also being made of the Mi-24VK-1, with similar avionics and sensors from Russkaya Avionika and the Ural Optical-Mechanical Plant (UOMZ).

Singapore

Longbow Apache progress

A major milestone in the development of the RSAF's attack helicopter capabilities was achieved on 9 April in the US, with official inauguration of the RSAF's Peace Vanguard Boeing AH-64D Apache programme detachment in Marana, Arizona. This detachment was initiated with assistance from the Arizona Army National Guard in October 2001, to give Singapore's Apache squadron valuable access to

firing ranges and training, and attain US levels of operational readiness.

The Peace Vanguard detachment received the first of eight AH-64D Longbow Apaches on 17 May 2002 at Marana. Twenty-one RSAF pilots have so far achieved US Army Mission Capable Aviator standard there, and deliveries of another 12 Singaporean AH-64Ds are scheduled from 2005.

Singapore also has four other training detachments in the US. These comprise Peace Carvin II and III, with Lockheed Martin F-16C/Ds at Luke AFB, Arizona, and Cannon AFB, New Mexico, respectively; Peace Prairie, with Boeing CH-47D Chinooks in Grand Prairie, Texas; and the Peace Guardian Boeing KC-135 jet tanker detachment at McConnell AFB, Kansas. Apart from providing essential training space, and improving inter-operability between the two armed forces, RSAF detachments in the US reinforce long-standing and close defence relations.

South Africa

New combat helicopter

The South African government formally approved the entry into service of Denel Aerospace's Rooivalk combat support helicopter in December 2002. Although the initial examples were assigned to Suid-Afrikaanse Lugmag (SALM), or South African Air Force's No. 16 Squadron at Bloemspruit Air Base in 1999, a further 2-3 years will pass before the aircraft enters operational service. A total of

12 helicopters was ordered and 10 have already been delivered. Developed using the powertrain of the Eurocopter Puma helicopter, the Rooivalk is equipped with a nose-mounted cannon and a range of underwing-mounted munitions.

United States

New USMC Osprey Squadron

The US Marine Corps will establish its first test and evaluation squadron at MCAS New River, North Carolina, during 2003. Designated VMX-22, the squadron will assume the MV-22B operational test mission that had been conducted by HMX-1. The new unit will report operationally to Commander, Operational Test & Evaluation Force (COTEF) at Naval Station Norfolk, Virginia.

C-130J environmental analysis

Indicating that it will likely become the first unit within the active air force to operate the C-130J, Pope AFB's 43d Airlift Wing recently announced that an environmental impact analysis has been completed. The findings of the analysis clear the way for the 43rd's conversion from C-130E to C-130J model as part of Air Mobility Command's C-130 modernisation plan.

Skyhawks to retire

As a direct result of the cessation of training on the Vieques Naval Training Range in Puerto Rico, the US Navy has announced plans to deactivate the commands which directly supported

USS Ronald Reagan (CVN 76), the penultimate 'Nimitz'-class carrier, is due to replace USS Constellation (CV 64) in 2003. Here it is seen during sea trials in the spring, testing the CMWDS (Counter-Measures Wash-Down System).

the range. Included are both the Atlantic Fleet Weapons Training Facility (AFWTF) and Fleet Composite Squadron (VC)-8. Both the AFWTF, which had been the command in charge of organising and executing the training exercises in Vieques, and VC-8 will stand down by 30 September 2003. VC-8 is based at Naval Station Roosevelt Roads and had provided search and rescue coverage and aggressor support. The squadron operates UH-3H helicopters and the last TA-4J Skyhawks in the Navy inventory and has been stationed at 'Rosey Roads' since 1959.

USMC Prowler pilots CQ
Six pilots and a number of electronic countermeasures officers (ECMO) assigned to the 'Moondogs' of Marine Tactical Electronic Warfare Squadron (VMAQ)-3 recently became the first Marine Corps personnel to conduct carrier qualifications (CQ) in the EA-6B since 1996 when they operated from the USS *John F. Kennedy* (CV 67). Although four Marine Corps Hornet squadrons make regular deployments as part of US Navy Carrier Air Wing (CVW), in recent years this has not been a requirement for Cherry Point's four Prowler squadrons. A recent agreement between Commandant of Marine Corps and the Secretary of the Navy, however, has committed the Corps to providing six additional squadrons to support CVW deployments.

Knighthawk deployed for SAR
The Sikorsky MH-60S Knighthawk recently made its first deployment as a search and rescue (SAR) platform aboard the USS *Essex* (LHD 2). Two aircraft, operated by Helicopter Combat Support Squadron Five, Detachment Six (HC-5, Det. 6), joined the 31st Marine Expeditionary Unit (MEU) aboard the amphibious assault ship, which was conducting training near Okinawa. The squadron, which is normally stationed at Andersen AFB, Guam, currently has 14 Knighthawks. Its six detachments are normally deployed aboard combat support ships in the vertical replenishment (VERTREP) but the squadron also provides SAR helicopters to the assault ship when it deploys.

Super Hornet first combat
The F/A-18E, which recently concluded its first deployment, conducted its first combat mission on 6 November 2002 when aircraft from USS *Abraham Lincoln* (CVN 72) fired on Iraqi targets in the southern 'No-fly' zone. The Super Hornets used precision-guided weapons to target two surface-to-air missile systems (SAM) near Al Kut, approximately 161 km (100 miles) southeast of Baghdad, and a command and control communications facility near Tallil, about 257 km (160 miles) southeast of Baghdad.

Hueys retired
Assigned to the Military District of Washington, and based at Davison Army Airfield on Fort Belvoir, Virginia, the 12th Aviation Battalion has retired the majority of its UH-1Hs and will soon complete a transition to the UH-60A. The unit, which comprises one general support and one command aviation company, currently operates 14 Blackhawks including 10 UH-60As and four VIP configured VH-60Ls. The unit recently transferred three UH-1Hs to the 1st US Army Support Battalion Aviation Company in the Egyptian Sinai and two were transferred to the Ronald Reagan Ballistic Missile Test Site, US Army Kwajalein Atoll in the Marshall Islands. The remaining 'Hueys' include just five VH-1Hs and these will be retired during the summer of 2003.

New facility for Nebraska NG
The Nebraska Army National Guard has announced plans to build a new army aviation support facility at Central Nebraska Regional Airport in Grand Island. The facility will eventually house 16 AH-64As that will be stationed in Lincoln from May 2004. Construction of the \$20 million facility should begin in 2005. Additional aviation units of the Nebraska ARNG will remain based at Lincoln.

D-M combat rescue squadrons
Air Combat Command has activated three Rescue Squadrons (RQS) Davis-Monthan AFB, Arizona as part of the 355th Wing. Although the units will operate as many as 12 HH-60Gs and 10 HC-130N/Ps by 2007, they will initially operate eight and five of these types, respectively. The initial aircraft were transferred from the Air Force Reserve Command's 939th Rescue Wing at Portland International Airport, Oregon, which has transitioned to an aerial refuelling mission. The third squadron at Davis Monthan is comprised of the pararescue jumpers (PJ) that support the combat search and rescue (CSAR) operations. The

command had looked at a number of other operating locations for the CSAR units before selecting Davis Monthan, including Edwards and Vandenberg AFBs in California. The units are respectively designated the 55th, 79th and 48th Rescue Squadrons – all have seen prior service as CSAR units.

Vipers for Vipers
The Minnesota Air National Guard's 148 FW/179 FS has converted from the Block 15 Air Defense Fighter variant of the F-16A/B to the Block 25 F-16C/D. The unit's aircraft were being transferred from the New York Air National Guard's 174 FW/138 FS, which in turn transitioned to the Block 30 model. The aircraft for the New York ANG came from Iowa's 185th FW/174th FS, which is converting to the KC-135E. The latter unit, based at Sioux City, Iowa, transferred its final five 'Vipers' to Syracuse on 17 January 2003 and welcomed a visiting KC-135E from the Washington Air National Guard's 141st Air Refueling Wing at Fairchild AFB. The 185th will not, however, receive its own tankers until October 2003 when the first aircraft are transferred from Fairchild.

US Navy carrier news
On 9 December 2003 the US Navy announced its intention to name CVN 77, the last aircraft carrier of the 'Nimitz'-class, to honour former President George H.W. Bush. Bush, the 41st President, is a former naval aviator who flew the Grumman TBF Avenger during World War II. Construction of the tenth ship of the class is underway and it will enter service in 2008.

US Navy and Marine Corps leaders have announced plans to proceed with an updated larger version of the

current 'Wasp'-class multi-purpose amphibious assault ship (LHD) as a replacement for the five 'Nassau'-class amphibious assault ships (LHA). Referred to as the 'LHD plug plus', the new class would be 23.47 m (77 ft) longer and feature a flight deck that is 3.048 m (10 ft) wider than the 'Wasp'-class. The new ships would be 22 percent larger than the 'Wasps' and would be capable of carrying about 20 F-35 Joint Strike Fighters, as well as helicopters and MV-22 tiltrotor aircraft. Other options include buying additional 'Wasp'-class vessels or designing a completely new class of ships.

Super Hornet transition
As deliveries of new F/A-18E and F/A-18F Super Hornets continue, the Navy has released preliminary transition dates that run from 2003 through 2007. Besides closing out Tomcat operations, the transition schedule includes dates for several F/A-18C squadrons, and the first of these will take place during 2003. According to the plan the last Tomcats will be phased out in 2007.

Year	Squadron	Current Type	New Type
2003	VFA-137	F/A-18C	F/A-18E
	VF-2	F-14D	F/A-18F
	VFA-22	F/A-18C	F/A-18E
2004	VF-154	F-14A	F/A-18F
2005	VFA-81	F/A-18C	F/A-18E
	VF-32	F-14B	F/A-18F
	VF-103	F-14B	F/A-18F
2006	VF-213	F-14D	F/A-18F
	VFA-86	F/A-18C	F/A-18E
	VF-211	F-14A	F/A-18F
2007	VF-11	F-14B	F/A-18E
	VF-143	F-14B	F/A-18F
	VFA-31	F-14D	F/A-18F
	VFA-105	F/A-18C	F/A-18E
	VFA-146	F/A-18C	F/A-18E

The last SRF-5A in Spanish service was given these special colours by Ala 23 at Talavera to celebrate the 50th anniversary of the Escuela de Caza y Ataque (fighter/attack school). Tragically, the aircraft was lost in a fatal crash in May.

Operation Iraqi Freedom

The long-awaited campaign against Saddam Hussein's Iraq commenced on 19 March with a failed attempt to kill the Iraqi leadership. While the result of the conflict was never in contention, it did demonstrate how far the West's air power has advanced since Desert Storm in 1991.

Iraq was listed as one of three countries forming part of the 'Axis of Evil' in a State of the Union address given by President George W. Bush on 29 January 2002. Increasing pressure to return weapons inspectors, thrown out of the country in the late 1990s, and rhetoric on the subject of Iraq's weapons of mass destruction programmes, led to United Nations Security Council Resolution (UNSCR) 1441 that held Iraq in material breach of previous resolutions. On 7 March the United Kingdom submitted a 17 March deadline for Iraq to comply with UNSCR 1441; it was followed by a 48-hour ultimatum when this deadline was reached. Two days later Operation Iraqi Freedom commenced.

American airpower had been taking up positions in the Middle East theatre since the beginning of 2003. While few countries in the Arab world wished to be openly associated with any operations against Iraq, in reality Bahrain,

Top: Somewhere over Iraq, Tornado F.Mk 3 ZE932/XC pulls away from a 117th ARW KC-135 fitted with a hose extension to the flying boom. On the nose of the Tornado is a depiction of the 'Dennis the Menace' comic character. Most of the in-theatre Tornados (including the GR.Mk 4s, right) received some form of nose art.

Below: The RAPTOR pod carried on the fuselage stores station has been in use since 2002 over the 'No-fly' zones in Iraq. This GR.Mk 4A also displays nine PGB mission markings under the cockpit. UK aircraft used 679 guided and 124 unguided munitions in Operation Telic.

Right: Considering the impact the type had in Desert Storm, the deployment of only 12 F-117As for Operation Iraqi Freedom came as a surprise to most. They were operated as the 8th Expeditionary Fighter Squadron based at Al Udeid in Qatar as part of the 379th AEW. During the campaign a total of 11 GBU-27s and 98 EGBU-27s (the same weapon but with a GPS-guidance system added) were used, with the majority (if not all) of them being employed by the 'stealth fighter'. The aircraft were one of the first to leave the theatre at the end of the war (above), departing Al Udeid on 14 April.

The RAAF deployed 14 F/A-18As to Al Udeid air base in Qatar, the aircraft coming from No. 75 Sqn based at Tindal, in Australia's Northern Territory. The aircraft flew a total of 302 sorties.

Jordan, Kuwait, Oman, Qatar, Saudi Arabia and the United Arab Emirates all hosted US aircraft.

In the expectation that a northern front would be opened up early in the conflict, the Turkish airbase at Incirlik housed around 50 US and British jets, while forward operating locations nearer the border with Iraq were available at Batman, Diyarbikar and Van. Turkish permission to use these bases for any attack was expected, especially as the United States and the United Kingdom had campaigned for NATO protection for Turkey should hostilities break out. Thus a blow was dealt to coalition plans on 1 March when Turkey voted against allowing foreign forces to conduct missions from Turkish bases, even after the offer of a substantial financial package.

Operation Telic

Apart from the United States, three countries were willing to provide some form of air power to further the implementation of United Nations resolutions concerning Iraq. From amongst the 'coalition of the willing', it was the United Kingdom (112 aircraft), Australia (22) and

Left: The RAF's deployed Harrier GR.Mk 7s, split between bases in Jordan and Kuwait, made extensive use of Maverick air-to-ground missiles, although this one is carrying a pair of dumb bombs and two Sidewinders. The US Marine Corps F/A-18 Hornets in the distance suggest this is Ahmed al Jaber Air Base in Kuwait.

Canada (three airlifters) that were to provide aircraft for what would become Operation Iraqi Freedom. The United Kingdom's contribution was known as Operation Telic. RAF aircraft had operated nearly continuously in the theatre since the end of Desert Storm, gaining an enormous amount of combat experience.

RAF Brize Norton, Oxfordshire, was used as the airhead for moving personnel and supplies to the Gulf. The Brize-based Air Movements Squadron saw its workload surpass that experienced for the first Gulf War and during February 2003 up to two chartered Antonov An-124s were being dispatched to the Gulf each day. Nos 10, 101 and 216 Squadrons increased their aircraft deployed to the region with tankers based at Muharraq in Bahrain (a long-term No. 216 Squadron deployment) and at Prince Sultan AB in Saudi Arabia (VC10s). A single aeromedical-configured VC10 was based at RAF Akrotiri on Cyprus to expedite casualties back to the UK. VC10s were also based at Incirlik in Turkey, where four Jaguar GR.Mk 3s were also based. Prince Sultan AB also played host to RAF Sentries, Nimrods (both a single R.Mk 1 and MR.Mk 2s) and some of the four No. 32(TR) Squadron BAe 125 CC.Mk 3s deployed to the area. The 14 Tornado F.Mk 3s drawn from Nos 11, 25, 43 and 111 Squadrons were also based at the Saudi airbase.

Tornado GR.Mk 4/4As were based at Ali Al

Salem in Kuwait and at Al Udeid in Qatar. As well as carrying the RAPTOR (Reconnaissance Pod for TORnado), the force also launched the first MDBA Storm Shadow conventionally-armed cruise missiles used in anger. No. 9 Squadron's Tornados were equipped to fire ALARM anti-radiation missiles at Iraqi air defence radars, each aircraft carrying up to five of the weapons. Tragedy was to strike the deployed Tornado force at 09:15 local (06:15 GMT) on 23 March, when a No. 9 Squadron crewed GR.Mk 4A was shot down by a US Army Patriot surface-to-air-missile, killing the two crew. The aircraft was returning to Al Salem after completing a mission over Iraq. The day after the accident a Patriot battery was attacked by an AGM-88 HARM fired by a F-16CJ approximately 30 miles (48km) south of Al-Najaf, damaging the radar of the battery. 'Blue-on-blue' engagements continued throughout the conflict, with a US Navy F/A-18 Hornet almost certainly having been shot down by another Patriot missile on 2 April.

As well as the combat debut of the MBDA Storm Shadow, Operation Telic also saw the first combat use of the AGM-65G Maverick by the RAF and the deployment of a low-collateral damage bomb, a Paveway II practice round with concrete replacing the explosive, the weapon relying on kinetic energy for its destructive power. The Maverick was used by

Operation Telic – aircraft deployed

The air component of Operation Telic peaked at around 100 aircraft, plus 77 helicopters of the Joint Helicopter Command. Numbers below reflect the total on station and not necessarily the total number of aircraft involved.

Aircraft	No.	Notes
BAe 125 CC.Mk 3	4	No. 32(TR) Sqn deployed aircraft to Al Udeid, Qatar and Prince Sultan AB, Saudi Arabia
Canberra PR.Mk 9	2	No. 39(No.1 PRU) Sqn aircraft based at Azraq, Jordan
Chinook HC.Mk 2	20	The SF Flight of No. 7 Sqn was deployed to Azraq, Jordan, while No. 18 Sqn was based initially at Ali al Salem, Kuwait (later Iraq) and on HMS *Ark Royal*
Gazelle AH.1	16	Based at Ali al Salem, Kuwait (3 Regt AAC, later to Iraq) and HMS *Ocean* (No. 847 Sqn)
Harrier GR.Mk 7	18	Azraq, Jordan (No. 3 Sqn) and Ahmed al Jaber, Kuwait (No. 4 Sqn)
Hercules C.Mk 4	4	LTW, based in the UAE
Jaguar GR.Mk 3	4	Incirlik, Turkey (No. 6 Sqn, possibly other units and personnel)
Lynx HAS.3SGM/HMA.8	8	HMS *Cardiff*, HMS *Chatham*, HMS *Edinburgh*, HMS *Liverpool*, HMS *Marlborough* HMS *Richmond* and HMS *York* embarked No. 815 Sqn Lynxes
Lynx AH.Mk 7	18	Ali Al Salem, Kuwait (3 Regt AAC, later into Iraq) and on HMS *Ocean* (No. 847 Sqn)
Merlin HM.Mk 1	4	No. 814 Sqn on RFA *Fort Victoria*
Nimrod R.Mk 1	1	A single No. 51 Squadron aircraft at Prince Sultan AB, Saudi Arabia
Nimrod MR.Mk 2	6	Based at Prince Sultan AB, Saudi Arabia, and Seeb, Oman. Aircraft and crews drawn from Nos 120, 201 and 206 Squadrons
Puma HC.Mk 1	7	No. 33 Sqn started at Ali al Salem, Kuwait, but moved on into Iraq
Sea King HC.4	10	HMS *Ocean* (No. 845 Sqn)
Sea King HAS.Mk 6	6	RFA *Argus* and RFA *Rosalie* embarked elements of No. 820 Sqn
Sea King AEW.Mk 7	4	HMS *Ark Royal* embarked a flight of No. 849 Sqn. Two tragically collided.
Sentry AEW.Mk 1	4	Prince Sultan AB, Saudi Arabia played host to Nos 8/23 Sqn E-3Ds and crews
Tornado F.Mk 3	14	Prince Sultan AB, Saudi Arabia (Nos 11, 25, 43, 111 Sqn)
Tornado GR.Mk 4	30	Ali Al Salem, Kuwait and Al Udeid, Qatar, from Nos II, 9, 12, 31, 617 Sqns
Tristar KC.Mk 1	4	Bahrain, Muharraq hosted No. 216 Sqn aircraft
VC 10 C.Mk 1K/K.Mk 3/K.Mk 4	9	aircraft were based at Incirlik, Turkey; Prince Sultan AB, Saudi Arabia and (for the aeromedical role) RAF Akrotiri, Cyprus

Top: A total of 42 F-15C/D Eagles was deployed to Prince Sultan AB, Saudi Arabia, as part of the 363rd AEW (67th FS aircraft, above left, operating as the 363rd EFS) and to Shaikh Isa AB in Bahrain (58th and 71st Fighter Squadrons). The 94th FS was stationed at Incirlik AB in Turkey. As in Desert Storm, coalition air superiority was soon won, being declared over all Iraq on 6 April, day 18 of the conflict.

Above: The 379th Air Expeditionary Wing based at Al Udeid in Qatar contained within its structure three F-15E Strike Eagle squadrons (the 333rd, 334th and 336th Fighter Squadrons) of the Seymour Johnson AFB, North Carolina-based 4th Fighter Wing. A total of 48 examples of the 'Beagle' was deployed for the conflict.

Left: Two 131st FS/104th FW A-10s on their way to the Gulf – the nearest wearing 11 September 'Lets Roll' artwork. The O/A-10A Thunderbolt II played a key role in providing close air support to coalition troops in the field. They were based at Ahmed al Jaber in Kuwait as part of the 332nd AEW (above left and below) and with the 410th AEW at Azraq in Jordan, but used the Tallil airfield Forward Arming and Refuelling Point in Iraq to decrease response times.

Above: A HARM-toting F-16CJ of the 157th Fighter Squadron of the South Carolina Air National Guard prepares to leave Al Udeid in Qatar on a mission in the dark. A total of 408 AGM-88 HARMs was fired by all types of aircraft during Iraqi Freedom.

Above: Four airbases in the Gulf played host to US Air Force F-16s, with a total of 60 F-16Cs and 71 F-16CJs being deployed to the region. These consisted of Ahmed Al Jaber in Kuwait; Al Udeid in Qatar; Azraq in Jordan and Prince Sultan AB, Saudi Arabia. At the latter base was the 363rd AEW, which included elements of the 'WW'-coded 35th Fighter Wing.

Right: The usual loadout for the defence suppression 'Viper' was two HARMS, three AMRAAMs and a Sidewinder. Just visible under the starboard front fuselage is the ASQ-213 HARM Targeting System.

Top: The 52nd Fighter Wing at Spangdahlem AB in Germany deployed F-16CJs of both the 22nd (red tail band) and 23rd FS (blue) to serve as part of the 379th AEW at Al Udeid in Qatar.

Right: Ground-attack configured F-16Cs of the 107th FS ('MI' tailcode) and 175th FS (Wolf's head on tail) are escorted by two F-16CJs of the 363rd EFS over northern Iraq. Clutching a LANTIRN II pod under the intake to allow precision drops of the GBU-12s carried, they have an ALQ-131 ECM pod under the fuselage and are armed with AIM-120Cs and AIM-9s for self defence.

Below right: A pair of F-16CJs of the 363rd Expeditionary Fighter Squadron sits outside at Prince Sultan Air Base in Saudi Arabia. The aircraft are from the 77th FS from Shaw AFB, South Carolina.

Below: OA-10A 78-0634 of the 190th FS/Idaho Air National Guard (392nd AEW) carries a mixture of Mavericks, 'dumb' bombs and a rocket pod containing marker rockets that release smoke on impact. There is no physical difference between the OA- and A-10A.

Above: Already wearing seven Operation Allied Freedom mission markings, an Operation Iraqi Freedom mission is added to B-2A Spirit of Arizona's wheel door. Iraq is the B-2A's third conflict.

Above: A total of 51 bombers flew 505 sorties over Iraq. Four B-2A Spirits from the 393rd Bomb Squadron were deployed to Diego Garcia, BIOT (left and top) as part of the 40th AEW, considerably shortening the 34-hour round trips to Iraq made by others flying from Whiteman AFB, Missouri. The conflict saw the first combat use of Mk 82 bombs by a B-2 and was also the first time all three US heavy bombers (B-1B, B-2A and B-52H) were involved in a single combat package.

Below and left: Thumrait in Oman hosted eleven B-1Bs with the 405th AEW, drawn from the 34th and 37th BSs. The B-1B dropped a large percentage of the weapons dropped in the conflict, and as a direct consequence of its usefulness over Iraq further retirements of the type to Davis-Monthan AFB, Arizona, have been postponed.

Above: Air-to-air refuelling was vital for the US bombers during Iraqi Freedom. This JDAM-carrying BUFF takes on fuel from a 931st ARG KC-135 while on its way to Iraq. A total of 28 B-52Hs was deployed during the conflict, operating from two forward locations.

Above right: B-52H-135-BW 60-0003 (of the 93rd BS 'Indian Outlaws') was assigned to the 40th EBS and was adorned with patriotic nose art and the shield of the Bossier City police department. At least 25 missions are recorded as axe markings under the '00' of the serial.

Top: The 23rd BS commander's B-52H (60-0023) gathers speed to depart RAF Fairford's runway, a clutch of JDAMs on the underwing pylons. A total of 14 BUFFs deployed to the base in Gloucestershire as the 457th AEG, arriving from 3 March 2003.

Harrier GR.Mk 7s extensively against armoured personnel carriers, artillery and surface-to-surface-missiles launchers. Harrier pilots also learned to use the infra-red version of the Maverick's seeker to find and track targets, the system reportedly having a greater resolution than the TIALD pod. RAF Harriers also occasionally used the Enhanced Paveway II, such as when destroying the Basra Ba'ath headquarters.

In addition to the aircraft-carrier HMS *Ark Royal*, the helicopter-carrier HMS *Ocean* was deployed to the area, both carrying assets of the Joint Helicopter Command (the 'Ark' leaving its Sea Harriers behind at RNAS Yeovilton, Somerset). Royal Fleet Auxiliaries *Argus* and *Rosalie* embarked Sea King HAS.Mk 6s of No. 820 Squadron, while four Merlin HM.Mk 1s of No. 814 Squadron made the type's conflict debut onboard RFA *Fort Victoria*.

Operation Falconer

The Australian armed forces were deployed using the code-name Operation Falconer. The RAAF deployed 14 F/A-18As of No. 75 Squadron to Al Udeid in Qatar, where the majority of the RAAF aircraft were based. Australian Hornets were used during the first 24 hours of the Operation Iraqi Freedom to protect tankers, AWACSs and other high-value assets but were also involved in strikes against Iraqi positions later in the conflict. The other Australian aircraft involved were three airlifters (C-130H and Js of Nos 36 and 37 Squadrons) and a pair of No. 92 Wing P-3Cs. Onboard LPA 51 HMAS *Kanimbla* was a Royal Australian Navy HS817 Sea King Mk 50, while based alongside their No. 7 Squadron Special Forces

Right centre: Two B-52Hs taxi down RAF Fairford's runways on 22 March devoid of external ordnance. B-52Hs used Litening pods for the first time during OIF.

Right: The 40th AEW controlled units on Diego Garcia, including the B-2As from the 393rd BS, B-52Hs from the 20th and 40th Bomb Squadrons, and tanker support in the form of KC-135Rs from the 462nd AEG's 28th ARS.

Coalition losses

Coalition losses in Operation Iraqi Freedom were on the whole the result of operational accidents and the harsh flying conditions that can occur in the region due to sand storms rather than to Iraqi defences. *Operation Iraqi Freedom – By The Numbers* produced by the US Air Force's Assessment and Analysis Division on 30 April 2003 listed four Class A, five Class B and 16 Class C aircraft mishaps excluding enemy action. Total losses to all causes account for 17 aircraft. The US Air Force is known to have lost one A-10A, an F-15E and one MH-53M, plus at least three RQ-1 Predators. It was reported that two stripped Predators were flown into Baghdad as 'bait' to test the air defences early in the campaign, eventually running out of fuel and crashing. Two AH-1Ws, one UH-1N, one CH-46E and one AV-8B were lost by the US Marine Corps, and one example each of the F/A-18C, F-14A, S-3B and HH-60H by the US Navy. The Hornet was shot down on 2 April near Kabala, southern Iraq, with US Central Command confirming on 14 April that

the most likely cause was an American Patriot. Several RQ-2A Pioneers were also destroyed. The US Army lost at least two AH-64s. In addition, five BQM-34 Firebees were expended laying chaff over Baghdad. RAF losses were confined to Tornado GR.Mk 4A ZD710/D flown by a No. 9 Squadron crew, which was shot down by a Patriot missile while returning from a mission on 23 March, tragically killing both the pilot and navigator. A pair of No. 849 Squadron Sea King

Three Thunderbolts were damaged on 8 April, with one of them being lost. OA-10A 80-0258 was reportedly hit by AAA, although the damage looks more consistent with a IR-guided MANPADS hit.

AEW.Mk 7s (XV650 and XV704) collided on 22 March, tragically killing all seven on board. At least two British Army Phoenix unmanned air vehicles were also shot down.

Flight RAF counterparts at Azraq in Jordan were CH-47Ds of the 5th Aviation Regiment, Australian Army Air Corps. Two other RAAF aircraft in the Gulf were believed to have been Boeing 707 tankers.

American forces

As was to be expected, the US military contribution to Operation Iraqi Freedom was dominant, accounting for over 92 percent of all the aircraft deployed. Excluding US Army helicopters embedded within Army formations, a total of 1,801 US aircraft was in the Gulf or in support of Iraqi Freedom outside the theatre.

Using the Air Expeditionary Wing (AEW) and Air Expeditionary Group (AEG) concepts, the US Air Force deployed packages to form the 320th AEG (Seeb, Oman), 322nd AEW (Ahmed Al Jaber, Kuwait), 363rd AEW (Prince Sultan Air Base, Saudi Arabia), 379th AEW (Al Udeid, Qatar), 386th AEW (Ali Al Salim, Kuwait) and 405th AEW (Thumrait, Oman). In addition, Shaikh Isa Air Base in Bahrain hosted an AEW, believed to be the 358th. The F-16C-equipped 120th EFS, operated from an offically unidentified base in Jordan that housed the 410th AEW, believed to be Azraq. AEWs and AEGs controlled the tactical fighters, tactical transports, command and control, tankers, rescue, intelligence, surveillance and reconnaissance assets required to prosecute the conflict. The US Air Force operated from 36 locations, most in countries that do not want to be identified.

Heavy bombers were located with the 40th AEW (B-2As and B-52Hs at Diego Garcia, British Indian Ocean Territories – BIOT), 405th AEW (B-1Bs at Thumrait, Oman) and the 457th

AEG at RAF Fairford, Gloucestershire (B-52Hs). Operation Iraqi Freedom was the first time the US heavy bomber triumvirate was used within the same combat package. At its height, the bombing effort involved six or seven sorties a day, with the aircraft loitering for up to eight hours over the battlefield, dropping weapons where requested. During the same period, they were also flying sorties over Afghanistan.

The B-1B, while flying only around 2 percent of all sorties, was responsible for dropping about one-third of all the JDAMs (Joint Direct Attack Munition) used – around 2,100 weapons – carrying 24 per mission. The B-1B's performance over Iraq was such that the current drawdown of the fleet to 60 aircraft has been halted. For the first time the aircraft used a Moving Target Indicator – an item reportedly cancelled from the Block D upgrade.

For the first time in combat the B-52H used a Litening II pod to allow it to drop laser-guided bombs. Early in April a pair of GBU-12s was dropped onto a radar and command complex on a northern Iraqi airfield. Litening II had been integrated onto the B-52H by mid-February, with work on system compatibility having starting in October 2002. It also allowed the aircraft's crew to visually confirm the nature of the target in the crosshairs from 35,000 ft (10668 m), a capability the B-52H did not previously have. A total of 12 B-52Hs will be modified with the capability to operate the system, but only two were available during the conflict. In a reversal of the overall trend, B-2As used Mk 82 'dumb' bombs in anger for the first time.

In order to support flights to the Gulf, tanker aircraft were deployed to several airfields

including Lajes on the Azores; Diego Garcia, BIOT; Bourgas, Bulgaria (409th AEG with KC-10As); Souda Bay, Crete (398th AEW) and RAF Akrotiri, Cyprus (401st AEW).

Specialised rescue forces were deployed to form the largest Joint Search and Rescue Center (JSRC) in history. The JSRC helped under take 55 missions, resulting in the rescue of 73 personnel during the operation.

Some unusual types were listed as having been deployed for Operation Iraqi Freedom, including a single CASA CN-235 (presumably 96-6049 associated with Special Operations Command) and a Pilatus PC-6. The Big Crow NKC-135E, operated by the 412th Test Wing, was based at Souda Bay on Crete from 17 March. The aircraft is used as an electronic warfare laboratory, able to recreate specific electronic threats and would have been of use to train units against specific Iraqi air defence systems, although the exact purpose of its contribution has not been disclosed.

The US Navy's new capabilites

A total of six carriers was involved in Operation Iraqi Freedom, two based in the Mediterranean and four in the Gulf. The USS *Harry S. Truman* with the embarked Carrier Air Wing 3 (CVW-3) and the USS *Theodore Roosevelt* with CVW-8 could not take part in strikes on Iraq for the first few days of the conflict because they did not have permission to overfly any of the countries between them and Iraq. On 22 March (day three) Turkey finally granted overflight permission and the two carriers joined the original three in the Gulf actively attacking targets. Without the consent

Some of the most colourful aircraft involved in Operation Iraqi Freedom were the vertical replenishment Sea Knights. Examples were based on – among other warships – the USS Camden (right) and the USS Rainier (above, with Det 5 HC-11).

Marine air power in the Gulf

In support of the 55,000 US Marines deployed initially in Kuwait as part of the 1st Marine Expeditionary Force (1st MEF), around 180 US Marine Corps aircraft were deployed on five amphibious ships. Unusually, to ease employment and maintenance of its air assets, the 3rd Marine Air Wing centralised some of its aircraft types, with all the AV-8Bs being grouped together instead of assigned to the medium helicopter squadrons as per standard. As a rough guide, AV-8Bs were focused on mobile targets, while Marine F/A-18s attacked fixed targets or those obscured by smoke or weather using the Nite Hawk designation pod. Two-seat F/A-18Ds were also used as airborne forward air controllers, marking targets with fire from rockets. Marine aviation undertook 4,948 sorties during Iraqi Freedom of which around 3,800 were by the fast jets and helicopter gunships.

Above: Marine Corp Harriers provided close air support to the troops on the ground, with 70 deployed to the theatre.

Right: The AH-1W crews developed new tactics in theatre to allow them to fire weapons while on the move, significantly reducing their chances of being hit by small arms fire.

Left: CH-46E BuNo 156418 of HMM-263 is seen on USS Nassau in early April – one of 67 of these Marine transports deployed for OIF.

to base tankers in Turkey, the volume of strikes from the Mediterranean-based carriers was down on what it could have been.

The USS *Abraham Lincoln* and CVW-14, USS *Constellation* (CVW-2) and USS *Kitty Hawk* (CVW-5) were all on station in the Arabian Gulf when hostilities commenced. They were able to engage targets from the start. The final American carrier involved was the USS *Nimitz* which had undertaken training in the eastern Pacific in early February before setting out for the Gulf. Onboard was CVW-11, the first air

An HC-5 MH-60S carries a JDAM between USS Abraham Lincoln and USS Nimitz in early April. Operation Iraqi Freedom was the type's first operational deployment.

wing to deploy without the F-14 Tomcat, having an all F/A-18 Hornet strike force that included the F/A-18F Super Hornets of VFA-41 'Black Aces' undertaking the variant's first operational cruise. Having missed out on the opening stages of the air campaign, CVW-11's VFA-14 and VFA-41 cross-decked F/A-18Es and F/A-18Fs to the USS *Abraham Lincoln* to fly alongside CVW-14's aircraft from 30 March. One system that limited the Super Hornet to attacking fixed targets was the Nite Hawk designation pod, which has some difficulties attacking mobile targets. The pod's replacement, the ATFLIR (Advanced Targeting FLIR), has yet to enter service in large numbers and the three examples deployed were found not

reliable enough for daily operational use. The F/A-18F also employed the SHARP reconnaissance pod, and was used in the tanking role for the first time during Iraqi Freedom. Overall, the Hornet was the type deployed in the greatest numbers for the conflict, with 250 on hand.

F-14D Tomcats were modified to be able to employ JDAMs in conjunction with their LANTIRN pods while onboard the carriers, the variant making its debut with this weapon over Iraq. Three squadrons of F-14Ds were located in the Persian Gulf and the type's long range

HM-14 Det II MH-53E Sea Dragon 'BJ/552' lands at Umm Qasr in late March. Tasked with mine clearance using towed sledges, the squadron maintained a detachment in Bahrain as well as afloat.

Above: OIF was the first time the F-14D Tomcat dropped a JDAM in anger, the modification being undertaken onboard the aircraft-carriers on station in the Gulf. All three operational marques of the Tomcat participated in OIF, these examples being F-14Bs of VF-32, deployed onboard the USS Harry S. Truman.

Above left: A US Marine Corps F/A-18D takes on fuel from a KC-10A. Land-based, the two-seat Hornet was used in the FAC-A (Forward Air Control – Airborne) role.

Left: A VS-22 S-3B refuels a VFA-105 F/A-18C – both units based on the USS Harry S. Truman. The S-3s topped up naval aircraft's tanks going to and from Iraq.

and the advantages of the LANTIRN over the Hornet's Nite Hawk pod in attacking mobile targets made the upgrade a viable endeavour.

As well as undertaking the suppression and destruction of enemy air defences roles, for the first time the EA-6B Prowler was also used in psychological operations. Over the last 12 years Iraqi radar systems operators had learnt the dangers of broadcasting any signals and were reluctant to turn on most radar systems. Thus most Iraqi surface-to-air-missiles were fired blind.

US Marine Corps airpower was also deployed to the Gulf on the helicopter assault ships USS *Nassau* (LHA 4) and the USS *Bonhomme Richard* (LHA 6) and land bases.

UAV and space assets

The single Northrop Grumman RQ-4A Global Hawk was based in the UAE for the duration of the conflict and used for the first time to provide real-time targeting data to strike aircraft. In some instances this diminished the target-to-shooter time to around 15 minutes. It was instrumental in locating the Medina Division of the Republican Guard, allowing strike planners to direct B-52Hs and B-2As to attack. The Global Hawk remained overhead to make a post-strike assessment of the damage inflicted on the Division.

Operation Iraqi Freedom Weapons

The MBDA Storm Shadow stand-off missile (left) was used in combat for the first time on the evening of 21 March 2003. A pair of missiles was carried by a Tornado GR.Mk 4 of No. 617 Squadron piloted by the squadron OC, Wg Cdr Dave Robertson, with Sqn Ldr Andy Myers in the navigator's seat. The aircraft was one of a pair assigned targets in and around the Baghdad area, the mission starting at 20:00 local (17:00 GMT). En route the second aircraft of the pair was fired at by a surface-to-air-missile and had to jettison its tanks while taking avoiding action to ensure that the missile missed. Around Naserihya the pair of aircraft came under fire from anti-aircraft artillery, which again missed. During the start of its attack run, the lead pair was targeted by a SA-2 'Guideline' surface-to-air-missile, but using on board counter measures they took avoiding action and continued their attack run and launched the Storm Shadows, which hit the intended target. Storm Shadow has not officially been accepted into RAF service and they were used by No. 617 Squadron – the unit designated to introduce the missile into service – under an initial operational capability.

During the 1991 Gulf War the vast majority of the munitions used were unguided – for Operation Iraqi Freedom this had decreased significantly, with only 38 percent of the ordnance dropped being unguided. Since the first Gulf War all of the carrier air wing's jet combat aircraft can now employ some form of guided munitions, as have all three of the US Air Force's heavy bombers and the majority of its tactical fighters. This change is best summed up by the change in planner's questions from how many aircraft do we need to destroy a target, to how many targets can we destroy per aircraft? A total of 19,948 guided and 9,251 unguided munitions was expended in Iraqi Freedom. Several US weapons saw their debut over Iraq. These were the CBU-105 Sensor Fuzed Weapon (the CBU-97/B fitted with a Wind Corrected Munitions Dispenser kit), and CBU-107 cluster bombs and the AGM-86D Conventional Air-Launched Cruise Missile (CALCM) hard target penetrator.

Well over one quarter of all the guided munitions used were members of the JDAM family, with the GBU-31 (5,086 used) accounting for the vast majority of them.

A total of 253 AGM-154A JSOWs was expended in Operation Iraqi Freedom, most launched from carrier-based F/A-18 Hornets.

A VMFA-115 Hornet drops a JDAM on 'ops' in late March. JDAM is a strap-on kit designed to produce a guided munition from a 'dumb bomb'. As a GPS-guided munition it can still be targeted through clouds.

A General Atomics MQ-1 Predator attempted to use a Stinger air-to-air missile to intercept an Iraqi Air Force MiG-25 that had attacked a second Predator, but the 'Foxbat' survived the encounter. More success was experienced with the Hellfire air-to-surface missile with, for example, a ZSU-23-4 mobile anti-aircraft-artillery piece being destroyed by an AGM-114K-firing MQ-1 outside of al Amarah on 22 March. Having started the war located in Jordan and Kuwait, the seven RQ-1 and nine MQ-1 Predators deployed later relocated to the captured airfields of H2 and H3.

Other UAVs operated in Operation Iraqi Freedom include two squadrons (VMU-1 and VMU-2) of AAI/IAI RQ-2A Pioneers working with the 1st Marine Division, each unit being issued with eight. During the first two nights over Baghdad the US Navy used ex-US Air Force BQM-34 Firebee UAVs to lay chaff to cover the approach of Tomahawk Land Attack Missiles. Three examples were air-launched from a DC-130 (presumably a former AVTEL-operated DC-130A) while the final pair was ground launched.

Commands for the control of the Global Hawk were routed via the Milstar 5 military communications satellite from the UAE. US space assets, including the Navstars on which the Global Positioning System – and thus GPS-guided munitions – keyed from, had a bigger impact on this conflict than any previous war. At least three KH-11 visible and infrared spectrum, and two or three Lacrosse radar imaging satellites were reported to be used for surveillance, making a total of 12 passes each day.

Helicopters

The landscape of Iraq is flat, giving helicopter pilots, used to 'nape of the Earth' low-level tactical flying, no way of masking their movements. US Army AH-64A and Ds were used as an independent manoeuvre force but proved to be vulnerable to small arms fire, with a large percentage of them damaged. The tactics of hovering to fire also increased the susceptibility of the Apaches to small arms fire, while the US Marine Corps AH-1W SuperCobras, that fired weapons on the move, suffered less. A tendency in the early days of the conflict for US Army and US Marine formations to call in attack helicopters rather than fixed-wing air support when faced with persistent resistance also added to the number of hits suffered by coalition helicopters.

David Willis

The Iraqi Air Defence's War

The Iraqi Air Force (Al Quwat Al Jawwiya Al Iraqiya) never recovered its strength after the 1991 Gulf War, the embargo hitting its ability to repair the damage inflicted or make good its losses. The poor performance of the service in the first war against the west resulted in the separation of the missile and anti-aircraft artillery component to command status, out of Air Force control. It was the Iraqi Air Defence Command that was to prove to be the more dangerous arm during the 1990s and also during Iraqi Freedom, damaging several aircraft and downing some UAVs. The Air Defence Command had had plenty of practice, regularly firing on allied aircraft in the 'No-fly' zones. In contrast, the Iraqi Air Force spent the war harbouring its assets with very few sorties being flown. Unlike the first conflict when Iraqi aircraft fell to allied fighters, it seems none was lost to this cause. Coalition attacks on Iraqi airbases destroyed most aircraft in the open, leaving the Air Force with aircraft secreted away off bases. The lack of meaningful opposition from the Air Force has led to some speculation of its tacit collusion with the coalition, but, even if this is true, it did not save it from being included in the organisations due to be disbanded post-conflict.

Above: Members of the Australian SAS patrol next to a well camouflaged MiG-25PU hidden among trees. After the 1991 conflict the Iraqi Air Force learned that camouflage gave a better chance of survival than putting its aircraft into HASs.

Right: A wrecked Iraqi Zlin Z326 sits forlornly as an earth-mover prepares an airfield for use by the allies.

Above: One of the most intriguing pictures to come out of Iraq was this photo of a destroyed (most likely museum-piece) Iraqi Mil Mi-4 at Tallil. Visible in the distance is a wrecked F-4 Phantom in Iraqi Air Force markings – a type not associated with that air arm. It is likely to be a captured Iranian example, but it is open to speculation as to whether the type ever flew in Iraqi markings. If the Phantom was preserved it would have been a brave base commander who would have displayed the aircraft in the markings of Iraq's bitter enemy, given the feelings towards Iran of the Iraqi leadership.

Above: With the inscription 'Give 'em hell' on its nose, the VFA-105 boss-bird awaits launch from the USS Harry S. Truman during daylight. A VFA-15 F/A-18C does the same at night on USS Theodore Roosevelt (below).

Above: VFA-15 F/A-18Es based on the USS Abraham Lincoln were used in both the strike role and as tankers, carrying four underwing fuel tanks under the wings (below) and a buddy pack under the fuselage.

A total of 15 U-2Ss undertook missions from Al Dhafra in the UAE, RAF Akrotiri in Cyprus and from Prince Sultan AB, Saudi Arabia, contributing to the 452 intelligence, surveillance and reconnaissance missions flown by the US Air Force in Operation Iraqi Freedom. The conflict was the first time six U-2s had been assigned missions on the same Air Tasking Order.

Above: EP-3E Aries II intelligence, surveillance and reconnaissance aircraft based with VQ-1 at Muharraq in Bahrain and at Souda Bay, Crete, in the Mediterranean with VQ-2 (with which '25' is in service) were used to help update the electronic order of battle of the Iraqi armed forces prior and during the conflict.

Above: The majority of the Sentries in the US and UK fleets were deployed to Prince Sultan AB in Saudi Arabia, the US aircraft forming part of the 363rd AEW. A total of 19 US and four British aircraft served in the conflict, with the USAF E-3B/Cs and E-8Cs undertaking 432 sorties and the RAF E-3Ds providing another 112.

Below: NKC-135E Big Crow test-bed 55-3132 deployed from Edwards AFB, California, via RAF Mildenhall to Souda Bay, Crete, as part of the based 398th AEW. The aircraft is used as an electronic warfare test bed, able to simulate a diverse range of electronic threats. Its exact role in OIF has not yet been revealed.

Below: Psyops accounted for 58 Commando Solo media broadcast missions flown by EC-130E Rivet Riders. The EC-130H Compass Call communications jammers (left) deployed as the 41st EECS to Al Dhafra in the UAE as part of the 60th AEW, alongside the Rivet Riders. Operation Iraqi Freedom was the first time the Compass Call aircraft were used in a psyop role.

Above: VAQ-135's 'boss' Prowler circles the USS Nimitz. The EA-6B provided the majority of the coalitions jamming capability and broadened its use to include psychological operations (psyops). 35 were deployed.

Below: HC-130P 64-14855 of the 39th RQW refuels one of the few desert-schemed MH-60G Pave Hawks, 90-26224 from the 39th RQS. The conflict saw the assembly of a large rescue organisation in the theatre.

Above: Elements of the 15th RS operated MQ-1 and RQ-1 Predator UAVs from Azraq in Jordan as part of the 410th AEW and Ali al Salim in Kuwait (386th AEW). 99-3059 is an RQ-1K, lacking the hardpoints of the armed MQ-1.

Above: A total of 16 HH-60G Pave Hawks was deployed to the theatre, being based at Ali al Salim in Kuwait and Azraq in Jordan. When the advance into Iraq started some were forward-based at the Tallil FARP. To these rescue helicopters could be added the HH-60Hs based onboard the aircraft-carriers, as well as US Army and US Air Force special operations MH-60s.

No. 847 Squadron initially started the conflict on HMS Ocean but moved to land bases as the conflict developed. Its Lynx AH.Mk 7s (below right) were equipped with TOW launchers. The unit's Gazelle AH.Mk 1s gained infra-red suppression gear on the engine exhausts to reduce the dangers of the significant Iraqi short-range air defence missile threat to the type.

Above: An element of No. 18 Squadron operated Chinook HC.Mk 2s from the deck of HMS Ark Royal, while HMS Ocean accommodated 10 Sea King HC.Mk 4s of No. 845 Squadron. The furthest Sea King along seen on Ocean's deck displays a temporary wash of white 'arctic' camouflage – the chances of coming across snow in the Gulf during March and April were not significant!

Above: US Air Force C-17A Globemaster IIIs were used to drop troops in northern Iraq, flying from Rhein-Main AB in Germany. The RAF's four examples (above left) were kept busy ferrying supplies to the Gulf.

Left: No. 32(TR) Squadron deployed four BAe 125 CC.Mk 3s to Al Udeid in Qatar and Prince Sultan AB, Saudi Arabia, for communications duties. It is next to an LTW Hercules C.Mk 4, a No. 99 Sqn C-17A and a C-5.

Below: Tallil airfield was soon (unofficially) renamed. Significant use was made by coalition aircraft of captured air bases.

Above: The bulk of the in-theatre fixed wing tactical transport duties were performed by the C-130 Hercules, with 124 deployed. This C-130E of the 37th AS/86th AW is seen departing Bashur in northern Iraq, while the example seen through NVGs (above far left) is delivering aid to Baghdad International in mid-April. In addition 22 US Marine Corps KC-130s (including KC-130T BuNo 163311 of VMGR-452 at Tallil FARP, left) were based at Ali al Salim and Camp Coyote in Kuwait, and at Shaikh Isa AB in Bahrain. Strategic airlift into the theatre was undertaken by a mix of US heavy-lifters (9th AS C-5, below) and civilian charter, with the Antonov An-124 (below left) being heavily utilised.

Above: A total of 149 KC-135s was deployed for Operation Iraqi Freedom. These examples formed a squadron within the 401st AEW based at RAF Akrotiri on Cyprus. The base hosted examples from both the 319th ARW (left, refuelling a VFA-105 F/A-18C) and 351st ARS.

Above right: The C-40B of General Tommy Franks, the Commander-in-Chief of Operation Iraqi Freedom, departs an airfield in the theatre – probably Al Udeid in Qatar – in front of a waiting KC-135R. The aircraft is officially assigned to the 1st AS, part of the 89th AW.

Above: A KC-10A refuels a B-1B of the 37th BS/28th BW while a pair of Tornado F.Mk 3s fly alongside. 33 KC-10s were based at Al Dhafra in the UAE (380th AEW), Al Udeid in Qatar (379th AEW) and at Bourgas in Bulgaria (409th AEG).

Right: Four VC10s of Nos 10 and 101 Squadrons line up next to KC-10As, probably at Prince Sultan AB in Saudi Arabia. Sand storms are a constant hazard in the region.

Below: Tristar KC.Mk 1 ZD951 of No. 216 Squadron tops up the tanks of US Navy EA-6B Prowler '523'. The 12 RAF tankers deployed to the region undertook 359 sorties during the conflict, proving to be particularly useful to US Navy and US Marine Corps aircraft.

Aerostar's MiG-21s

Romanian upgrade updates

Romania's Aerostar is an aerospace and defence concern that has established a solid international reputation for its engineering expertise. At its home in Bacău, the company is the overhaul and maintenance specialist for the jet fleet of the Romanian air force (the Fortele Áeriene Române, or FAR). It is also an aircraft builder in its own right, producing over 1,800 Yak-52 trainers – as the Iak-52 – and still evolving that basic design with new and improved versions. Aerostar has proven its ability to develop, produce and integrate advanced systems upgrades for Eastern (*i.e.* Russian) combat aircraft. This reputation is founded upon the Lancer MiG-21 programme for the FAR and the related MiG-29 Sniper technology demonstrator. Still unequalled in its scope, the Lancer programme has seen the upgrade of 96 MiG-21M/MFs and 14 MiG-21UMs to an advanced common standard based around a radar, avionics and weapons fit developed by Aerostar and Elbit. In April 2003 Aerostar marked the 50th anniversary of its foundation with the handover of the last upgraded Lancer for the FAR, but it was accompanied by the roll-out of the first aircraft in a new upgrade programme for the Croatian air force (Hrvatsko Ratno Zrakoplovstvo, HRZ).

The Croatian MiG-21 programme is the first such work to be undertaken by Aerostar for a foreign customer, although not as far-reaching as the Romanian Lancer upgrade. Previous reports have associated Croatia with MiG's own MiG-21-93 upgrade project and the Aerostar Lancer, but the HRZ did not have the funding

available to support such an ambitious plan. Instead, Aerostar is providing an upgrade and overhaul package that is essentially a 'return-to-service' effort with a modest new avionics fit. By 2002 the serviceability of Croatia's MiG-21 force had decayed to such an extent that its two squadrons of (approximately) 20 aircraft were essentially non-functional. Of paramount importance to the HRZ was to get its front-line fighter force back into the air and Aerostar's ability to deliver a fast-paced programme that would return an effective core of aircraft to operations was key in Croatia's decision to work with Aerostar.

In a new departure for the Romanian firm, Aerostar is providing pilot and technical training to the HRZ, along with in-service support, spares and logistics. Even more significantly, Aerostar has acquired four 'new' aircraft for Croatia. Mystery surrounds the provenance of Croatia's MiG-21bis 'Fishbed-Ls'. Only three former Yugoslav air force aircraft fell into Croatian hands during the early 1990s, but by the middle of the decade the HRZ had at least 20 in service, with as many as 30 in its inventory. The origins of these MiGs remain a closely-guarded secret in Croatia – some

This view shows the MiG-21 overhaul and upgrade line at Aerostar's Bacău facility. Every one of Croatia's mysterious MiG-21bis aircraft was free of any identifying marks, save for a scribbled-on tactical code.

sources have suggested they are former East German aircraft but this is not the case. Adding to the mystery, Aerostar has now procured four additional MiG-21UMs for the HRZ as part of the first tranche of 12 aircraft upgrades. Aerostar will say only that these aircraft were sourced in 'an Eastern European country'.

Aerostar is also involved in the transfer of former-FAR MiG-21s to other third-party customers and, during April 2003, two MiG-21UMs were being prepared for transfer to a new user. By establishing such an independent chain of supply – using either Romanian stocks or aircraft sourced from elsewhere – Aerostar now has the ability to deliver a total 'capability package' to MiG-21 users around the world. This is underlined by the Croatian deal. Such an arrangement cuts out the need to deal with sometimes unreliable suppliers in Russia, but equally runs the risk of incurring the wrath of MiG which has already directed much venom against Aerostar and every other company offering independent overhauls or upgrades for MiG aircraft.

In total, the Croatian upgrade deal covers eight MiG-21bis and four MiG-21UM aircraft (which become MiG-21UMD after the upgrade). The contract was signed in September 2002 and all eight single-seat aircraft

Left: Prior to its formal handover, Croatian MiG-21UMD '86' was painted with temporary Romanian roundels to cover the time spent test flying after its rework.

Below: The (official) first upgraded MiG-21UMD for Croatia was presented alongside the last Lancer for Romania during Aerostar's 50th anniversary ceremonies.

Operational MiG-21 Lancer As (ground-attack), Lancer Bs (conversion training, seen above) and Lancer Cs (air defence, seen right) are flown by the FAR's fast-jet training unit, the 95th Fighter Aviation Group, which is co-located at Bacău alongside the Aerostar plant. The FAR has four operational groups equipped with the Lancer, including a rapid-reaction unit on standby for international deployments.

were delivered by road soon afterwards. Once in the Bacau plant, the MiGs were stripped and disassembled for Aerostar's standard schedule of airframe and engine inspection and overhaul. Aerostar has the facilities to deal with all of the aircraft's mechanical, hydraulic and electrical systems and also maintains a full overhaul, repair and test capability for the MiG-21's Soyuz (Tumanskiy) turbojet engine. An important point to note is that Romania never operated the MiG-21bis, and that the R-25-300 engine of this variant is quite different to the R-13-300 that powers the MiG-21M/MF (and by extension the Lancer). Nevertheless, Aerostar has acquired the resources to work on the R-25 and is currently overhauling a separate batch of about 20 such engines for Egypt.

The new cockpit avionics fit for the Croatian MiGs incorporates a Honeywell VOR/ILS/DME, a Garmin GPS, a Rockwell Collins VHF/UHF radio, and a Thales IFF. The fully NATO-compatible IFF (Mark X, Mode IV) is part of Romania's national IRIS IFF system, and is produced as part of a wider joint venture with Aerostar. The transponder and control box unit for every Thales IFF set is now built in Romania and shipped to France for final assembly. Around each MiG's exterior several new antennas have been fitted, chiefly for the GPS, IFF and new radio systems but, by-and-large, the upgraded HRZ jets are indistinguishable from standard aircraft. One obvious change is the new matt-finish camouflage scheme of two-tone green and dark grey, with blue undersides. This replaces the green and brown colours seen previously on Croatia's MiG-21s.

Handover to the HRZ of the first two upgraded aircraft for the 21. ELZ took place at Pleso airbase, near Zagreb, on 14 May 2003. Photographs of this event show a MiG-21UMD in the new 'upgraded' colour scheme wearing a serial (164) that was not carried by any of the 12 aircraft present at Aerostar a few weeks earlier (MiG-21bis: 108, 110, 115, 116, 117, 120, 121, 122. MiG-21UM: 22, 26, 56, 86). The modified aircraft are being recoded prior to delivery and are being painted in full Croatian colours, including the Sahovinca crest.

The last of the upgraded MiG-21 Lancers for Romania – Lancer C '9611' – made its final test and acceptance flights in mid-April 2003, prior to its handover.

Aerostar is optimistic that a second batch of Croatian MiG-21s will follow the current aircraft under a new contract to be agreed, perhaps, later this year. While the work for the HRZ is nothing like that demanded by the Lancer, it is still welcome revenue. Aerostar is diversifying into airliner maintenance, and has a growing ultralight aircraft manufacturing business, although its military competence remains paramount. It continues to seek opportunities with MiG-21 and MiG-29 operators (although the MiG-29 is no longer the priority it once was) and a recent contract with the Bangladesh air force to return its six Aero L-39ZAs to operational service opens another new market.

The hand-over of the last Lancer upgrade, an air defence Lancer C, to the FAR on 17 April was a poignant occasion for Aerostar. The company knows it will never see a such a large and well-funded programme again, but equally the Lancer will be the FAR's only combat aircraft until at least the end of this decade and so will remain important to Aerostar. Already the first Lancer has returned to Bacău for depot-level maintenance after 800 flying hours.

While recent sales of Aerostar's Iak-52 have been to civil customers, the sale, support and upgrade of military aircraft – like this Hungarian example – continues.

Aerostar's president and general director, Grigore Filip, acknowledges these changing times saying, "while the market for Western aircraft is dynamic, the situation with Eastern types is quite different and much worse than four or five years ago. By-and-large the countries that still operate these aircraft have no money for maintenance, let alone upgrades. The aircraft themselves are becoming less and less airworthy as the operational infrastructure of pilots and technicians disappears. There is a continuous, understandable, pressure to buy new aircraft and so, as time goes by, the probabilities decrease – but there is a still a market and Croatia is an example of that."

For the future, Aerostar is looking to Romania's emerging next-generation combat requirement which should see a competition for 30 to 40 new aircraft opening in the 2008 timeframe. As the only all-round military airframer in the country, Aerostar hopes to play an important part in the supply of aerostructures and perhaps even assembly work that will come with the new fighter.

Robert Hewson

Turtmann and Buochs

Swiss AF wartime base closures

Under the Armee XXI programme, which comes into effect on 1 January 2004, the Swiss Army is reducing its forces to 119,000 'active' servicemen, with a reserve of 80,000. This compares to the 600,000 men of the Armee 61 programme and 400,000 of Armee 95. The new profile for the 21st Century has a post-Cold War posture, massively reducing the size of the country's army and restructuring its civil protection scheme. It will have a major impact on the traditional militia army, and on the Schweizer Luftwaffe (Swiss Air Force).

Since the F-5 Tiger is no longer considered to be fully capable of fulfilling its roles of air policing (Luftpolizeidienst) and air defence in a modern air war environment, the veteran Tigers will be gradually phased out. Part of Armee XXI is the disbandment of two Tiger squadrons during 2003, reducing the fleet from 85 to 54 aircraft. The 12 F-5F two-seaters will once more be used exclusively in the advanced training and jet conversion role, following the premature retirement of the Hawk T.Mk 66 (last operational flight on 13 December 2002). By 1 January 2004 the Alouette III force will have

been reduced from 60 to 35. In the meantime, the F/A-18 fleet will undergo an intensive upgrade programme (Kampfwertsteigerungsprogram) including the incorporation of the Raytheon AIM-9X, helmet-mounted visor and MIDS datalink, a process planned to be completed before 2009.

As a consequence of the reduction from five to three Tiger squadrons, it was announced on 11 July 2002 that the war base (Kriegsflugplatz) at Turtmann would close by 1 April 2004. At the same time, it was announced that the war base at Stans-Buochs would also close in line with the retirement of the Mirage IIIRS/DS fleet in December 2003. By closing Turtmann the Swiss treasury will save SFr 1.2 million in maintenance and operations each year, and SFr 9

Two F-5s taxi to the runway at Turtmann. The base's mountainous surroundings provided a challenge for pilots, even in good weather conditions.

million in planned investments. By comparison with 1994, when the Swiss Air Force operated some 350 fighters and could rely on some 12 war bases, from 1 January 2004 onwards the Schweizer Luftwaffe will have three F/A-18 and three F-5 Tiger squadrons. These will be operating from only three war bases – Payerne, Sion and Meiringen, of which only the latter has a cavern hangar structure. The remaining training airfields will be Emmen, Locarno and Dübendorf, although the latter is planned to end Tiger operations by 1 January 2005 and Hornet operations around 1 January 2007.

Goodbye Turtmann

Turtmann lies in the region of Oberwallis and is situated at an elevation of 2,060 ft (628 m). A Fliegerkompanie was based there as far back as 1929, and in 1943 a concrete runway was built. Between 1951 and 1958 the airfield was turned into a cavern-airfield (Kavernenflugplatz), when three caverns were constructed in the mountain on the south side of the airfield – two for aircraft and one for troop lodging.

From 1958 the airfield was occupied twice a year by two squadrons conducting two-week

F-5Es are seen during the last Wiederhohlungs Kurs (WK) at Turtmann. The aircraft at left carries a suitable message on its fuselage, while the badge on the nose is for Flugplatz Abteilung 3 and Fliegerkompanie 8, which maintained the base and aircraft, respectively.

For the WK period Turtmann housed Fliegergeschwader 3, a wing which would have been activated in wartime to control FISt 1 and FISt 6 (6ème Esc Av) and their assigned ground/airfield units. Here the traffic is held up while an F-5 taxis to the runway through the village.

refresher camps (Wiederhohlungs Kurses or WKs), the last such camp being WK'03. In 1967, one of the caverns was modified to operate the Mirage, and between 1986 and 1994 the Hunter operated from Turtmann. It then became an F-5 base. A regular user was the pilot recruitment school (Flieger Rekrutenschule) which used the airfield for its off-base training courses.

Turtmann was memorable for all pilots who flew from there. The airfield is located in the middle of a narrow valley, surrounded by some very striking mountains, requiring special pilot skills to counter the typical but unpredictable winds. The runway is short and narrow.

To land on runway 08 the pilots had to fly 12° off the centreline at an 8° incidence. To land on runway 26, the pilots have to approach the runway 4° off centreline, with an angle of 11°, and cannot land as a pair. At the west end of the airfield the runway is situated amid Turtmann village, with several public roads crossing the runway.

Turtmann is a typically Swiss air base, with the runway on one side of a main road and the caverns on the other. Offering a fair deal of protection, the cavern concept was originally aimed to minimise the period the aircraft were exposed before take-off and after landing, although it is now considered somewhat outdated given the capabilities of today's precision weapons. For operations the F-5s were towed outside the cavern (theoretically each

As well as cavern accommodation, Turtmann also had two hardened taxi-through shelters at each end of the runway, well camouflaged with grass. These were used by aircraft on quick-reaction alert, and for 'last-chance' weapons arming.

squadron had its own cavern) with the pilots already strapped in, only requiring the external supply of compressed air in order to be able to start engines. Military reservists would have stopped traffic and over-enthusiastic spectators from getting too close, allowing the Tigers to cross the road to the runway. Tigers returning from a mission were towed back inside the caverns immediately after engine shut-down with the pilots still inside their cockpits.

During the two-week exercise period of a WK, it is the only time that both the ground troops and the air squadrons are simultaneously in service. The purpose of the F-5 deployment at Turtmann was therefore aimed at the combined training of both the pilots and the airfield personnel attached to the Flugplatz Abteilung 3 (air base battalion 3), working together in operations from the caverns.

Ground troops operated the whole airfield (including airfield security and air defence), while the personnel of Fliegerkompanie 8 (FlKp 8) serviced the aircraft.

Between 3 and 14 March 2003 the 'wartime' airfield at Turtmann was activated for the last time to accommodate two F-5 Fliegerstaffeln, or air squadrons: Fliegerstaffel 1 (the last F-5 unit with professional pilots) and Fliegerstaffel 6 (a reserve unit). On 5 March, some 22 different F-5E Tiger IIs were flown in from different air bases around the country for the very last time. For the exercise Fliegerstaffel 1 and 6 deployed 17 Profi-pilots (professionals) and 11 Miliz-pilots (reservists), between them accumulating 190 flying hours during the WK. The missions flown were planned by the Swiss Air Operations Centre (AOC) and covered a wide spectrum, including air-to-air combat scenarios,

intercepts, air policing, and air-to-ground gunnery at the Axalp range. No distinction was made between the type of missions the professional and reserve pilots flew, although the professionals were usually tasked as lead pilots. A few years ago Swiss F-5s stopped flying night missions, something also no longer practised during WKs. Only limited night flying operations are allowed due to the noise nuisance they cause, and only the F/A-18s fly at night.

Some 600 soldiers were stationed at Turtmann for the duration of this last WK of Flugplatz Abteilung 3. Two aircraft received a text just behind the cockpit reading 'Bye bye airbase Turtmann, FlKp 8 14.03.2003' to commemorate the last WK at Turtmann by Fliegerkompanie 8. Many other aircraft received the non-standard air base badge on the right-hand side of the nose. At the end of WK'03 the airfield was closed and the seven military personnel affected by the closure were either transferred to Sion or retired early.

Flugplatz Abteilung 3 based at Turtmann, as well as Fliegerstaffel 1, will be dissolved before the end of 2003. Since the latter is formed with professional pilots, the pilots will be either reassigned to Fliegerstaffel 16 or will convert to the F/A-18. Six pilots were selected to go to the USA at the end of May to begin the ground

The sight of a Mirage taxiing along a public road as troops hold up the traffic was once commonplace in Switzerland, but is now consigned to history. Buochs will no longer host operational deployments, and by the end of 2003 the Mirage will have retired. Until then, Fliegerstaffel 10 at Dübendorf remains Europe's last front-line Mirage III user.

course at Lackland AFB, after which they move to NAS Lemoore for the flying course. After some 60 flying hours they are cleared to fly the Hornet and upon their return to Switzerland two pilots will be assigned to each of FlSt 11, 17 and 18. The other Turtmann-assigned squadron – FlSt/Esc Av 6, a reservist unit with pilots from the Romandie region – will move to Payerne to replace Fliegerstaffel 13, which will be the second Tiger squadron to be disbanded by the end of 2003 under the Armee XXI plan.

Goodbye Buochs

Buochs lies in the region of Kanton Nidwalden at an elevation of 1,473 ft (449 m). With its alternate name of Stans-Buochs, the airfield is located near to Luzern, close to the shore of the Vierwaldstättersee. The airfield was built in the 1940s as a Militärflugplatz (military airfield) for the Schweizer Armee. Besides its use by the military, the airfield has played host to a growing aviation industry complex. For more than 60 years, Pilatus Flugzeugwerke (now Pilatus Aircraft Ltd.) has had its main

plant at Flugplatz Buochs, and has been joined by Ferner, RUAG Aerospace (previously known as Schweizerische Unternehmung für Flugzeuge und Systeme/SF) and Aerolite. The joint civil-military airfield has two parallel runways, one measuring 2000 m (6,560 ft) in length.

Between 24 March and 4 April 2003 the 'wartime' airfield at Buochs was reactivated for the last time for a full WK by Fliegerstaffel 10, the Swiss Air Force's last Mirage IIIRS reconnaissance unit, and Fliegerstaffel 19, a reserve unit flying F-5E Tiger IIs. Each squadron oper-

ated from its own cavern, and four Tigers fitted with single ACMI pods operated from a ramp on the south side of the airfield. Flugplatz Abteilung 10 operated the airfield during the exercise, Fliegerkompanie 16 (FlKp 16) serviced the Mirage IIIRS aircraft, and Fliegerkompanie 20 serviced the Tigers.

A FlSt 10 Mirage IIIRS is towed from the squadron's cavern at Buochs. The aircraft's engine is running and the tow truck is about to be disconnected. At the 'last chance' before the runway the protective covers for the Sidewinder missile will be removed.

FlSt 10 has marked the final year of Mirage IIIRS service by painting two of its aircraft in special schemes based on the unit's badge. The schemes, known as 'Black' and 'White', were revealed during the last WK at Buochs in March 2003.

The final Buochs WK also featured FlSt 19 flying F-5Es. This aircraft carries special markings behind the cockpit for the ultimate Buochs deployment, the legend reading 'Farewell Buochs, 4 April 2003, FlKp 20/FlSt 19'. It also carries the FlSt 13 badge on its nose, this unit being one of two F-5 squadrons to be disbanded in 2003.

Fliegerstaffel 10 was formed on 1 September 1963 by redesignation of the Aufklärer-Gruppe, which had been formed in 1954. It received Mirage IIIs from 1968. In 1992 the Mirage IIIRS fleet was distributed among three squadrons – Fliegerstaffel 3, 4 and 10, operating in peacetime from the bases of Sion, Payerne and Dübendorf, respectively. Fliegerstaffel 4 undertook its last WK in April 1999 at its war-base in Payerne, and at the end of 1999 was disbanded, its personnel and aircraft integrating with those of Fliegerstaffel 10. On 31 December 2001, Fliegerstaffel 3 underwent the same fate. From 1 January 2002 there was again only one reconnaissance squadron in the Swiss Air Force,

Fliegerstaffel 10 flying all the remaining Mirage IIIRS and DS aircraft.

For the WK exercise, Fliegerstaffel 10 brought six Mirages to Buochs and 10 pilots, plus two former pilots to help in operations. Fliegerstaffel 19 could count on 12 aircraft, the majority having been flown in from Turtmann from the WK exercise there finished just one week earlier. Until 1993 FlSt 19 underwent WKs at nearby Alpnach, moving to Mollis the following year. In 2000 the unit had its first WK at Buochs.

Most recce missions during the WK were flown over France to an area near Paris. Unusually, a few night missions were flown, the Mirages being equipped with an infra-red reconnaissance pod fitted on the centreline station. The squadron still had three pilots qualified for flying night reconnaissance missions.

The departure of a Mirage III from the cavern is spectacular. Just before the mission starts, the

departure is announced to the groundcrew through loudspeakers inside and the doors of the cavern are then quickly opened. The pilot initiates start-up while still being towed inside the cavern. The moment the aircraft is just outside, the engine is fully ignited, and a few seconds later it is released from the towbar while still moving. The tow-cart then speeds away, while the Mirage crosses the public road, heading for the runway.

On 26 March, two specially painted recce Mirages were officially presented. Affectionately dubbed 'Black' and 'White', they were adorned with their unit badges superimposed over the total aircraft surface, with one aircraft painted in a blueish dark grey and the other white. Additionally, a text on the central fuel tank read 'nil non videmus' ('nothing we do not see').

At the end of this last WK at Buochs, Flugplatz Abteilung 10 was dissolved, although the base hosted one more training camp in May, a lower-key one-week TK exercise undertaken by Fliegerstaffel 1's F-5s (their last) involving mainly the less experienced pilots from the squadron.

Buochs airfield will continue to be operated by Airport-Buochs AG for civilian use. The air force will also continue to use it for its Ranger ADS-95 reconnaissance drones, while the base has been placed on the reserve list.

Meanwhile, the Mirage is destined for a busy year until its retirement. In May FlSt 10 held its last WK at Sion, and was due to take part in Exercise Elite at Dübendorf, as well as the Axalp firepower demonstration. The last flight of the Mirage is planned for 17 December. Until then 'Black' and 'White' were scheduled to make several airshow appearances.

Marnix Sap/MIAS

With part of the Pilatus plant forming a backdrop, a Mirage IIIRS makes one of the type's last landings at Buochs. The Mirage was originally slated to stay in service until 2007.

NATO peacekeepers over Afghanistan

Operation Enduring Freedom – EPAF

Although major combat operations in Afghanistan came to an end in late 2001, many areas of the country have remained volatile, and attacks against the international stabilisation forces have been recorded on a regular basis since. To provide an on-call air support capability, continuous patrols are mounted over the country. B-1s flying from Thumrait in Oman have been used daily – even at the height of Operation Iraqi Freedom they were still operating over Afghanistan. Another important asset has been the EPAF (European Participating Air

Forces), a detachment based at Ganci air base, Manas, Kyrgyzstan. Between 1 October 2002 and 31 March 2003 the EPAF consisted of six F-16A MLUs each from Denmark (below), the Netherlands and Norway (right). The EPAF was supported by a single KDC-10 tanker from the KLu's No. 334 Squadron (above), which joined USAF tankers in refuelling coalition aircraft over Afghanistan. For routine CAS patrols the F-16s typically carried two GBU-12 laser-guided bombs with a LANTIRN targeting pod, together with AIM-120 AMRAAMs on the wingtips.

Dassault Mirage 2000

'Deux-Mille' comes of age

In production since 1984, the Dassault Aviation Mirage 2000 is still one of the finest and most versatile fighters in service anywhere. Once considered as an outsider in a very competitive market, the design has over the years become a real commercial success, with seven export customers totalling 286 orders to date, on top of 315 French deliveries. Dassault designers have continuously updated the aircraft, and the latest variants are among the best equipped jets in the world today. The latest technological evolutions of this proven fighter – the Mirage 2000-5 Mk 2 and the Mirage 2000-9 – are now being produced for Greece and the United Arab Emirates.

Dassault produced 124 Mirage 2000Cs and 30 2000Bs for the Armée de l'Air. Today the survivors equip four 20-aircraft squadrons at Orange (EC 5, above) and Cambrai (EC 12, top), plus a five-aircraft detachment in Djibouti (EC 4/33). EC 2/5 acts as the type OCU, and has a correspondingly large proportion of two-seaters. Mirage 2000B/Cs in current service are all to either S4 or S5 standard, with RDI radar. They are receiving some upgrades, notably to the radar.

During Operation Allied Force eight Mirage 2000Cs were deployed first to Istrana, and then to Grosseto. They provided air defence cover for the NATO forces attacking targets in Serbia and Kosovo. This EC 2/12 aircraft, being prepared for a mission from Istrana, has a standard air-to-air loadout of two Magic 2s, two Super 530Ds and a centreline 1300-litre (286-Imp gal) tank.

After a successful development programme, the first Mirage 2000C went into service with the French Armée de l'Air at Dijon in 1984. The first aircraft were then equipped with the interim Snecma M53-5 engine, rated at 88.30 kN (19,842 lb), and with the fairly basic RDM (Radar Doppler Multifunction) coherent monopulse multi-mode radar that featured only limited look-down capability. The RDM was later upgraded with a continuous-wave illuminator which permitted the firing of the semi-active Super 530F missile. The RDM radar has now been withdrawn from service in France since the introduc-

tion of the Mirage 2000-5F and the new RDY radar. For autonomous navigation, the Mirage 2000C/B is equipped with the Uliss 52 inertial navigation system with a 63-waypoint capability.

Single-seat Mirage 2000Cs are armed with two internal DEFA 554 30-mm cannons, each with 125 rounds. Produced by GIAT Industries, these guns offer an excellent compromise between onboard weight/volume and fire-power, and their wide range of ammunition is adapted to varied targets, allowing the pilot to handle different air-to-air and air-to-ground situations. Their reliability is excellent, with a mean rounds between failures of more than 5,000. Each single-barrelled weapon is capable of firing at rates of 1,200 (for air-to-ground attacks) or 1,800 (for the air-to-air role) rounds per minute in either limited – 0.5- or 1-second duration – or unlimited bursts.

However, two experienced Armée de l'Air pilots have admitted choosing the air-to-ground mode for air combat and vice-versa. "With the 1,800 rounds/minute rate, the gun vibration level is less important, and the accuracy is better against a pinpoint target," explains one of them. "With the 1,200 rounds/minute firing rate, the two cannon vibrate slightly, and the shells are dispersed in a wider envelope, increasing the hit probability against a fast-moving hostile fighter." The two-seaters are not fitted with internal guns, but can nevertheless carry a CC630 pod armed with two DEFA 554 cannon with 300 rounds each.

Peru's single squadron (Escuadrón de Caza-Bombardeo 412) of 10 Mirage 2000EPs (left) and two DPs (below) is the jewel in the FAP's crown. The aircraft are the most capable air defenders in Latin America, and are also PGM-capable through the use of ATLIS pods, AS 30L missiles and BGL 1000 laser-guided bombs. Intertechnique 231-300 'buddy' refuelling pods were also supplied, allowing the aircraft to act as tankers. The Mirages have certainly seen action in border clashes with Ecuador, although no details have been released. It is likely that they have undertaken bombing sorties, using unguided weapons.

Delivery in 1987 of the definitive Mirage 2000C RDI, powered by the higher rated M53-P2 (95.16 kN; 21,385 lb) and fitted with the RDI (Radar Doppler d'Interception), was a decisive boost for French air defence capabilities, as the new radar set optimised for air-to-air work was fully capable of detecting and tracking very low-flying targets. Adopted from the 38th production aircraft onwards, the X-band RDI is used in conjunction with the semi-active Matra Super 530D medium-range missile, an upgraded variant of the Super 530F then used by French Mirage F1Cs and Mirage 2000 RDMs. The Super 530D offers huge improvements over the previous Super 530F, as it can engage targets at altitudes ranging from 50 to 80,000 ft (15 to 24384 m), with a 40,000-ft (12192-m) snap-up capability. It is worth noting that the pulse-Doppler RDI radar has never been cleared for export by the French authorities.

The RDI radar is currently being improved with an NCTR (Non-Cooperative Target Recognition) mode that allows the pilot to identify targets at long range. The development of this mode was initiated in 1993, when it appeared that the IFF had a number of limitations. The NCTR, which relies on identifying specific patterns in the radar pulse caused by rotating engine compressor blades, is a new way of positively identifying a target to minimise the risk of 'blue-on-blue' incidents.

Although mostly used in the air defence/air superiority role, Armée de l'Air Mirage 2000C/Bs are occasionally utilised in the air-to-surface role, and SAMP 250-kg (551-lb) and Mk 82 500-lb (227-kg) bombs were carried on forward fuselage pylons during missions over Bosnia in the 1990s. Pictures of Mirage 2000Cs carrying Matra 68-mm rocket pods have also been released. Interestingly, the Mirage 2000C and 2000B are collectively known as Mirage 2000DA, for Défense Aérienne (Air Defence).

Export variants

Dassault products have long been considered as good alternatives to US and Russian fighters, and the French company has created its own market niche. For export customers, Dassault developed the Mirage 2000E fitted with the substantially improved multi-function RDM+ radar that introduced real look-down/shoot-down capabilities, coupled with continuous-wave illumination of aerial targets for the Matra Super 530D. The RDM+ also offers air-to-surface modes such as ground-mapping and obstacle avoidance. This variant met with considerable success, and its derivatives were selected by no fewer than five countries: Greece (40 Mirage 2000EG/DGs), India (59 Mirage 2000H5/H/TH5/THs), Peru (12 Mirage 2000EP/DPs), Egypt (20 Mirage 2000EM/BMs) and the United Arab Emirates (36 Mirage 2000EAD/RAD/DADs), while a contract signed with Jordan for 12 Mirage 2000s was eventually cancelled.

These fighters, although all fitted with the M53-P2 engine and the RDM+ radar, vary in details. For example, UAE Mirages were delivered with a radar warning receiver (RWR) and a jammer devised by Elettronica of Italy, whereas Greek aircraft are equipped with an extensive internal self-defence suite called ICMS Mk 1 (Integrated Counter Measures System Mk 1), with another pair of superheterodyne antennas on the fin-top, additional wingtip antennas and Spirale chaff/flare dispensers in the rear of the Karman fairings. The ICMS Mk 1 is fully integrated, and its RWR can detect a 50-km (31-mile) range air-to-air radar at a distance of about 100 km (62 miles), leaving ample time for the pilot to either avoid the threat or adopt convenient offensive tactics. The ICMS Mk 1 jammer is capable of both noise and deception jamming tech-

Below: Like the Mirages supplied to India and Peru, the 2000EMs for Egypt were delivered with ATLIS laser designation pods. They also reportedly carry Matra ARMAT anti-radiation missiles, and are fitted with RDM+ radar offering Super 530D capability. Twenty aircraft were delivered in total, including four Mirage 2000BM two-seaters, and they serve with a single squadron (No. 82) based at Bir Ket, with air defence as the primary tasking.

niques, and its response time is extremely short (less than one second).

Export Mirage 2000s are all capable of carrying ATLIS laser designation pods, although Dassault would not discuss which countries have selected and purchased that system. French, Indian and Peruvian Mirage 2000s have been used operationally, and the type has proved very effective during the numerous combat missions carried out by the three air forces, especially in India where the type was used to deliver PGMs during the 1999 Kargil war.

Ground attack versions

The Mirage 2000 was an obvious contender for the replacement of the Mirage III and of the Jaguar in the ground attack role. However, the electronic suite of the 'Deux Mille' had to be extensively modified for that new role, and Dassault developed two dedicated variants of its new delta fighter. In sharp contrast with the older Jaguar and Mirage III, equipped only with basic navigation kits and rudimentary stand-alone self-defence systems, the ensuing Mirage 2000N and 2000D are among the most modern attack fighters currently in service in the world. Their admission into service has brought the Armée de l'Air into the era of automatic terrain-following, all-weather, day and night precision attacks, and integrated self-protection systems.

Initially designated Mirage 2000P (for Pénétration), the Mirage 2000N (N standing for Nucléaire) specialises in the nuclear deterrence role. Its origins can be traced back to an Armée de l'Air standing requirement for a nuclear strike aircraft to replace the ageing Mirage IIIE and Jaguar A, two types armed with the 25-kT AN-52 pre-strategic free-fall bomb. Dassault decided to design a specific variant of the Mirage 2000, taking advantage of the outstanding qualities of the tailless delta to produce a two-seater optimised for low-level flight. It became the first fighter in service in France with automatic terrain-following capabilities, thanks to the new Antilope 5 radar, which also offers air-to-surface modes, such as searching and tracking of ground targets, and air-to-air modes, such as combat and air-to-air search. The weapon system of the Mirage 2000N was developed around the ASMP (Air-Sol Moyenne Portée, Air-to-Surface Medium Range) nuclear missile, a ramjet-propelled weapon which is so fast – Mach 3 at high-level and Mach 2 at low altitude – that it is nearly impossible to shoot down. The first Mirage 2000N prototype made its maiden flight in February 1983, while the first flight of a production aircraft (s/n 301) took place in March 1986.

Mirage takes a dip

In 1997, the Hellenic Air Force contacted Dassault to recover Mirage 2000EG 210, which had crashed at low speed during a final approach. The aircraft was lying in 5 m (16 ft) of sea water, and it was felt that a rebuild could be attempted provided that corrosion had not set in. After three days in the sea, the Mirage was successfully brought to the surface and immediately washed with fresh water and Ardrox 6345 while still lying on the rescue barge. The airframe was taken to *terra firma*, where it was dismantled, and the largest components – fuselage and wings – were subsequently immersed in a purpose-built pool filled with water and Ardrox 6345, before being dried and covered with Ardrox 396-1E28. All these components were then stored while the experts assessed the feasibility of a rebuild. Eventually, the wings and fuselage were sent to France for a major rebuilding effort, and it is anticipated that the fighter will be brought back into service in mid-2003. A successful first flight was carried out by Dassault test pilot Eric Gérard at Tanagra in March 2003 (left).

A total of 75 fighters was ordered, and the first Armée de l'Air squadron, EC 1/4 'Dauphiné', re-equipped with the type in 1988. Today, 60 Mirage 2000Ns are in front-line service with Escadrons de Chasse 1/4 'Dauphiné' and 2/4 'La Fayette' at Luxeuil, in the east of France, and EC 3/4 'Limousin' at Istres, in the south-east of the country. In Armée de l'Air service, operational records have shown that the Mirage 2000Ns very rarely fly at supersonic speeds (especially as the 2000-litre/440-Imp gal RPL 541/542 drop tanks regularly carried are not cleared to exceed Mach 1 anyway) and, as a consequence, the moving air-intake half-cone centrebodies have now been locked in a fixed position, reducing the maximum speed to Mach 1.4 but noticeably minimising maintenance requirements.

Conventional weapons

Over the years, the range of weapons carried by the Mirage 2000N has progressively evolved. The type, initially delivered in the N-K1 standard, was limited to nuclear strike missions as it was then only capable of firing ASMPs and Magic IR-guided self-defence missiles. In order to boost their operational efficiency, the capabilities of the Mirage 2000Ns were increased step by step, and the N-K2 standard was introduced to deliver unguided conventional weapons such as 250-kg (551-lb) and 400-kg (882-lb) bombs, BAP100 anti-runway weapons, BAT120 area-denial bombs, and 68-mm rockets. The first N-K2 aircraft was Mirage 2000N s/n 322, and the variant initially became operational with EC 1/4 before entering service with all three squadrons. Noteworthy is the fact that neither the Mirage 2000N nor the subsequent Mirage 2000D are fitted with internal guns.

Designed for a specific mission, the highly-specialised Mirage 2000N cannot deliver precision conventional weapons, and is therefore currently not used in Kosovo- or Afghanistan-type scenarios where collateral damage has to be avoided. The adoption of laser-guided weapons for this variant is nevertheless seriously considered but, as the aircraft would still not be fitted with a laser designator, the aircrews would have to opt for the buddy-lasing technique that was so effectively used in Kosovo by the Mirage F1CTs and the Super Etendard Modernisés. It is anticipated that

the 2000N will eventually be fitted with two GBU-12 LGBs fitted to a Rafaut AUF-2 dual-station carrier under the centreline. Nonetheless, the nuclear strike role remains the primary mission of the 2000N, and about 70 percent of sorties are still devoted to nuclear attack training.

Apart from the fast disappearing Mirage IV, the Mirage 2000N was the first modern tactical aircraft with a crew of two in service in France. This was a real cultural shock for Mirage III and Jaguar pilots converting to the type, but the advantages of a two-man crew quickly became evident, especially at very high speed and very low level in adverse weather. Thanks to its advanced terrain-following radar, the Mirage 2000N can be safely and automatically flown at night and in bad weather at altitudes as low as 300 ft (91 m) in peacetime, and much lower in case of necessity. Over the years, the Mirage 2000N has proved very effective and, when it first participated in Red Flag exercises, the efficiency of its jamming system came as a nasty surprise to all participants: the fighter proved very hard to track and 'shoot down', even though the crews used only part of the capabilities available to them in an effort not to compromise the real efficiency of the system.

Unique to the UAEAF's Mirages is the Alenia Marconi Al Hakim family of PGMs. The family covers three types of seeker – PGM-1 (semi-active laser), PGM-2 (TV) and PGM-3 (imaging infra-red) – and two warhead weights (500-lb/227-kg and 2,000-lb/907-kg). The latter are denoted by A or B suffixes, respectively. Thus, a PGM-3B is a 2,000-lb IR-guided weapon. This Mirage 2000 SAD8 carries a 500-lb weapon on the wing, probably a PGM-2A. On the centreline is a datalink pod which relays imagery from the missile, and guidance commands to it.

Upgrading the Mirage 2000N

The withdrawal of its 18 Intermediate Range Ballistic Missiles (IRBMs) from the Plateau d'Albion and the specialisation of the Mirage IVP in the reconnaissance role has left the Armée de l'Air with only one nuclear vehicle, the Mirage 2000N. However, the type has now been in operational service for over 15 years, and is in need of a modernisation programme. As a result, the Armée de l'Air has funded a series of improvements to ensure that it remains fully effective, even when pitted against the latest threats in a demanding combat environment.

The self-defence suite has benefited from a series of upgrades to enhance its ELINT (Electronic Intelligence) capabilities: it can now record more effectively enemy radar emission parameters for later addition to electronic warfare threat libraries, thus complementing dedicated intelligence platforms. Since mid-2002, a SAT Samir DDM (Détecteur de Départ Missile, or missile launch detector) has been fitted to the rear of the Magic wing-pylons: when the flash of a missile firing is detected, the DDM provides

a visual and audio warning to the crew and, depending on the dispensing mode selected, a decoying sequence is automatically initiated. The DDM considerably enhances the chance of survival against attacks by beam-riding and IR-guided missiles, and its introduction was eagerly awaited by aircrews who vividly remembered the loss of a Mirage 2000N downed by a short-range missile over Bosnia on 31 August 1995.

During its development programme, the DDM had encountered higher than expected false-alarm rates. This problem took considerable time and effort to solve, but the system is now fully compliant with the Armée de l'Air's requirements. Additionally, the decoying technique will be progressively ameliorated from late 2002, the system now taking into account the aircraft's altitude to optimise the decoy release. At an undisclosed date, decoy dispensers mounted on the fighter's spine will be adopted, as already fitted to the Mirage 2000D fleet. From 2007/08, it is expected that the self-defence suite will be further improved, and will jam individual threats in a dense environment with greater efficiency. The adoption of a towed decoy is also under consideration.

The ASMP missile has now been in service for 15 years, and it will be replaced by the end of the decade. The development contract for the ASMP-A (Air-Sol Moyenne Portée-Amélioré, Air-to-Surface Medium Range-Improved) nuclear missile was recently confirmed, and the first is due to enter operational service in 2008. This 850- to 900-kg (1,874- to 1,984-lb) weapon, which will have enhanced precision, will boast a much longer range than its predecessor.

Over the years, the Mirage 2000N fleet had been more or less restricted to its nuclear deterrence role, and many people within the Armée de l'Air and the French Ministry of Defence thought that it had been a waste of valuable

Operation Vijay – Mirage 2000s over Kargil

In late 1998 the mild winter in Jammu and Kashmir provided Pakistani Army regulars and mercenaries with the means to infiltrate across the Line of Control (LOC) and set up camps on the Indian side. The build-up of forces, unknown to the Indian authorities at the time, was to have a major impact on the Indian Air Force (IAF) the following year, as it fought to dislodge the enemy in Operation Vijay, its part of the overall Operation Safed Sagar.

What became known as the Kargil war was to take place over some of the highest terrain in the world, where aircraft and weapons were limited in their effectiveness. Conventional targets – like airfields, command and control centres, and convoys – did not exist. Instead, the IAF was to be confronted with tents and fortified bunkers, the single biggest structure being a hangar capable of taking a helicopter. The advance along the LOC was eventually to penetrate up to 10 km (6 miles) in some areas along a 200-km (125-mile) line. All the ridges in the area were offshoots of K2 (8611 m/28,251 ft), the world's second tallest mountain, or Nanga Parbat (8126 m/26,660 ft), both located on the Pakistan side. Owing to the prevailing climatic conditions and wind direction, much of the region comprises sheer cliffs facing the Indian side with gradual gradients on the other. Elevation varies along the line, with a low of

2700 m (8,858 ft) at Kargil, going up to 3400 m (11,115 ft) at Dras and 5000 m (16,405 ft) at Tiger Hill. The area has no habitation and is covered in snow for most of the year. From a strategic perspective, the area cannot be used as a launch pad for any major offensive, its sole importance being control of the heights and the threat to National Highway 1A.

At such demanding altitudes the IAF was limited in what it could use. Attack helicopters were initially preferred, with Mi-17s being employed but, after early experiences, it was deemed that high-altitude bombing by fighter aircraft, and in particular the Mirage 2000, was the best option. IAF Mirage 2000s were delivered from 1985, and serve with two squadrons: Nos 1 'Tigers' and 7 'Battleaxes' at Gwalior. For 14 years they had been highly regarded in the IAF, being a stable weapons platform and having a good safety record. The Kargil Conflict was to push that high esteem to even greater heights.

Thomson-CSF ATLIS laser designator pods had been delivered with the Mirages, enabling them to deliver Matra BGL 1000 1000-kg (2,205-lb) LGBs, which were purpose-built for the destruction of reinforced targets. These weapons were highly capable but were very expensive. It was decided, therefore, to augment laser-bombing capability with a 1,000-lb (454-kg) bomb coupled with a Paveway II

laser-guided bomb kit. The IAF had ordered a number of these, but they had been supplied with an incorrect part. Because of the arms embargo placed in reaction to India's nuclear tests, the correct parts could not be obtained. Consequently, IAF technicians had to remanufacture this part in order to make the Paveway II serviceable for use on the Mirage.

Escalation along the highway

On 9 May, Pakistani shelling along the Line of Control intensified and appeared to become more accurate. This began to threaten the strategically important National Highway 1A, which ran from Srinagar to Leh, and which was only open from May until October because of the weather. It became apparent that artillery spotters were entrenched along the peaks on the Indian side of the LOC. Initially, the Indian Army suffered badly due to the terrain and having to attack forces that were well dug-in to defensive positions high up in the peaks. On 26 May the Army Chief formally requested IAF support after losses had begun to exceed acceptable levels.

Until that time the IAF had only flown photo-reconnaissance missions. This had resulted in a Canberra from No. 106 Squadron being hit by a Stinger missile on the 21st while over Batalik. Luckily, the aircraft was able to make a safe recovery to Srinagar. PR missions were subsequently flown at higher altitude by MiG-25RBTs from No. 102 Squadron.

Following the commitment of combat resources by the IAF, initial operations involved low-level air strikes in the Tololing Sector using Mi-17s and MiG-21, MiG-23 and MiG-27 fighters. Two fighters were lost on 27 May and one Mi-17 was lost on the 28th. The loss of the helicopter proved to be a turning point in IAF thinking. The aircraft had been on standby for an attack mission to Tololing and was not fitted with adequate self-protection in the form of chaff and flares. However, a helicopter which was fitted with the necessary protection had aborted its mission. As a result, four crew were killed when the Mi-17 was attacked by three Stingers.

A Mirage 2000H prepares to launch from Gwalior during an exercise, followed by a Jaguar IT. The Jaguar was the only other aircraft type to drop an LGB during the Kargil conflict.

resources. New roles for the under-used fleet were thoroughly examined, and enhancements will be introduced to increase the Mirage 2000N's multi-role capabilities. After a very long study, the French Air Force has now officially decided that the Mirage 2000Ns will eventually carry out tactical reconnaissance missions with a dedicated pod. The anticipated reduction in the number of available Mirage F1CRs from 40 to 30 by 2005 had made this choice necessary.

From 2006, the Pod de Reconnaissance Nouvelle Génération (Pod Reco NG, or New Generation Reconnaissance Pod) will enter service. Designed and produced by Thales, this system will be first fielded by suitably modified Mirage 2000Ns. Mounted on the centreline pylon, the Pod Reco NG is able to operate at very low altitude/very high speed, or very high altitude. The Pod Reco NG will make good use of electro-optics technology, and its CCD optronic sensor will allow the fighter to operate at stand-off ranges (up to 50 km/31 miles from the target). The pod is fitted with two sensors: a bi-spectral (infrared and near-infrared) sensor for day and night medium/high-altitude reconnaissance, and a mono-spectral infrared sensor for high speed and low-level (resolution: 0.15 m at 250 m/6 in at 820 ft). The imagery provided will be exceptionally good, and will allow high-resolution shots to be taken at stand-off distances using the 1100-mm (43.3-in) focal length optics. Daylight resolution is 1 m (39 in) at 90 km (56 miles), while at night the same resolution can be achieved from 45 km (28 miles). The pod will be equipped with a datalink to relay data in a timely and accessible way for real-time interpretation at long-range (up to 350 km/217 miles). The datalink will provide field commanders with real-time imagery, notably enhancing overall combat efficiency. Additionally, all imagery will be stored on a digital recorder.

From 2008 onwards, Dassault Aviation Standard F3 Rafale tactical fighters will be equipped with the new reconnaissance pod to supplement the Mirage 2000Ns, and to supersede Armée de l'Air Mirage F1CRs and Aéronavale Super Etendard Modernisés. A total of 23 Pod Reco NGs will be acquired, including eight for the Marine Nationale. Six ground stations will also be ordered, including two for the Aéronavale, with one due to be mounted on board the aircraft-carrier *Charles de Gaulle*. Aerodynamic compatibility testing has been carried out on board a Rafale in late 2002, and the first fully functioning Pod Reco NG prototype will be delivered in 2004 for flight testing under a Mirage 2000. Deliveries of production pods are planned for the 2006-2010 timeframe, with initial operational capability in 2007.

The upgraded Mirage 2000N – with the improved electronic warfare suite, the ASMP-A and the reconnaissance capabilities – will be known as the Mirage 2000N-K3.

A new attack variant

Development of laser-guided weapons promised a huge improvement in attack accuracy and, consequently, Armée de l'Air planners asked Dassault to design an improved, precision strike variant of the Mirage 2000N capable of performing attack missions with a very wide range of conventional guided weapons. In 1986, the Mirage N'

In 2000 the UAEAF dispatched Mirage 2000s to take part in the ODAX exercise at Dijon-Longvic, where they flew alongside the Mirage 2000-5Fs of EC 2, one of which is depicted here escorting a 2000EAD and a 2000DAD two-seater. Abu Dhabi's first-generation SAD8 Mirages incorporate a number of non-French systems, including Elettronica RWR and ECM, and are the only Mirage 2000s compatible with the AIM-9 Sidewinder missile. Thirty are being upgraded to SAD91/92 (2000-9) standard to complement 32 new-build aircraft.

A bombed up No. 7 Squadron Mirage 2000 is seen between missions during the Kargil conflict. The Mirage's reliability held up superbly throughout the two-month campaign.

Immediately, the IAF reassessed the situation. Of the attack helicopters only the Mi-17 could operate at this altitude, whereas the more appropriate Mi-24/35 could not. However, the loss of the Mi-17 showed that the environment was evidently awash with man-portable SAMs and was deemed too hostile to commit further helicopter resources. Planners at IAF HQ began to re-think their offensive strategy. They thought about committing the Mirage 2000 to the conflict to augment the other jet fighters. This aircraft could fly at this altitude with no problem, but it had no high-altitude attack capability.

On 30 May, IAF HQ decided to commit the aircraft, which had already moved to forward operating locations to perform the air defence role, to the offensive. Now, the work of the back room staff and pilots was to intensify greatly. As of early June, aircraft, pilots and technicians were spread around various Western Air Command bases and Gwalior. The Mirage 2000 had always been regarded within the IAF as an air defence fighter with only a limited ground attack capability. Consequently, when the call came to adopt the air-to-ground role, resources such as bombs, pylons, tooling, testers and ground crew experience were either lacking or widely dispersed. A big push was instigated at Gwalior to round up sufficient equipment and expertise to get the platform prepared

for its new task. By 12 June, IAF personnel had ironed out most of the faults.

Enough equipment was found to make 12 aircraft capable of delivering bombs at any given time. However, bombs were not so readily available, so a rapid search of the IAF inventory was instigated. Vintage 250-kg (551-lb) bombs that had been held in storage were found and made available. Originally they had been made in the 1970s in Spain, for use by the HAL Ajeet. On 1 June the weapon was cleared for use in a one-off trial, carried out from Jaisalmer over the Porkoran Range. The bombs were immediately rushed into service with the Mirages of No. 7 Squadron.

Initial missions were flown using dumb bombs only. Each aircraft would be configured with 12 bombs, one fuel tank and two Magic 2 air-to-air missiles. A typical bombing mission would involve four Mirages from No. 7 Squadron loaded with dumb bombs leaving a base in Punjab, together with a two-seat Mirage loaded with a laser-guided bomb and ATLIS pod. This five-ship would rendezvous with three fighter escort Mirages (armed with Matra Super 530D BVR missiles) from No. 1 Squadron, which were operating from elsewhere. The rendezvous point would change from mission to mission and, once the formation had joined up, one escort aircraft would

A French 2000C lands behind this Super 530D-carrying 2000H during a joint IAF/AdA exercise at Gwalior in February 2003. During Operation Vijay the air defence-configured Mirages, working with MiG-29s, kept Pakistan Air Force F-16s – which were sometimes just a few miles away – from interfering with bombing operations on the Indian side of the Line of Control.

return. As the formation arrived over Jammu and Kashmir, it would be joined by MiG-29 fighters giving top cover. These only had a 20-minute duration in the area and would usually be supplemented by another pair. Over the target the Mirages with the dumb bombs would visually acquire the target and drop their bombs. The two-seater, which would be filming the events from behind, would only use the LGB if required to do so.

During the war only nine LGBs were dropped, eight by the Mirage fleet and one by a Jaguar. Normal procedure employed during the dumb bomb attacks was for the aircraft to commence a dive at about 30,000 ft (9144 m) and designate the target at a distance of about 15 km (9.3 miles). At 8 km (5 miles) distance, anything from six to 12 bombs would be despatched towards the target. Procedure for an LGB attack differed in that the target would be acquired at

(N *Prime*) programme was announced, and full-scale development was launched in 1988. To minimise the risks of confusion with the 2000N, the aircraft was re-designated Mirage 2000D (for Diversifié, multi-role) in 1990. The programme progressed rapidly, and the Gulf War confirmed the usefulness of the requirement: Armée de l'Air Jaguars, although very effective with their laser-guided armament, were unable to operate in bad weather or at night.

The first Mirage 2000D prototype (a reworked Mirage 2000N prototype) made its maiden flight from the Dassault Flight Test Centre at Istres on 19 February 1991, while the first production aircraft (s/n 601) first flew on 31 March 1993. In April 1995, EC 1/3 'Navarre' was declared fully operational with the type. Today, four squadrons are equipped with the variant: EC 1/3 'Navarre', EC 2/3 'Champagne' and EC 3/3 'Ardennes', each with 20 aircraft at Nancy-Ochey in the east of France, plus EC 4/33 'Vexin' with three Mirage 2000Ds at Djibouti. The last example was delivered in June 2001 from the Dassault plant at Mérignac, bringing to an end the production of the Mirage 2000 for the Armée de l'Air.

To this day, only two Mirage 2000Ds have been lost in two separate accidents (in 1997 and 2000) with no loss of life, quite an achievement in flight safety statistics compared with older types such as the Mirage III, for example. This can be attributed to better flight safety awareness, and to considerably enhanced aircraft reliability.

Staged improvements

As is the case with the Mirage 2000N, a stepped modernisation approach has been chosen for the Mirage 2000D. It was initially planned to clear the 2000D to first drop/fire unguided weapons. Instead, the aircraft was first certified to deliver BGL 1000 (Bombe Guidée Laser 1000-kg/2,204-lb) laser-guided bombs and AS30L missiles, two weapon types used in conjunction with the PDL-CT pod. Only six aircraft (601 to 606) were delivered in this interim standard – called R1N1L – and they formed a combat-ready flight known as the Cellule Rapace (bird-of-prey cell), ready to deploy anywhere in case of necessity. The next 18 Mirage 2000Ds (607 to 624) were also delivered in an interim standard (R1N1), cleared to drop

a 20-km (12.4-mile) distance, designation would occur at 15 km (9.3 miles) with release of the weapon at 8.5 km (5.3 miles). As the LGB travelled towards the target the Mirage would turn away, still illuminating the target. At the time of impact the aircraft would be approximately 6 km (3.7 miles) away.

No. 7 Squadron Mirages were heavily committed during the Kargil conflict. The first bombing attack took place on Point 5140, near Tololing in the Dras Sector, and four strikes took place against the same area over three days. The Indian Army re-took this position on 20 June after fighting in conjunction with

continuous IAF strikes. Other notable airstrikes were those carried out on Muntho Dhalo, Tiger Hill and Point 4388 in the Dras Sector.

On 16 June, the major enemy supply depot at Muntho Dhalo in the Batalik Sector was sighted by a Mirage on the LDP. The following day the target was hit and destroyed by aircraft from No. 7 Squadron using dumb bombs. This camp was the major re-supply base in the Batalik Sector, and the devastating attack left over 100 dead and 50 structures destroyed.

On 24 June, the enemy's Battalion HQ on top of Tiger Hill was hit by two Mirage 2000s employing Paveway II LGBs. This was the first operational use of an LGB by the IAF. In another mission on the same day, Mirages struck the same target using dumb bombs. This strike proved to be particularly effective, causing severe damage to the enemy while boosting

the morale of Indian Army troops who watched from nearby.

Also witnessing this raid was Air Chief Marshal Tipnis, then commanding officer of the IAF, who was flying in the back seat of a Mirage 2000TH. The day after, at a press conference, an IAF spokesperson stated that: "New weapons delivery techniques have been developed by Western Air Command, that have proved very accurate and have caused considerable damage to the enemy positions. They have been effective in achieving the desired results." Attacks continued on Tiger Hill, which was heavily defended with man-portable Stinger missiles, for several days and nights. After continuous air attacks, the Indian Army recaptured Tiger Hill on 4 July after an 11-hour night battle.

Also on 4 July, a strike with dumb bombs was made against gun positions and a supply camp at Point 4388 in the Dras Sector. These attacks proved to be highly successful, and culminated in a serious degradation of the enemy supply chain. The series of attacks on Point 4388 was an excellent example of how lethal airstrikes could be when combined with good reconnaissance. Enemy plans to shift to alternative supply routes were quickly detected and then attacked, quickly strangling the supply arteries. Follow-up attacks were made by Mirages on 6 July,

conventional unguided weapons only. The definitive R1 Standard allowed the whole range of conventional weaponry/laser-guided armaments in French service to be carried. Over the years, R1N1L and R1N1 aircraft were progressively brought to R1 Standard.

The next logical step was to increase even further the jet's capabilities with an updated self-defence system and new weapons. The R2 variant was first declared operational in the summer of 2001, and R2 Standard aircraft are now being delivered at the rate of four per month. With the advent of this version, a host of modifications are introduced (detailed below). Further improvements, such as the integration of the Scalp EG stand-off missile, are already being implemented while others are still being discussed. For instance, a limited reconnaissance capability is considered desirable, but funding is likely to be a problem as the Mirage 2000N fleet is already planned to receive a reconnaissance system. The introduction at a later date, from 2005/06, of a secure Link 16 Multifunction Information Distribution System (MIDS) and encrypted SATURN (Second-generation Anti-jam Tactical UHF Radio for Nato) radios is envisioned to boost interoperability with NATO fighters, AWACS and command centres.

Under the skin

Mirage 2000Ns and 2000Ds may look similar but these two variants are very different under the skin. At first glance, the Mirage 2000D immediately betrays its Mirage 2000N origins, albeit with a green-painted radome without the pitot tube of its predecessor, a wrap-around camouflage, and a new fin top. The gold-covered cockpit transparencies, adopted to protect aircrews against laser emissions which might bounce back into their eyes, have been progressively replaced by standard canopies as the risk was assessed as negligible. Compared with that of the Mirage 2000N, the weapon system/navigation suite of the 2000D had to be enhanced because the conventional attack role of the new variant required more precision than the ASMP delivery profile.

When developing the Mirage 2000D, Dassault introduced a host of improvements: upgraded Uliss 52P navigation kit with embedded GPS, upgraded terrain-following

Above: Although the 2000N has no self-designation capability, it can drop laser-guided bombs for designation from other sources. This aircraft, on the strength of the CEAM test centre at Mont-de-Marsan, carries two GBU-12 LGBs on the centreline.

Above left: Mirage 2000N crews routinely practise the delivery of unguided weapons. Here Mk 82 bombs are loaded on to an EC 3/4 'Limousin' machine at Eielson AFB, Alaska, during the Cope Thunder exercise in May 2001.

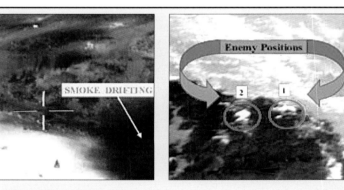

The Muntho Dhalo supply camp (above) was the largest target hit by Mirages in the war. The image at right is from an ATLIS pod during the dumb bomb attack.

Tiger Hill was a small but important target (left). Precision attacks were carried out by the Mirages (ATLIS imagery, above), the first time that a laser-guided bomb had been used.

four aircraft dropping 24 dumb bombs, and on 10 July, when three aircraft dropped 15 dumb bombs. These attacks broke enemy resistance and, because of the high casualty rate among officers, seriously degraded command and control, in turn leading to the overrunning of their position by the Indian Army. In fact, by 9 July the Indian Army had recaptured 99 percent of the Batalik Sector and 90 percent of the Dras Sector.

both squadron commanders: Wing Commander S. Neelakantan (No. 1) and Wing Commander Sandeep Chhabra (No. 7). Wing Commander R. Nambiar received the Vayu Sena Medal, as did fellow No. 7 Squadron pilots Squadron Leaders M. Rao,

D.K Patanaik and Flt Lt Tokekar. The technical officer of No. 1 Squadron, Squadron Leader K. Ravi, received the Vayu Sena Medal for his work in adapting bombs. Several other aircrew were Mentioned in Dispatches.

Simon Watson and Phil Camp

The final tally

During Operation Vijay, the total of IAF fighter sorties reached 1,199, of which the Mirage fleet accounted for over 500. No. 1 Squadron was deployed to Ambala, from where it flew 274 operational sorties, the majority of which were air defence and strike escort missions. (The unit's history on the IAF website also refers to Elint sorties, and lists the Ambala sortie number as 234. Furthermore, it notes that some aircraft were detached to Jodhpur, from where 153 sorties were flown.) No. 7 Squadron completed over 240 strike missions during the conflict, dropping over 55000 kg (121,252 lb) of ordnance. Several Gwalior-based personnel received decorations following the war, including the award of the Yudh Sena Medal to

A two-seat 2000TH 'laser-bomber' accompanied many of the Kargil missions, its ATLIS pod and extra crew member being crucial to the ability to spot small targets in difficult terrain.

back navigates and handles the self-defence suite. In the Mirage 2000D, the role of the weapon system operator is much more complex as it also includes manipulating the laser designator." For the pilot, the management of the weapon system and flying functions has been substantially simplified by the adoption of the HOTAS (Hands On Throttle And Stick) concept (the navigator only has basic throttle and stick for flight safety). The pilot now has three displays at his or her disposal: a lateral multi-function screen, the HUD and a central radar screen. The weapon system operator relies on three colour displays, a grip with quick-access functions, and various controls on the throttle. It is worth noting that both the Mirage 2000N and 2000D can be flown with NVGs, although only the pilot actually wears the goggles as the navigator is mainly head-down to take care of the navigation and weapon-aiming kit.

Electronic warfare suite

To defeat the latest air and ground threats, the Mirage 2000D is fitted with an extremely efficient, internally-mounted self-protection suite, which incorporates a Serval radar-warning receiver, a Caméléon jammer and a Spirale decoying system. The Serval and Caméléon are closely integrated, and the reaction time to deal with any arising threat is very short. The two Spirale twin-tube dispensers have a combined capacity of 112 bundles of chaff, whereas the two flare dispensers in the lower fuselage carry 16 60-mm IR cartridges (or up to 40 with the optional Eclair dispenser).

The decoying system has four operating modes: automatic, semi-automatic and manual, all selected by the navigator, plus a survival mode which can be triggered by either crew member, although the pilot can select it much quicker thanks to a large paddle on the side of the cockpit, above the throttle. "Usually, the crew will choose a fully automatic dispensing sequence for the attack run, and a semi-automatic programme for the rest of the sortie," says Commandant Louis 'Ptilouis' P. The self-protection system proved its worth over Kosovo, and the Mirage 2000Ds escaped unscathed even though there were a number of close calls. An aircraft even had to jettison its drop tanks before diving away to safety to avoid a SAM.

The compact Mirage 2000 is very difficult to detect due to its low radar cross-section, and small visual and thermal signatures. The aircraft is very agile too, even at high altitude. "During Operation Allied Force, we always egressed

While nuclear strike remains the 2000N's primary role, the importance of this has diminished in an era when single-role types are an expensive luxury. Accordingly, the 2000N is being updated to K3 standard to carry the Pod Reco NG multi-sensor pod for reconnaissance duties, as well as being updated to carry the forthcoming ASMP-A.

EC 3/4 'Limousin' flies the 2000N from Istres. Long-range strategic strike sorties are regularly practised by the 2000N units, but about 30 percent of the training is devoted to conventional attacks, by both day and night.

system, new glass cockpit, improved self-protection suite and wider weapons-carrying capability. The aircraft is designed to fly low – down to 200 ft (61 m) – and fast – up to 600 kt (1111 km/h) – thanks to its Antilope 50 automatic terrain-following radar. For precise navigation, the Mirage 2000D is fitted with two Uliss 52P inertial navigation units with hybridised GPS. Up to 64 waypoints can be stored, and another 10 can be created in flight, for example when overflying targets of opportunity. The terrain-following radar has been improved and, contrary to the Mirage 2000N, the new variant can deliver weapons automatically (in loft and level flight) with the terrain-following system activated.

Compared with those of the earlier Mirage 2000N, both cockpits have been extensively modified in the new version. "The Mirage 2000D is much more than a traditional two-seat aircraft: it is a combat two-seater," says Commandant Louis 'Ptilouis' P., Mirage 2000N/D Project Officer (the Armée de l'Air's policy is to keep secret the names of its aircrews). "What I want to stress here is the fact that the two aircrew each have a well-defined role within the aircraft, and that the mission cannot be carried out by the pilot alone. In the Mirage 2000N, the guy in the

targets in the track-while-scan mode (out of 24 displayed), irrespective of their aspects and flying altitudes. Interception data are calculated for the four priority targets, allowing firing of four thrust-vectoring MICA (Missile d'Interception, de Combat et d'Autodéfense, interception, combat and self-defence missile) missiles in quick succession in their full range envelopes. In the air-to-surface role, it is capable of accurately identifying fixed and mobile targets, and of designating them to the weapon system. To increase maintainability, the new radar is divided into six line replaceable units only: exciter, transmitter, antenna, receiver, signal processor and data processor.

The first full-scale airborne tests of the RDY radar were conducted onboard a Mystère 20 in July 1987. In 1989-1990, RDY prototypes were installed in Mirage 2000-04 (the fourth Mirage 2000 prototype) and in the Mirage 2000X-7 (production Mirage 2000C No. 7 which had been damaged in 1985).

New systems

The key to modern air combat is situational awareness, and the Mirage 2000-5 pilot-system interface has been designed to significantly reduce aircrew workload. If one compares the cockpit of a Mirage 2000C RDM with that of a Mirage 2000-5, there is considerable difference: the older

variant only had a radar screen and conventional instruments, while the new-generation aircraft is equipped with a glass cockpit composed of three multi-function multichromatic displays and a combined wide-angle HUD/head-level display. The three multi-function screens are generally used to display navigation, electronic warfare, tactical situation and system information, whereas the head-level screen is used for the radar picture.

Four screens offer worthwhile improvements over most other designs, which have only three. The added benefit of the new head-level display is that it is collimated to infinity, allowing the pilot to shift instantly from head-up flying to radar monitoring without a need to refocus, and its field of view is larger than that of a traditional screen. The combined HUD/head-level display provides short-term information, while the other three convey medium-term information. It should be noted that the rear cockpit of the two-seater is not equipped with the head-level display.

The Mirage 2000-5's HOTAS system has been optimised with simplicity in mind. The end result is an advanced cockpit which is highly automated and very easy to use. For example, the eight most menacing targets are automatically tracked without any pilot action. Then, the system designates the four priority targets, and the aircrew only has to press one HOTAS button to validate this choice before shooting four MICAs.

The quickest identifying feature on the Mirage 2000-5 compared with earlier variants is the lack of the traditional pitot tube, replaced by a side probe air data system. This choice was dictated by the need to increase radar perfor-

Aircraft 77 was one of the two 'prototype' conversions of 2000Cs to 2000-5F standard, undertaken at Istres and first flying in early 1996. They were followed by 35 'production' conversions undertaken at Bordeaux (with airframe work performed at Argenteuil), the first of which was handed over to the CEAM in April 1998. The EC 2/12 markings reflected 77's former assignment prior to conversion.

Left: In May 2001 the 2000-5F undertook its first major overseas exercise deployment, travelling to Eielson AFB in Alaska to participate in the Cope Thunder exercise with US aircraft.

Below: 2000-5F 77 is seen again in 1998, by now wearing the badge of its new assignment: the Centre des Expérimentations Aériennes Militaires at Mont-de-Marsan.

Advanced missile

At the heart of the Mirage 2000-5's weapon system is the new MBDA MICA, which boasts true fire-and-forget capabilities. When compared with the latest active BVR-type weapons that were being developed in the USA and in Russia, the heavy and bulky Matra Super 530D of the Mirage 2000C/E was looking increasingly outdated. The French answer to these new active weapons was the thrust-vectoring MICA, which is the primary air-to-air missile fielded by the Mirage 2000-5.

Development of the new weapon started in 1978, and the first ground test firing was carried out as early as 1982. The multi-target multi-mission MICA is capable of both interceptions beyond visual range and close-up dogfight combats. This lightweight (112-kg/247-lb) weapon is fitted with a high-impulse Celerg powerplant which has an extremely short combustion time – approximately 2.5 seconds – to minimise its visual footprint and reduce firing detection (no telltale smoke trail). A jet deviation system, combined with aerodynamic control surfaces and its long fins, provide MICA with eye-watering agility (load factors of up to 50 g). The MICA also features a fighter/missile data link capability which gives good fire-control capabilities in adverse environment and at long range, thus massively increasing the overall lethality of the Mirage's weapon system. The wing-mounted MICAs can be rail-launched at up to 9 g during a dogfight, whereas the ejected fuselage missiles can be fired at up to 4 g. Two variants of the MICA will be used by the Rafale – the radar-guided MICA EM (Electromagnétique) and the infrared-guided MICA IR, but only the first one has so far been approved for the Dash 5.

Thanks to its Thales AD4 active radar seeker, the MICA EM is fully autonomous after launch, so that a pilot can either engage several targets simultaneously or turn away after a shot, reducing the time spent in a potentially dangerous area or denying the enemy interceptor any firing possibility. When fired at a distant target, the missile climbs to very high altitude to minimise drag and boost range. This climb/high-altitude cruise also allows it to dive at the target during the terminal stage of the interception and, as such, to retain enough energy to engage and destroy a hard manoeuvring aircraft. The MICA EM has been tested in very demanding environments, and numerous test firings have confirmed its excellent efficiency, even when faced with the latest countermeasures. The MICA EM is currently in service on Dassault Mirage 2000-5s with the Armée de l'Air, the Republic of China Air Force (Taiwan) and the Qatar Emiri Air Force, and on French Navy Dassault Rafale fighters. It is on order for the United Arab Emirates Air Force for its Mirage 2000-9s and for the Hellenic Air Force for its Mirage 2000-5 Mk 2s.

MICAs are fitted to four hardpoints under the sides of the fuselage, and it should be noted that, although these pylons existed on previous variants, the ones of the Dash 5 can 'talk' to the missiles via a databus connection. With its added pylons, the capacity of the fighter has grown to

2 Escadre de Chasse at Dijon-Longvic was the first wing to receive the Mirage 2000, converting its three squadrons from 1984. Today's successors – EC 1/2 'Cigognes' and EC 2/2 'Côte d'Or' at the same base – were chosen to be the lucky recipients of the limited number of 2000-5F conversions, allocated 15 aircraft each (the remainder forming a trials fleet and attrition reserve). EC 1/2 aircraft from the 'Stork' squadron wear the badges of two famous World War I aviators: Guynemer's stork (above) on the port side of the fin and Fonck's stork (below) to starboard.

mance, the nose-mounted pitot tube being an obstacle to radar wave propagation.

Survivability is of paramount importance for air supremacy, and the Mirage 2000 was designed with that aspect in mind from the outset. The export Dash 5s are equipped with the very modern ICMS Mk 2 jamming/electronic warfare system which, including the chaff and flare dispensers, is entirely mounted internally, leaving all nine pylons available for weapons and drop tanks. According to Thales, the ICMS Mk 2 is capable of detecting and jamming all known radar systems. For redundancy, the jammers are equipped with a basic built-in radar detector so that the fighter does not lose all self-defence capabilities if the radar warning receiver fails. The two aforementioned Spirale twin-tube dispensers have a combined capacity of 112 bundles of chaff. The two flare dispensers in the lower fuselage carry 16 IR cartridges (or up to 64 with the optional Eclair M dispenser). The decoying system has five operating modes: automatic, semi-automatic, navigation, manual and survival.

Right: At the time of its procurement it was suggested by some that the 2000-5F acquisition was a politically motivated decision, driven by the need to demonstrate French confidence in the product and therefore reinforce efforts to export it. Whether this was the case or not, the Armée de l'Air was happy to receive the new fighter, which provided a state-of-the-art air defence capability while the service waited for the Rafale. Here MICA EMs wait to be loaded on to a 2000-5F of EC 2/2, resplendent in the 'Grim Reaper' (Mort qui fauche) badge of SPA 94.

six missiles (four BVR air-to-air MICAs and two combat/self-defence Magic IIs), and the Mirage's combat persistence has been sharply increased. Furthermore, the two wing pylons that were previously used to carry Super 530D missiles are now available for either 1700-litre (374-Imp gal) external tanks as used by Taiwan, or 2000-litre (440-Imp gal) RPL 541/542 drop tanks as those in service on French Mirage 2000-5Fs. These new configurations nearly double the operational range of the fighter. The single-seat Mirage 2000-5 retains the two DEFA 554 30-mm guns of the earlier variants, while the two-seaters can still carry a CC630 pod.

Export contracts

With the advent of the Mirage 2000-5, Dassault salesmen had a new product to offer on the export market, and the first success was scored in 1992 when the Republic of China Air Force (Taiwan) announced an order for 48 single-seat Mirage 2000-5EIs and 12 two-seat Mirage 2000-5DIs as part of a major revamping of its combat fleet. The contract also included MICA and Magic air-to-air missiles. The 60 fighters have now all been delivered, and are used as high-altitude interceptors.

The second export order came in 1994 when Qatar selected the Mirage 2000-5 to supersede its long-serving Mirage F1EDAs. The order was split between nine single-seaters and three two-seaters, and the first batch of four Mirage 2000-5EDA/DDAs was delivered in December 1997.

The new Mirages are regarded as a quantum leap forward in operational viability compared with the older Dassault Mirage F1EDA/DDA.

At the beginning of the 1990s, the French Air Force and the French Ministry of Defence elected to have 37 earlier Mirage 2000Cs brought up to the new Mirage 2000-5F standard. The Air Force commanders had realised that the new variant represented a quantum leap when compared with the earlier Mirage 2000DA, and they agreed that their air defence assets could be considerably updated at a modest cost. These air defence-optimised fighters were taken from the Mirage 2000C RDI fleet because they had a long remaining structural life. The modernisation process was in fact quite complicated, as the RDI radar sets were transferred to older RDM-equipped Mirage 2000Cs, paving the way to the complete withdrawal of the RDM radar from French service. However, the programme was completed on time and within budget.

Armée de l'Air Mirage 2000-5Fs are not equipped with the latest EW suite fitted to the export Dash 5s but retain the systems of the Mirage 2000C: Serval (Système d'Ecoute Radar et de Visualisation de l'ALerte) radar warning receiver, Sabre (Système d'Autoprotection par BRouillage Electromagnétique) self-defence jammer and the widely used Spirale. The Mirage 2000-5F will remain the most important air defence asset of the Armée de l'Air until the arrival of the Rafale at Saint-Dizier Air Base in 2005, and the CEAM (Centre d'Expérimentations Aériennes Militaires) has

Qatar's limited airpower needs have been amply met by the single squadron of Mirage 2000-5s it ordered in 1994. The fleet is split between nine single-seat 2000-5EDAs and three two-seat 2000-5DDAs (illustrated). This aircraft was the first of the 12 to fly, in 1995, and was among the first four handed over to the customer on 8 September 1997. Delivery to Qatar was accomplished in three batches of four, the first on 18 December 1997. Between completion and handover the aircraft were retained in France at Mont-de-Marsan to train Qatari pilots. In QEAF service the 2000-5 has been used for both air defence, with Magic 2 and MICA EM missiles, and for air-to-ground use. In 2003 unconfirmed reports suggested that the fleet was up for sale.

The undercarriage of Mirage 2000-5EI no. 28 is cycled during a pre-delivery acceptance test flight. Taiwan ordered the Dash 5 in 1992, the first of which – a two-seat 2000-5DI – flew in October 1995, so becoming the first new-build second-generation aircraft to take to the air. The first Taiwanese aircraft were handed over to the ROCAF on 9 May 1996, but were retained in France to allow training to be undertaken (with Escadron de Transformation Temporaire 85.330, part of the CEAM) before the first delivery to Taiwan in May 1997.

This Mirage 2000-5EI carries a typical air-to-air loadout of four MICA EMs, two Magic 2s, two 1700-litre (374-Imp gal) wing tanks and a 1300-litre (286-Imp gal) tank on the centreline. The Mirage 2000 force at Hsinchu forms Taiwan's principal high-altitude air defence against intruders from the mainland, the nearest point being about 150 km (93 miles) from Hsinchu across the Formosa Strait. A number of fully armed aircraft are maintained on quick reaction alert at all times.

developed new, revolutionary air combat tactics centred around its expanded capabilities. The first operational unit to receive the Dash 5F, in March 1999, was EC 1/2 'Cigognes', based at Dijon-Longvic, and its sister-unit EC 2/2 'Côte d'Or' was declared fully operational in 2000. The Armée de l'Air is extremely satisfied with its aircraft, and a limited upgrade programme, with a secure Multi-function Information Distribution System (MIDS) datalink and a helmet-mounted sight, coupled with MICA IR missiles, is seriously envisioned.

Series production

Although the fighter has evolved considerably over the last two decades, production is still carried out in the same four factories: Argenteuil, where the fuselage is built, Martignas (wings) and Biarritz (fin), with final assembly at Bordeaux-Mérignac in the south-west of France. The acceptance programme of a Mirage at Mérignac first starts with a ground run to ensure that the systems work correctly

and that the engine delivers its normal maximum thrust. Three flights are then performed. The first sortie is called a Domaine (flight envelope) flight during which the aerodynamic performance and behaviour of the Mirage will be thoroughly checked. The second mission, a further Domaine flight, also includes the first radar checks. The combat capabilities of the fighter will be fully tested in the course of the third sortie, dedicated to the weapon system.

During this last mission of the acceptance programme, the Mirage will be pitted against another Dassault fast jet (usually a Mirage F1), which will act as an exercise target to help to check the performance of the weapon system suite. Furthermore, targets of opportunity – generally Armée de l'Air fighters operating out of Cazaux – will be acquired to test the Track-While-Scan mode. When all these tasks have been completed, the fighter will be ready for delivery to the customer. French Mirage 2000-5Fs were not newly built and, as such, needed only two flights before being taken into account by the Centre d'Essais en Vol for further acceptance testing by Ministry of Defence personnel.

Introducing the Mirage 2000-5 Mk 2

The success of the Mirage 2000-5 did not deter Dassault from improving the type and, although the Mirage 2000-5 was already much more potent than earlier versions in the air defence/air superiority role, areas where further enhancements could be introduced were soon identified by the design team. Following close co-ordination between operational users and industrial partners, new improvements were brought to the Mirage 2000-5 family, such as an advanced PGM capability for night attacks.

A low-cost/low-risk path was chosen, with Dassault and Thales progressively improving mature systems to boost combat efficiency. The end result is the new Mirage 2000-5

Flying the Mirage 2000-5

To experience a Mirage 2000-5 mission at first-hand, the author was privileged to join Philippe Rebourg, Dassault Chief Test Pilot for Military Aircraft, for a 45-minute familiarisation sortie from the Dassault Flight Test Centre at Istres.

A former Armée de l'Air Mirage IIIE flyer, Philippe Rebourg was selected in 1989 for test pilot training. In 1990, he graduated from the US Air Force Test Pilot School, at Edwards, before joining the Centre d'Essais en Vol – the French Flight Test Centre – where he worked on the Mirage 2000D, Mirage 2000C and Rafale programmes. He was recruited by Dassault in 1995, and has now logged over 5,700 flying hours, including over 700 in Mirage 2000s and 700 in Rafales.

Walk-out

For the sortie, our allocated aircraft was Mirage 2000-5 B01, a two-seat demonstrator which has been extensively used for the development of the new Mirage 2000-5 variant. As such, it is fitted with a full 'glass' cockpit and took part in the RDY radar development programme. Its current standard is broadly equivalent to that of a Qatari Mirage 2000-5DDA, with Thales RDY radar and Uliss 52 Inertial Navigation System (INS). This prototype is unusual in that it can be started from the rear seat. However, it is not fitted with an IFF interrogator. At the time of writing, the Mirage 2000-5 B01 had logged only 1,170 flying hours, and it was being modified for a helmet-mounted display trials campaign. For our sortie, the aircraft was flown clean with 3100 kg (6,834 lb) of internal fuel, and our take-off weight amounted to 11360 kg (25,044 lb), including 300 kg (661 lb) of test instrumentation.

Walking towards the waiting Mirage 2000, one can only be impressed with its beautiful lines, especially with no external stores as it is today. "If they look right, they fly right" once said the late Marcel Dassault, and the Mirage 2000 is no exception. The initial impression when sitting in the cockpit is that it is roomy and that the Martin-Baker Mk 10F ejection seat is very comfortable. The state-of-the-art instrument panel in the rear cockpit is almost identical to that of the front seat, and is dominated by a HUD repeater and three multi-function screens (instead of four in the front). The cockpit is extremely well laid out, and all the controls and instruments come easily to hand.

In the front seat, Philippe Rebourg rapidly went through the pre-flight checks and, with a loud rumble, the M53-P2 crackled into life. For the M53 turbofan, Snecma has adopted the 'start and forget' concept, and engine parameters are automatically monitored to reduce pilot workload. For instance, the afterburner would be automatically switched off in the unlikely case of a thermal lock in the exhaust at high altitude and very low speed, and would be automatically switched back on again when the phenomenon has disappeared. As a result, pilots can devote all their attention to combat and tactics.

Monitoring the Jx

Our sortie was conducted as a real flight test, and a test controller at Istres provided an air traffic advisory service. Moreover, two flight test engineers from Dassault and from Thales monitored the mission from a dedicated facility. A special radio system allowed them to listen in real time to what was said in the cockpit, and to easily communicate with us. Their first task was to determine the Jx, that is the planned longitudinal acceleration, which is calculated on gross weight, temperature, airfield elevation and atmospheric pressure. For this flight, the scheduled Jx was +0.70 *g* with a 'no-go' at +0.63 *g*. For take-off, Philippe Rebourg had selected the HSI on the left lateral display, and an artificial horizon format on the right one.

Vision forward from the back seat is not good, but excellent to the sides and behind. The engine intakes, and small strakes mounted on the sides, impair downwards vision slightly.

Mk 2/2000-9 family. "The Mirage 2000-5 Mk 2 is a fully multi-role fighter," says Paul Mustacchi, Chief Test Engineer in charge of the Mirage 2000-5/2000-5 Mk 2/2000-9 programmes. "The modifications implemented, such as the introduction of the Damoclès laser designation pod or the pylon-mounted FLIR, give unprecedented strike efficiency to the fighter, even at night." It is often said that the Mirage 2000-5 Mk 2 combines the capabilities of the Mirage 2000D attack variant and of the Mirage 2000-5 air defence version into one aircraft.

"The core of the enhanced capabilities of the Mirage 2000-5 Mk 2 lies in a new Modular Data Processing Unit (MDPU) borrowed from the Rafale," explains Paul Mustacchi. "It replaces two mission computers and two symbol-generation units, and is composed of up to 18 line replaceable modules, each with a processing power 50 times higher than that of the 2084 XRI-type main computer which equipped the Mirage 2000-5. In fact, only 12

modules have been used so far, giving the new variant six hundred times more computing power than the Mirage 2000-5, and the other six available module slots provide a 50 percent future growth capacity to integrate new weapons or electronic warfare systems." The MDPU, composed of commercial off-the-shelf elements to drive down costs, substantially enhances the avionics/weapon integration capability. Thanks to its modular architecture, the system is highly adaptable, and new avionics and new ordnance now under development can be easily adopted. Enough growth potential has been built into the Mirage 2000-5 Mk 2 to ensure that the design has warfighting relevance beyond 2030.

Empty weight of the Mirage 2000 has not changed much since the first deliveries as the electronic suite is getting lighter for a given power output, but Dassault nevertheless took advantage of the Mk 2 programme to augment the maximum take-off weight from 16500 kg (36,376 lb) to

This is the first of Taiwan's 48 single-seat 2000-5EIs. By December 1997 the first squadron at Hsinchu had been declared operational, and the 2nd/499th TFW force was rapidly built up to three squadrons. One of these (41st TFS) has a type conversion function, and consequently operates most of the 2000DIs. In common with Qatar's aircraft, the Mirage 2000-5EI/DI batch was delivered with full ICMS Mk 2 electronic warfare suite. For Elint sorties they can carry the ASTAC pod on the centreline.

Istres is a massive air base with considerable air traffic, and although it was early in the morning, other aircraft were taxiing towards runway 33. We were nevertheless cleared to take off before an Armée de l'Air DC-8 which was already waiting at the holding point. Simulating a quick reaction alert, Philippe Rebourg expedited our departure: after lining up, he pushed the throttle to full dry power and released the brakes before selecting maximum afterburner. The acceleration was brisk, to say the least. In the HUD, the Jx readout peaked at +0.69 g, well over the calculated +0.63 'no-go'. The obvious benefit of the Jx is that it eliminates the need to perform a time/distance check using runway markers to confirm adequate acceleration and thus engine performance. At 125 kt (231 km/h), Philippe Rebourg pulled the stick fully back, and the wheels left the concrete after an estimated 350-m (1,150-ft) ground run. The gear came up quickly and, at 250 kt (463 km/h), the afterburner was cancelled while we turned left towards the Mediterranean Sea.

The fighter's rear seat is raised slightly, but this is still not enough to provide any good forward visibility. Thankfully, two-seat Mirage 2000 fighter variants are fitted with a HUD repeater. Unfortunately, our aircraft was still equipped with the old-generation black and white system, which tended to white-out when the aircraft pointed directly at the sun. The system worked reasonably well in other conditions however, and it should be noted that a new colour LCD HUD repeater is now being fitted to Mirage 2000-5 Mk 2/2000-9s to offer much improved forward vision.

Aerobatics

The Mirage 2000 is highly praised for the efficiency of its fly-by-wire flight controls, which make the

aircraft safe and easy to fly, even at very low speed where it excels. There are two sets of limits, according to the configuration of the fighter: heavy loaded (+6 g, 20° angle of attack, 150° per second roll-rate) and air combat (-3.2/+9 g, -12/+29° AOA, 270° per second roll-rate). However, these are 'soft limits' that can be overridden in some conditions. If needs be, to avoid hitting the ground for example, the pilot can apply more pressure on the stick to pull up to 11 g and 31° AOA. This outstanding safety feature has already saved many lives in various countries.

As briefed before departure, Philippe Rebourg offered to begin the demonstration with a few aerobatics. He handed over the jet to me as we were still in the climb over the Rhône river estuary, and first suggested trying a loop. Entering the manoeuvre from a 20° nose-up attitude at 400 kt (741 km/h) IAS at 7,300 ft (2225 m) was no problem and, still climbing, I eased the stick backward to pull 4 g. At the top of the manoeuvre we reached 12,400 ft (3780 m) and the speed decayed to 153 kt (283 km/h), while the

AOA peaked at 24° during the descent. We ended up at 7,900 ft (2408 m) and 224 kt (415 km/h), and we were still climbing slightly although the throttle was, by this time, nearly closed to idle.

Military power was then selected, and full left stick at 410 kt (759 km/h) IAS at 8,300 ft (2530 m) resulted in a roll-rate of about 250° per second. We concluded our short but exciting aerobatic session with a hard turn, accelerating to 510 kt (944 km/h) in full afterburner before turning hard right towards the shore. Pulling up to 8 g was no problem up to the normal operating detent. The AOA reached the 13.6° mark and, at 432 kt (800 km/h), I released the stick.

The Mirage 2000-5 controls proved extremely light and crisp, and the eye-watering roll-rate has to be experienced at least once to fully understand what a modern fighter's agility really is. Energy management is considerably facilitated by the 'energy chevrons' displayed in the HUD. Their up and down movement is linked to both throttle position and aircraft attitude. When these chevrons are above the velocity vector,

The two-seat 2000-5 Mk 2 can be used for conversion training, but also has application as a two-seat combat aircraft for missions where a second crew member is appropriate. These could include complex attack sorties in heavily defended airspace, large-scale air defence missions in which the back-seater acts as an on-scene fighter controller, and Fast-FAC duties. This view highlights the backwards rake of the ejection seats, which provides additional g tolerance compared to an upright seat.

between flights. For the Mirage 2000-5 Mk 2, Dassault has so far retained the proven and reliable Snecma M53-P2 rated at 95.16 kN (21,385 lb) thrust with reheat.

Two countries have already selected the Mirage 2000-5 Mk 2: the United Arab Emirates and Greece. The variant ordered by the UAE is known as the Mirage 2000-9, and is basically a 2000-5 Mk 2 with a few modifications to suit the UAEAF requirements. In May 1999, the UAE signed a contract for a total of 63 Mirage 2000-9s, including 33 upgrades of the older Mirage 2000EAD/RAD/DADs (this total was later amended to 32 new-build fighters, plus 30 update kits). Greece has ordered 15 new Mirage 2000-5 Mk 2s and 10 update kits for older Mirage 2000EGs.

At the time of writing, the aircraft was being aggressively marketed in Brazil and India. In Brazil, the Força Aérea Brasileira has outlined plans for 48 new fighters, and the Dassault/Embraer Mirage 2000BR, a derivative of the 2000-5 Mk 2, is a leading contender. Dassault has teamed with Embraer and, if the Mirage 2000BR is selected by the FAB, up to 32 aircraft could be built under licence locally. In India Dassault is pursuing an order for a reported 126 aircraft, most of which would be built locally by HAL Bangalore to follow on from licensed production of the SEPECAT Jaguar.

An even better radar

Thales has developed the upgraded RDY 2 radar for the Mirage 2000-5 Mk 2/2000-9. In order to implement the full multi-role capabilities, improvements have been introduced to the original RDY radar. New air-to-ground modes, such as Doppler Beam Sharpening for mapping (with a resolution said to be 15 m/50 ft) and multi-target GMTT/GMTI (ground moving target tracking/indicating), are now available. Greece had the RDM radars fitted to some of its Mirage 2000EG/DGs modified for ship attacks with the

The 2000-5 Mk 2 introduces improved multi-role capabilities, including additional air-to-ground radar modes thanks to the RDY 2. Here the demonstrator carries 250-kg bombs and MICA EM/IRs.

17500 kg (38,580 lb) thanks to the adoption of new, stronger Michelin tyres and modified oleo strut settings. An OBOGS (onboard oxygen generation system) has been developed, giving infinite oxygen endurance, and helping reduce costs through a diminution of necessary ground equipment and maintenance. It also contributes to the reduction of the already very short turn-around time

---

the pilot instantly knows he can either accelerate or climb. When they are under the velocity vector, he is slowing down. At this stage, with the throttle fully forward and the afterburner lit, the only way to accelerate again would be to descend.

Radar work

Although the RDY in the track-while-scan mode is fully capable of tracking up to eight targets, irrespective of their aspects and flying altitudes (out of 24 displayed), we had to make do with a few airliners and a lone French Test Pilot School Jaguar. With the introduction of the Thales RDY multi-mode radar, Mirage 2000-5 operators benefit from a first-class radar which offers excellent performance. Obviously, no radar detection range or missile-firing ranges can be discussed as all such data is classified. However, it can be said that the RDY had no difficulty tracking targets of opportunity at the selected display format range/scale, that is 80 nm/148 km (out of a 320 nm/ 592 km maximum scale). As soon as an aircraft is detected by the RDY, a 'V' shape symbology – called 'gull' by Dassault pilots – is first displayed on the radar screen. Without any pilot action, the system will automatically start tracking this 'raw' contact after a few seconds. When a target is tracked, the 'V' symbology changes into a square blip. For aircraft flying in close formation, a raid assessment mode is provided.

RDY waveform management is entirely automatic to optimise detection range, but the pilot can override that manually if required. Similarly, depending upon the tactical situation, aircrews may choose the number of bars (that is radar sweeps in the horizontal plane) from one to four. A bar is in fact the angle of the beam generated by the radar. According to the tactical situation, the pilot can also choose the minimum Doppler speed above which the target will be tracked: high (120 kt; 222 km/h), medium (90 kt; 167 km/h) and low (60 kt; 111 km/h). If an enemy fighter attempts a 'beam manoeuvre' to disappear in the Doppler notch, it may be advisable to select the lowest speed option in an effort not to lose the contact. However, in normal operations, the highest option might be better suited because light aircraft or cars speeding on motorways will appear on the screen if their velocity is higher than the target rejection speed of the medium and low options.

Similarly, chaff expendables which slow down after release will be more easily filtered by the high option algorithms. Helicopters will be detected in any case, even when stationary, as their rotor blade tip speed largely exceeds the RDY Doppler rejection velocity. "In fact, the radar is so good, that in real life pilots seldom override the system," noted Philippe Rebourg.

Fictitious engagements

Then acting as a fighter controller, the test controller in Istres used his all-round radar coverage to vector us towards aircraft in the vicinity. Without any specific threats pitted against us, interceptions of targets of opportunity had to be simulated, and the French Test Pilot School Jaguar was the first aircraft to be 'engaged'. Philippe Rebourg quickly locked the RDY on the hostile 'fighter', and the 'V' shape symbology turned into a square blip with an associated vector giving a graphical illustration of its speed and heading. Alongside the target track on the radar screen was a figure indicating the height at which the Jaguar was flying. On the lower side of the screen, additional information, such as its speed, heading, altitude and distance appear, and the target data relative to the selected 'bullseye' – a fixed position, common to the air defence controllers and the pilots, around which target range and bearing are referred to – is automatically shown on the upper side of the display.

If, for any reason, the pilot decides to engage another target, he can move a controller on the throttle to overlay a cursor on the contact, then press the designation switch once to allocate the new target to the weapons computer. Nice and easy. Alternatively, the pilot can use the 'pinkie switch' to instantly jump from one target to another, or from one group of aircraft to another. The 'pinkie switch' is also utilised to tell the system that the pilot wants to keep a specific aircraft as the primary target. All these procedures proved very intuitive to use.

When the target is tracked by the system, the firing parameters for the MICA missiles are automatically calculated. Interception data is computed for the four priority targets, allowing ripple-firing of four thrust-vectoring MICAs. The weapon system automatically selects the nearest or most threatening hostile fighter, chosen by the system according to a number of criteria such as distance and

closing speed, and the pilot only has to accept the proposal and shoot (or instantly switch to another target if the tactical situation or orders received dictate another choice). The pilot may instantly select various combat modes by using a rocker switch on the stick. In a combat mode, the RDY scans at a distance of 10 nm (18.5 km) whereas, with the improved RDY-2 of the Mirage 2000-5 Mk 2/2000-9, this range has been extended to 15 nm (28 km). A target designation aid is also available to predict the enemy aircraft flightpath if it moves out of the radar's scanning envelope, considerably easing target reacquisition.

By this time, the Armée de l'Air DC-8 which had taken off after us was conveniently placed for an interception, and the test controller vectored us towards this new prey. Having the radar firmly locked on to the four-engined aircraft, Philippe Rebourg decided to simulate the firing of a MICA. For lock-on-after-launch firings, the fighter-to-missile datalink may be used to update the missile in flight. Obviously, short-range lock-on-before-launch operations are also possible. The MICA EM is equipped with an AD4 Thales-designed radar seeker which boasts an excellent look-down capability, even in a high-clutter/ECM environment. In the HUD, two lines representing the MICA's maximum range and its no-escape zone clearly showed when to shoot. Thanks to this symbology, aircrews can instantly choose the most adequate tactic to achieve immediate superiority: engage the target at long range or wait to maximise hit probability. During a within-visual-range dogfight, the minimum firing range would be displayed instead. After firing at the DC-8, specific data appeared in the HUD, and the remaining fighter/missile link times were shown graphically while the remaining missile flight-times were displayed numerically.

The wing-mounted MICAs can be rail-launched at up to 9 g, whereas the ejected fuselage missiles can be fired at up to 4 g. Using the low-probability-of-intercept, unjammable, encrypted fighter-to-missile datalink, two missile flight paths can be simultaneously updated in flight. In the Mirage 2000-5 Mk 2/2000-9, this capability has been boosted to four missiles at a time. For a long-range interception, missile trajectory is specifically shaped to maximise endurance, and the MICA climbs to very high altitude to minimise drag before diving at very high speed

AM 39 Exocet Block 2 missile, and Dassault decided to offer that option on the RDY 2 radar. The new multi-target air-to-sea function now allows the pilot to fire two missiles in quick succession at two widely separated ships. A Synthetic Aperture Radar mode has been introduced and is currently being tested, allowing the detection of small surface targets from stand-off distances, even in adverse weather, and massively expanding the ground attack efficiency of the fighter (resolution quoted at 1 m/3 ft). The detection range of the RDY 2 is said to be 15 percent greater than that of the earlier RDY thanks to a slightly larger antenna and, although neither Thales nor Dassault would confirm it, an NCTR mode is thought to have been introduced.

RDY 2 air-to-air modes
  Velocity Search (VS)
  Track While Scan – Manual (TWS-M)
  Track While Scan – Automatic (TWS-A)
  Single Target Track (STT)
  Raid Assessment (Zoom)
RDY 2 combat modes
  Boresight Aiming
  HUD Search
  Vertical Search
  Azimuth Acquisition Scanning
  Slaving on other sensors
RDY 2 air-to-ground modes
  Ground Mapping (GM)
  Doppler Beam Sharpening (DBS)
  Air-to-Ground Ranging (AGR)
  Contour Mapping (CM)
  Ground Moving Target Tracking/Indicating (GMTT/GMTI)
  Synthetic Aperture Radar (SAR)
RDY 2 air-to-sea modes
  Search While Scan (SWS)
  Track While Scan – Manual (TWS-M)

The RDY 2 boasts inherent growth potential, and other modes and functions could be introduced at a later stage.

Improved avionics

When sitting in a Mirage 2000-5 Mk 2, most casual observers would think that the cockpit is unchanged compared with the earlier Mirage 2000-5. In fact, new larger multi-function displays supplied by Thales Avionics (previously known as Sextant) have been adopted for the family, giving the pilot a much better view of the tactical environment. The two LCD lateral displays are now 4 x 5-in (10 x 12.7-cm) – compared to 3.5 x 4.5-in (8.9 x 11.4-cm) for the Mirage 2000-5, while the LCD head-down display used to display the tactical situation is even larger at 5 x 5 in (12.7 x 12.7 cm). Enhancements have not been limited to the front cockpit: the back-seater now benefits from a new colour CCD HUD repeater with the same 18° x 24° field of view as the HUD itself. Both front and rear cockpits are fully NVG-compatible.

Network-centric warfare is now firmly on the agenda of most air forces, and the Mirage 2000-5 Mk 2/2000-9 family fully complies with the latest requirements. The aircraft is able to exchange sensor data via a secure, jamming-resistant digital datalink, improving even further pilot situational awareness. The advantages of such a system are varied: it helps to avoid incidents of fratricide, enhances mutual support, prevents overkill and allows faster target sorting. The Time Division Multiple Access of the system gives a classified number of slots on the data exchange network

Modern air warfare is all about the precision guidance of weapons, and the Mirage 2000 has been LGB-capable since the 2000D was introduced. The Dash 5 Mk 2 continues to offer this capability, at the expense of one MICA, through a designator pod mounted on the forward starboard pylon. Here the pod is the tried and trusted ATLIS (with GBU-12 Paveway II bombs), but the newer and much improved Damoclès will be the primary system. Although accurate, laser guidance is adversely affected by weather and other factors, and a new family of French weapons (AASM) with GPS guidance is under development. The accuracy of these weapons can be enhanced by adding laser seeker heads.

---

onto its target. After motor burn-out, the smokeless missile is virtually undetectable, and its warhead is powerful enough to take out a hostile aircraft, even if a direct hit is not scored.

IDF

To simulate an airliner identification, Philippe Rebourg switched to the identification (IDF) mode, and a specific symbology appeared in the HUD. This IDF mode is considered essential nowadays, especially since the 11 September attack. Airliners are now regularly intercepted, and this mode enables safe interceptions of targets down to close formation, even in bad weather. At night, the specific mode is designed to guide the pilot within the operating envelope of the interception light, which is powerful enough to illuminate and identify a target at up to 300 m (985 ft). During our flight, however, we had to maintain a safe separation with the DC-8, and Philippe accurately and easily remained at precisely 2.8 km (1.5 nm) behind the four-engined aircraft.

During the whole interception, the autopilot was engaged, and Philippe Rebourg precisely flew the Mirage with the 'coolie-hat' on top of the stick, turning left or right, or climbing or descending with his thumb instead of moving the stick itself. The Mirage 2000-5's autopilot is more advanced than that of earlier air defence and export versions, and it incorporates new modes borrowed from the Mirage 2000D attack fighter. It is fully integrated into the fly-by-wire system, and can be easily disengaged by pressing a dedicated trigger on the front of the stick. By releasing this trigger, the autopilot would be switched on again. The basic Dash 5 is not fitted with an autothrottle, but both the Mirage 2000-5 Mk 2 and the 2000-9 have been equipped with this extremely useful system. Additionally, an automatic, hands-off terrain-following mode has been integrated in the Mirage 2000-9, and is on offer for the 2000-5 Mk 2. "The Mirage 2000's autopilot is so good, that automatic ILS approaches can be performed down to 200 ft [61 m]," explained Philippe Rebourg.

Impressive tail-slide

At that time, Philippe Rebourg invited me to execute a split-S manoeuvre. At 307 kt (568 km/h) IAS and 19,100 ft (5822 m), I rolled the aircraft inverted and pulled the stick aft to the normal operating detent

while selecting full afterburner. The angle of attack peaked at 27° while the *g*-load reached 5.5, and I stabilised at 14,700 ft (4480 m). Final airspeed was approximately 220 kt (407 km/h), and we lost only 4,400 ft (1341 m) in the process. At 246 kt (455 km/h) IAS and 14,700 ft, I initiated another split-S and, this time, the angle of attack peaked at 26° while the *g*-load reached 4.6. I stabilised at 11,200 ft (3414 m) after losing only 3,500 ft (1067 m). Except for a slight buffeting, the aircraft remained very stable throughout the manoeuvres.

The rear cockpit of the 2000-5 is dominated by the large HUD repeater which offers the same field of view as the HUD itself. The MFD screens in the Mk 2 are slightly larger than those of the 2000-5.

Then, Philippe Rebourg took over the controls to demonstrate the docile low-speed behaviour of the aircraft. At 11,800 ft (3597 m) and 305 kt (565 km/h) IAS, he closed the throttle to 70 percent rpm and raised the nose to 75° climb-angle above the horizon to perform a tail-slide. The aircraft zoomed to 15,000 ft

Above: In the days when warfare becomes ever more clinical, and weapon/target matching more critical, the GBU-22 is an important addition to the arsenal. It combines the relatively low blast effect of the 500-lb Mk 82 bomb with the highly accurate and versatile Paveway III seeker.

Above right: Damoclès is the next-generation laser designation pod for the Mirage 2000, offering improved performance and range. It is seen here on the 2000-5 Mk 2 demonstrator, complete with the pylon which houses a nav FLIR. The UAE has ordered the Damoclès pod (known as the Shehab) and FLIR pylon (known as the Nahar) for its 2000-9s.

Opposite page: MBDA's Scalp EG (or Black Shaheen for the UAE) is a potent stand-off weapon with precision capability against hardened targets.

and several of these slots could be attributed to an AWACS. Typically, with a combat patrol of eight Mirage 2000-5 Mk 2s operating with an E-3 Sentry, combat information is updated every 2.5 seconds. Sensor and datalink data are fused into a single tactical picture on the lower display.

Accurate navigation

To minimise collateral damage during peacekeeping operations, air strikes have to be conducted with utmost precision. New-generation fighters are all equipped with highly accurate navigation systems. For the Mirage 2000-5 Mk 2/2000-9 family, precise navigation is provided by a new Thales Avionics Totem 3000 ring laser gyro INS with hybridised GPS which has been chosen to replace the Uliss 52 INS, in service since the introduction of the Mirage 2000C. The new design is technologically very advanced and features INS and embedded GPS – with code P for NATO countries – in one single unit, with man-machine interface (MMI) through full-colour LCD Control Display Unit or instrument panel lateral displays. Its technology is based on the acclaimed Pixyz three-axis ring gyro laser. Inertial Navigation Systems of the Pixyz/Totem series have been selected for the Indian MiG-21 upgrade, South African C-130s, the Tiger attack helicopters and the Scalp EG/Storm Shadow missile. It is worth noting that the Totem 3000's MMI is also used to control the IFF interrogator. Up to 10 satellites can be tracked simultaneously, thus widening reception coverage of the GPS, and the system can also work with the Russian GLONASS navsat.

This new-generation INS represents a radical improvement over the Uliss 52 of the previous Mirage 2000 variants: it is more accurate, its alignment time is greatly reduced (4 minutes for a normal alignment compared with 10 minutes with the older design), and the number of waypoints that can be stored has been substantially increased, now reaching 59, plus 59 offsets. If necessary, the Totem 3000 can even be aligned in flight with a high level of accuracy. Finally, its 1,200-hour Mean Time Between Failure is remarkable.

As well as the Totem 3000, a Digital Terrain System (DTS) has been included in the avionics package. Based on the TERPROM system, it features ground proximity warning, obstacle warning/cueing and database terrain-following (coupled with the autopilot) modes. The added advantage of this system is that it dispenses with radar emissions, dramatically increasing the stealthiness of the fighter. Finally, a better data transfer system with greater capacity cartridges has been adopted. Up to four missions can now be recorded on the same 'brick', giving the possibility to shift from one mission to another while in flight.

Digital technology

Today's Mirage 2000 bristles with electronic warfare equipment from nose to tail. While designing the Mirage 2000-5 Mk 2, advantage has been taken to enhance the electronic warfare suite of the aircraft. The ICMS Mk 3 selected for the Mirage 2000-5 Mk 2 is fully digital, enhancing the efficiency of the system compared with the earlier ICMS Mk 1 and Mk 2 variants. It is capable of high sensitivity detection/identification of enemy radar, and the eight highest priority threats are displayed to the pilot. As expected, the exact capabilities of the system remain clas-

Elements of the 2000-5 Mk 2 cockpit are tested by Dassault at Istres using this avionics bench, which has the main displays (three head-down MFDs, head-level display and head-up display), and a throttle and stick to allow HOTAS inputs.

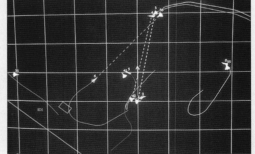

An option for the 2000-5 is the Serpam (Système d'enregistrement et de restitution des paramètres avions-missiles), which allows multi-aircraft missions to be replayed for debriefing. The system displays aircraft and weapon parameters.

(4572 m) and the speed rapidly decayed to zero, although the speed indicator in the HUD did not go below 30 kt (55 km/h) as the airflow around the pitot tubes is then too disturbed to give accurate readings. This was accomplished without any impact on engine performance, and it should be noted that the M53-P2 is fitted with a digital engine regulation which allows the throttle to be 'abused' in the most aggressive way without any risk of surge, compressor stall or engine flame-out. During the recovery, Philippe Rebourg moved the stick fully right, and the aircraft gently rolled, with the roll-rate rapidly increasing as the speed built up. "The aircraft was totally safe during the whole manoeuvre," said the test pilot. "Only a few fighters have such benign low-speed

characteristics. This is a big plus for flight safety." The FBW control system features spin, deep stall and 'dead leaf' protection. The Mirage 2000 family has now logged over one million flying hours and no crash has ever resulted from a flight control system malfunction.

Air-to-ground attack

Philippe Rebourg next offered to perform a few attack runs on the Faraman lighthouse, in the Rhône river estuary. With the Mirage 2000-5, the days of the map and stopwatch have been definitely overtaken by technology, and up to five digital flight plans can be stored in the navigation computer. In the Mirage 2000-5 Mk 2/2000-9, the number of stored flight plans

remains unchanged, but the number of waypoints for each has been doubled. With no external laser designation pod fitted to our aircraft, the radar was the main sensor for the attack. Although the first-generation RDY radars were optimised for air-to-air work, the variant fitted to Qatari Mirage 2000-5s can still operate in air-to-surface modes for mapping and telemetry, and the mapping mode showed a reasonably good image of the coastline. However, we had to move out of this mode back to radar telemetry to carry out the attack.

When a Damoclès laser designation pod is fitted to their Mirage 2000-5 Mk 2/2000-9, aircrews can use the RDY-2 radar in the air-to-air mode while utilising the pod for an attack, thereby keeping an eye on hostile fighters while prosecuting their intended target. Mirage 2000-5 Mk 2/2000-9 pilots can then choose to call the radar format on the right screen, while at the same time displaying a targeting pod image on the head-level display. A Synthetic Aperture Radar (SAR) mapping mode is now actively being developed for RDY-2 radars, and it will be fitted to Mirage 2000-5 Mk 2/2000-9s, allowing aircrews to 'paint' high-resolution maps of surface targets from stand-off distances in any weather, day and night, and to designate a precision aiming point to the fighter's weapon system.

At under 10 nm (18.5 km) from our intended target, a prediction of its position appeared in the HUD. Using the Beauduc lighthouse as an initial point, Philippe Rebourg placed the designation diamond on the building to update the INS. The attack was conducted in level-flight at 495 kt (917 km/h) and 700 ft (213 m), and simulated retarded bombs were 'released' in quick succession. He then offered to 'toss' bombs from low-level at the lighthouse, but the presence of a Test Pilot School Fennec helicopter flying at 1,600 ft (488 m) prevented us from carrying sified but it is capable of very smart and effective pre-emptive and reactive jamming modes that can be totally automatic, totally manual or can be programmed by the pilot according to the situation. The ICMS Mk 3 enables the Mirage 2000-5 Mk 2 to detect and jam advanced radar associated with the latest air-to-air and surface-to-air threats (SA-10, SA-11, SA-15 and SA-20).

Although neither Dassault nor Thales officials would confirm it, the ICMS Mk 3 is also said to be capable of jamming AMRAAM-class fighter-to-missile datalinks. Jammer coverage ranges from 2 GHz to 18 GHz. The system boasts an integrated ESM capability which records data for later analysis by electronic intelligence specialists. An optional SAT Samir DDM missile launch detector is available for the Mirage 2000-5 Mk 2.

Although the ICMS Mk 3 is the basic choice for the Mirage 2000-5 Mk 2, the United Arab Emirates Air Force has elected to purchase a similar digital system called IMEWS (Integrated Modular Electronic Warfare System), designed by Thales Airborne Systems and Elettronica. It is thought that the IMEWS offers a real Suppression of Enemy Air Defences (SEAD) capability thanks to a very accurate localisation of hostile radar emitters. However, neither Dassault nor Thales would discuss IMEWS capabilities in detail.

New weapons

To increase combat effectiveness and offer true multi-role capabilities, Dassault has selected a new range of air-to-air and air-to-ground weapons to arm the Mirage 2000-5 Mk 2 family. Originally envisioned for the Rafale only, the MICA IR is now also offered on the Mirage 2000-5 Mk 2 and has been selected by the UAE Air Force for its Dash 9 variant. The MICA IR is a huge improvement over the MBDA Magic 2, which has now been produced for over 20 years. Both IR and EM heads can be fitted to the same MICA airframe. This innovative concept ensures a massive reduction in direct costs as the airframes, warheads and motors are the same for both variants, the only difference being the IR seeker, which has many advantages for such a long-range missile. Easily distinguished by its blunt nose, it offers true first-shot/first-kill capabilities and greatly increased acquisition range in blue sky and clutter. It has

out the planned pull-up. Instead, he pressed on for another level run, squeezing the trigger to authorise the attack when within range, and the bombs were fictitiously dropped at the helpless building. For a toss bombing, pull-up cues for a 6-g pull up automatically appear in the HUD. With the Damoclès pod, the Mirage 2000-5 Mk 2 will be able to self-designate, track and illuminate a target to deliver guided ordnance with metric accuracy.

Extremely short landing

With only 900 kg (1,984 lb) of fuel remaining, it was time to head back to Istres. After the traditional break over the 4975-m (16,322-ft) runway, the test pilot handed over the fighter to me to fly the approach. Thanks to the HUD repeater, I was fully capable of following a precise flightpath towards the selected aiming point on the runway. Clearly, with a little experience, accurate touchdown would be achieved consistently and without difficulty. Final approach was flown at 150 kt (278 km/h) with 14° AOA. On short finals, Philippe Rebourg took over the controls, flared, and the Mirage 2000 gently touched down at 135 kt (250 km/h).

Full afterburner was immediately selected to perform a very tight circuit for an even more impressive demonstration: when back on the runway, Philippe initially kept the nose high and, as soon as the speed dropped below the 105-kt (194-km/h) mark, he moved the stick fully forward to bring the nosewheel back down. The high-performance carbon brakes ensured a remarkably short landing run, and I

estimated that we came to a full stop in less than 700 m (2,300 ft) without brake chute. I was thrown forward into the straps, certainly the most powerful braking ever experienced by this author. "There is no point designing a fighter which can take off in 300 m [984 ft] when clean if its landing distance is more than 1500 m [4,920 ft] in the same configuration," said Philippe Rebourg. "Dassault engineers are always very careful to ensure that landing rolls remain within acceptable tactical limits for short runway operations."

At the end of the 45-minute sortie, Philippe Rebourg pointed at the 0.2-nm (0.37-km) position error calculated by the Uliss 52 INS. This is quite a performance for a mechanical INS and, needless to say, the Totem 3000 ring laser gyro INS with hybridised GPS which now equips the Mirage 2000-5 Mk 2 would have given an even more accurate

reading, the GPS eliminating the drift associated with a traditional INS.

Every minute of that flight was superb, and I found the whole experience electrifying. The Mirage 2000-5 certainly is one of the most modern and efficient fighters in the world, and its man-machine interface is setting new standards. I want to stress the fact that I was immediately comfortable with the controls and displays, and I experienced no difficulties at all accurately flying the aircraft and tracking radar contacts. If a journalist with no experience at all on-type can automatically operate the correct switches without problem, an experienced aviator will obviously master the whole system in just a few days. As a direct consequence, conversion courses are expected to be remarkably short and straightforward.

Henri-Pierre Grolleau

Another view shows the Mk 2 demonstrator carrying GBU-22 bombs. The 2000-5 Mk 2 is fitted with the ICMS Mk 3 EW suite, although the improvements are internal. Abu Dhabi's 2000-9s have a similar IMEWS system with some additional capabilities.

Opera and Aramis – mission-planning aids

In modern air warfare, thorough preparation is the key to success, and fighter pilots spend a lot of time preparing the sorties in detail, making sure that everything is perfectly understood by everybody. Thanks to the Dassault Aviation Opera mission planning aid/restitution tool, Mirage 2000-5 Mk 2 pilots can easily load a data transfer cartridge with all the information that they will need during the flight: waypoints/target co-ordinates, threats (surface-to-air missile engagement zones), radio frequencies, IFF codes, weapon programmes, and friendly intelligence data (friendly troops and surface-to-air missile systems positions).

Prior to the mission, the pilots can visualise on the computer screen the approach to the target, and choose weapon release points according to the selected armament. The information programmed is then transferred to the 'brick', ready to be downloaded into the fighters' navigation systems. Courtesy of the Opera's open architecture, customers can easily modify the parameters for other aircraft types. Its man-machine interface has been tuned to be user-friendly.

The Hellenic Air Force has now adopted the Opera, and the system can be used in conjunction with the Thales Aramis electronic countermeasures data preparation tool. Aramis is utilised to programme the Integrated Counter Measures System in order to optimise the reaction of the defensive suite to match the threat. After the mission, Aramis will allow the aircrew and intelligence officers to visualise the various threats encountered, helping to update the intelligence picture/threat libraries and refine the jamming and decoying techniques/responses.

With a wide swathe of sea to defend, a key requirement for Greece is to retain MBDA AM 39 Exocet anti-ship missile capability in its new Mirage 2000-5 Mk 2s, as displayed by the two-seat demonstrator. Some of the current 2000EG fleet have been upgraded to 2000EG-S3 standard with the ability to support the missile, and all the new machines will have it.

excellent angular resolution – thanks to dual-band imagery – and is totally stealthy: the passive homing head enables absolute 'silent' interceptions.

As expected, both IR and EM seekers have very low false alarm rates and superior counter-countermeasures performance. Another major advantage is the detection capability of the IR missile: the seeker can act as an IRST and give a detection indication in the HUD. The missile can even be locked on a target, and the radar on another one if need be. These modes were also available with the Magic 2, but the considerably better all-aspect seeker of the MICA enhances detection probability and range. The MICA IR can be launched in two different ways: in the lock after launch mode the missile is fired at a distant target outside seeker

Noteworthy in these views of the Dash 5 Mk 2 demonstrator are the MICA IR missiles carried on the outer wing rails. These have replaced Magic 2s as the 2000-5 Mk 2/-9's main heat-seeking weapon.

range and relies on the fighter/missile datalink until it is close enough to the target. Alternatively, it can be locked first and then fired.

As on the Rafale, an optional Thales Avionics Topsight E helmet-mounted display (HMD) is on offer for the new Mirage, and could be used to cue the MICA IR for off-boresight interceptions in close-combat situations. Topsight also features a reverse cueing mode in which the displayed symbology guides the pilot's eyes to the target tracked by the aircraft sensors. The adoption of the MICA IR cued by HMD for the Armée de l'Air Mirage 2000-5Fs to supersede Magic missiles is a distinct possibility which would boost the aircraft's capability, and it is expected that this integration will go ahead.

The weapon system of the Mirage 2000-5 Mk 2 configures itself according to the weapon selected or to the flight phase. Besides the six MICA IR/EM missiles, the aircraft can carry a wide range of air-to-ground ordnance. The main long-range air-to-ground weapon offered for the Mirage 2000-5 Mk 2 is the MBDA Apache/Storm Shadow family of stealthy long-range stand-off weapons already in service or about to enter service on Armée de l'Air Mirage 2000Ds. Up to 500 are planned to be ordered by France for its Mirage 2000Ds and Rafales, including 50 for the Aéronavale. The Black Shaheen variant has been chosen by the UAE for its Mirage 2000-9s, and the Scalp has been selected by Greece for its Mirage 2000-5 Mk 2s (56 missiles ordered). To complement the Apache/Scalp EG, the AASM (Armement Air-Sol Modulaire, Modular air-to-ground armament) is also on offer.

Greece has ordered the AASM for its Mirage 2000-5 Mk 2s, and will also carry the proven AM 39 Exocet anti-ship missile. Over the years, the AM 39 Exocet has progressively been updated, and the Block 2 variant is now in service with various air forces and navies. The successive improvement programmes have mainly concentrated on radar seeker counter-countermeasures capabilities, limiting vulnerability to jamming and decoying while increasing lethality. The 655-kg (1,444-lb) missile has a range of between 50 and 70 km (31 and 43 miles), depending on launch speed and altitude, and its 165-kg (364-lb) blast fragmentation warhead is powerful enough to sink a frigate

Air-to-air weapons

Right: Until MICA IR becomes available, the Magic 2 remains in widespread use for short-range engagements, including self-defence of the attack versions.

Above: As well as moveable rear fins, MICA has a thrust-vectoring nozzle.

Below: MICA IR is superseding the Magic 2 as the Mirage's heat-seeking missile.

Above: The Mirage 2000's cannon is the Giat 30-550 F4, previously known as the DEFA 554.

MICA EM is the Mirage 2000-5's main weapon, with fire-and-forget capabilities.

Defensive systems

Right: Of the Armée de l'Air versions, the Mirage 2000D is fitted with the best EW suite, akin to ICMS Mk 1. The fin has two small antennas (like ICMS) and a forward-facing Caméléon jammer. The large aft-facing antenna serves the Serval RWR system.

Right: For dispensing chaff several Mirage 2000 variants have the Spirale dispenser mounted in the aft end of the Karman fairing. Each twin-tube dispenser carries 112 bundles of chaff, released either automatically or manually.

First-generation Mirages, and AdA 2000-5Fs, have the Sabre jammer at the base of the fin.

Mirage 2000Ns and Ds have the aft-facing antenna of the Caméléon jammer in place of the Sabre.

The latest 2000D-R2 aircraft have two 24-shot flare dispensers either side of the spine.

Targeting pods

This is the PDL-CTS laser designation pod, recognisable thanks to a new air scoop at the back. It offers improved imaging compared with the PDL-CT.

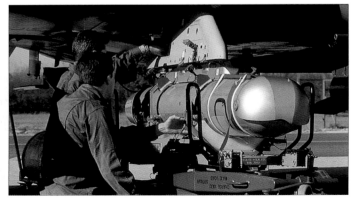

Combat-proven in several conflicts, the day-only ATLIS designation pod remains in service with the 2000D, and popular with the crews.

ROCAF
The ROC
Combat
Each cor
which ha
role-spec
but whe
they are
The ROC
systems
(2nd TFW
TFW). Th
originally
the stren
disparate
there wa
digit of t
(4+9+9=
In today'
sophistic
system l
it remair
and wing
ROCAF
Mirages

Cannon
Mirage 2
ammuni
traces it:
Post-wai
engineer
became
was inst
F1/Jagua
2.01 m (
mph). Th

Mirage 2000 operators

EGYPT — Al Quwwat Al Jawwiya Il Misriya

In January 1982, Dassault won its first export contract for its latest fighter when Egypt announced it had ordered 20 Mirage 2000EM/BMs. The order was split between 16 single-seat Mirage 2000EMs (101 to 116) and four two-seat Mirage 2000BMs (201 to 204). Deliveries took place from June 1986 to December 1988. A standing requirement for another batch of 20 Mirages will never be met. The Egyptian Mirage 2000s are thought to be mainly used for aggressor training for F-4 and F-16 units.

252 Tactical Fighter Brigade
82 Squadron   Bir Ket

FRANCE — Armée de l'Air

The Armée de l'Air is the largest Mirage 2000 operator. No fewer than 124 Mirage 2000Cs (serial numbers/construction numbers 1 to 124), 30 Mirage 2000Bs (501 to 530), 75 Mirage 2000Ns (301 to 375) and 86 Mirage 2000Ds (601 to 686) have been delivered to the French Air Force, with deliveries stretching from 1984 to 2001. Initially, the number of aircraft was planned to be higher, but the collapse of the Soviet Union and the ensuing reduction in defence spending have progressively led to cuts in procurements. A typical example is the Mirage 2000D fleet: when the 2000D programme was launched, a total of 105 aircraft was to be funded, but this was later reduced to 90 and eventually to 86 as the defence budget receded.

Mirage 2000D-R2 607 taxis in at Cazaux, carrying a PDL-CTS pod. This aircraft is permanently assigned to Dassault and the CEV for trials programmes.

Individual AdA squadrons maintain the traditions of two or even three historic units. This EC 3/4 Mirage 2000N wears a shield with badges of its three escadrilles (GC I/9-1e – Aigle, GC I/9-2e – Fennec, and SPA96 – Le Gaulois).

The Mirage 2000 has been successfully utilised by the Armée de l'Air in numerous conflicts and crises, stretching from the 1991 Gulf War to operations over Afghanistan. During the Gulf War, Mirage 2000Cs flew CAPs and escort missions from Al Ahsa Air Base in Saudi Arabia. French Mirage 2000s were also very active over Bosnia, Serbia and Kosovo, taking part in countless combat missions. During Operation Deliberate Force, in August 1995, a Mirage 2000N was lost to enemy fire when it was struck by a shoulder-launched SA-7 missile near Pale. This event prompted the adoption of the Eclair chaff/flare dispenser to boost the number of decoys carried by

Squadron	Base	Codes	Equipment
EC 1/2 'Cigognes'	Dijon-Longvic	'2-Ex'	15 x Mirage 2000-5F
EC 2/2 'Côte d'Or'	Dijon-Longvic	'2-Fx'	15 x Mirage 2000-5F
EC 1/5 'Vendée'	Orange-Caritat	'5-Nx'	20 x Mirage 2000C/B
EC 2/5 'Ile de France'	Orange-Caritat	'5-Ox'	20 x Mirage 2000C/B
EC 1/12 'Cambrésis'	Cambrai-Epinoy	'12-Yx'	20 x Mirage 2000C/B
EC 2/12 'Picardie'	Cambrai-Epinoy	'12-Kx'	20 x Mirage 2000C/B
EC 1/3 'Navarre'	Nancy-Ochey	'3-Ix'	20 x Mirage 2000D
EC 2/3 'Champagne'	Nancy-Ochey	'3-Jx'	20 x Mirage 2000D
EC 3/3 'Ardennes'	Nancy-Ochey	'3-Xx'	20 x Mirage 2000D
EC 1/4 'Dauphiné'	Luxeuil-St Sauveur	'4-Ax'	20 x Mirage 2000N
EC 2/4 'La Fayette'	Luxeuil-St Sauveur	'4-Bx'	20 x Mirage 2000N
EC 3/4 'Limousin'	Istres-Le Tubé	'4-Cx'	20 x Mirage 2000N
EC 4/33 'Vexin'	Djibouti (detachment)		5 x M2000C, 3 x M2000D
CEAM	Mont-de-Marsan	'330-Ax'	Mirage 2000-5F/B/C/D/N

French Mirage 2000s. The Mirage 2000D was barely in service when it first flew in anger in September 1995, when an AS 30L was fired at a Serbian target. In 1999, the Armée de l'Air deployed 23 Mirage 2000C/Ds to Italy for Operation Allied Force over Kosovo. They carried out 865 sorties totalling 3,574 flying hours. From 27 February 2002 to 2 October 2002, Mirage 2000Ds operating over Afghanistan from Manas Air Base, in Kyrgyzstan, logged more than 1,200 flying hours. In all, 47 GBU-12 Paveway II laser-guided bombs were dropped (plus 12 BLU-111s by French Navy Super Etendards).

Today, the Mirage 2000 still serves with 14 Armée de l'Air units and, according to the latest information, the type should soldier on until 2031, by which time the oldest aircraft, a Mirage 2000D, will be 30 years old. One of the latest modifications in the French order of battle is the withdrawal of the Mirage F1C from Djibouti and the introduction of a mixed fleet of Mirage 2000Cs and 2000Ds. Accordingly, the unit could be renamed EC 3/12 'Vexin' in the near future.

Moreover, the Mirage 2000 is used in several variants for various test and evaluation programmes by the Centre d'Essais en Vol, from Istres and Cazaux, and by Dassault, from Istres.

The two air defence squadrons (EC 1/2 right and EC 2/2 above) at Dijon-Longvic operate the AdA's small fleet of Mirage 2000-5Fs, the four other air defence units flying unmodified Mirage 2000B/C-S4/S5s. Although it is removeable, the refuelling probe is always fitted for use with C-135FR and Transall tankers, and for refuelling from other Mirage 2000s equipped with the 231-300 buddy pod. Although developed for the attack Mirages, the 2000-litre tanks (above) are routinely used by fighter variants.

GREECE — Elliniki Polemiki Aeroporia

Already operating Mirage F1s, Greece signed a contract in July 1985 for 40 Mirage 2000EG/DGs. Code-named Talos, this programme included 36 single-seat Mirage 2000EGs (210 to 245) and four two-seat Mirage 2000DGs (201 to 204). Greek Mirage 2000s are fitted with the ICMS Mk 1 (Integrated Counter Measures System Mk 1) electronic warfare/self-defence suite, and an undisclosed number of Greek Mirage 2000s have been modified to the 2000EG-S3 standard which allows the carriage and firing of the AM 39 Exocet anti-ship missile. Reports that a Greek Mirage 2000 shot down a Turkish F-16D during an encounter over the Aegean Sea cannot be confirmed.

The EPA (Hellenic Air Force, HAF) was extremely satisfied with its first-generation Mirage 2000EG/DGs and decided to order 15 new-build Mirage 2000-5 Mk 2s, plus update kits to bring 10 older aircraft to the same standard. Announced in May 1999, the contract

also includes technology transfer and the 10 Mirage 2000EG/DGs will be modernised by Hellenic Aircraft Industries at Tanagra. However, the Greek development aircraft, a two-seater, is expected to fly at Istres in mid-2003, and the first new-build production Mirage 2000-5 Mk 2s will be delivered at the rate of two per month from early 2004.

The EPA's Mirage 2000-5 Mk 2s will be equipped with NATO-compatible SATURN (Second-generation Anti-jam Tactical UHF Radio for Nato) frequency-hopping encrypted radios and an ICMS Mk 3 self-defence suite. The deal also includes the delivery of the Opera mission planning system. It was initially announced that the other Mirages would not be modernised so that the type would still be capable of using the Super 530D missiles, which are not compatible with the RDY 2 radar. However, it is possible that all remaining Mirages will be updated in batches.

114th Pteriga Mahis (Combat Wing)
No. 331 Mira (Squadron) 'Aigeas'
Tanagra
No. 332 Mira (Squadron) 'Geraki'
Tanagra

Greece's two-seaters fly with 332 Mira, which acts as the type OCU, as well as having a front-line commitment. Some aircraft wear the 114 Pteriga Mahis badge on the fin.

INDIA — Indian Air Force

The Indian Air Force is one of the largest Dassault customers, with no fewer than 59 Mirage 2000H5/H/TH5/THs in service or on order. The Indian Mirages were purchased in three batches of, respectively, 40, nine and 10 aircraft. In

Although they have a primary air defence role, with secondary conventional bombing duties, unconfirmed reports suggest that the Mirage 2000s were the IAF's first aircraft capable of delivering nuclear weapons.

Indian service, the Mirage 2000 is known as the Vajra (Thunderbolt).

The first batch of 40 Mirages was ordered in October 1982, with deliveries starting in June 1985. The first 26 single-seaters and first four two-seaters, respectively designated Mirage 2000H5 and Mirage 2000TH5, were powered by interim M53-5 engines, but the last 10 single-seaters of the first batch were delivered fresh from the factory with the higher-rated M53-P2 turbofan. The older aircraft were all eventually retrofitted, becoming 2000Hs and 2000THs in the process. The second batch, of nine aircraft, was ordered in

March 1986, with deliveries from April 1987 to October 1988. In all, the first and second batches comprised 42 single-seat Mirage 2000s (KF101 to KF142) and seven two-seaters (serial numbers KT201 to KT207). It should be noted that eight aircraft were extensively damaged when a hangar roof collapsed. They were all eventually repaired.

The last batch of 10 Mirage 2000s will be delivered between early 2003 and early 2004. These aircraft differ from their predecessors in having the RDM 7 radar instead of the older RDM 4, plus upgraded mission computers

and an improved navigation suite. The RDM 7 offers increased look-down performance, plus air-to-sea surveillance modes.

It has often been stated that some of the Indian Mirage 2000s had been equipped with the Antilope terrain-following radar coupled with two Uliss 52 inertial navigation systems (effectively turning them into ground

In 2002/03 the IAF tested some experimental camouflage schemes. This 2000TH wears an AdA 2000D/N-style green-grey scheme, and also shows evidence of a desert pattern.

attack aircraft), but informed sources within Dassault and Thales confirmed that all IAF Mirages were in fact equipped with the RDM radar.

Indian Mirage 2000s are equipped with the ATLIS laser designation pod, used in conjunction with various laser-guided munitions such as BGL 1000 and Paveway II. They have been used operationally against Pakistan, destroying several targets in the Siachen Glacier area, the highest battlefield in the world. The last 10

Mirage 2000Hs due to be delivered from 2003 will also carry the ATLIS laser designation pod, The IAF has also purchased the Litening II pod for the Mirage 2000 fleet, this system providing night-time designation and improved reconnaissance capability.

At the time of writing, a major contract was thought to be under negotiation for between 100 and 150 Mirage 2000-5 Mk 2s for the IAF, possibly to replace MiG-23 'Floggers'. Persistent rumours have mentioned the

possibility of a purchase of 126 aircraft: 36 to be built in France and 90 to be assembled in India by HAL, effectively supplanting the SEPECAT Jaguar on the Bangalore assembly line. These aircraft could be armed with a Russian air-to-air missile, possibly the R-77 'Amraamski' which would offer a certain level of armament commonality with the new IAF Su-30MKI 'Flanker'. With the signature of such a contract, the IAF would become the largest Mirage 2000 operator outside France.

Although based at Gwalior in Central Air Command, the IAF's Vajra fleet often deploys to Western Air Command bases for exercises and operations, notably during times of tension with Pakistan.

Central Air Command
No. 1 Sqn 'Tigers' — Gwalior
No. 7 Sqn 'Battleaxes' — Gwalior

Southern Air Command
ADA — Bangalore

PERU – Fuerza Aérea del Peru

The Peruvian Air Force received 12 Mirage 2000EP/DPs. Initially, Peru had planned to order 26 Mirage 2000s (including four two-seaters) but, due to financial constraints, this order was eventually cut to 10 single-seat Mirage 2000Ps (serial numbers 050-054, 060-064) and two two-seat Mirage 2000DPs (serial numbers 193 and 195). Deliveries to Peru started in December 1986 after

the first pilots had been trained in France. The role played by these aircraft during the border skirmishes with Ecuador is unclear. As far as is known, none has been lost in FAP service.

Grupo Aereo de Caza 4
Escuadrón de Caza-Bombardeo 412
    Base Aérea 'Mariano Melgar', La Joya

QATAR – Qatar Emiri Air Force

To replace outdated Mirage F1s then in service, Qatar placed an order in August 1994 for 12 Mirage 2000-5s, split into nine single-seaters (QA90 to QA98) and three two-seaters (QA86 to QA88). The first Qatari pilots were trained by French instructors at Mont-de-Marsan. A dedicated unit, Escadron de Transformation Temporaire 85.330 (Temporary Training Squadron 85.330)

had been set up in September 1996 to train Taiwanese pilots, and the Qataris also took advantage of that squadron/programme. Qatari Mirage 2000-5EDA/DDAs were delivered in three batches of four.

No. 1 Fighter Wing
No. 7 Air Superiority Squadron
    Al Udeid, Doha

A Mirage 2000-5EDA is seen at Bordeaux-Mérignac in December 1997 prior to delivery. It has yet to receive Qatari national insignia.

TAIWAN – Republic of China Air Force

With the ever-present threat of mainland China, the Republic of China Air Force (ROCAF) needed a high-tech interceptor which could simultaneously engage large numbers of opposing fighters and bombers. While the numerical superiority of the continental Chinese Air Force could not be disputed by Taiwan, the ROCAF realised that an

Taiwan's Mirage 2000s carry a ROCAF serial, in this case 2026, which is based on the Dassault build-number (EI26).

interceptor armed with new active weapons was the answer to its stringent requirement. Dassault offered the Mirage 2000-5 and the fire-and-forget radar-guided MICA missile which represented a quantum leap over the outdated F-104 Starfighter interceptors armed with IR-guided Sidewinder missiles then in service.

Announced in 1992, the contract for 60 Mirage 2000-5s also included an undisclosed number of MICA and Magic missiles, plus a full-mission dome flight

simulator produced by Sogitec and two cockpit procedure trainers. The order was split between 48 single-seat Mirage 2000-5EIs (serial numbers 2001 to 2048, construction numbers EI01 to EI48) and 12 two-seat Mirage 2000-5DIs (2051 to 2062, DI01 to DI12). The 60 Mirage 2000-5EI/DIs were delivered between May 1997 and November 1998 to replace F-104s which were suffering from serious flight-safety problems.

Taiwan's Mirage 2000-5s serve with a dedicated air defence unit, the 2nd (499th) Tactical Fighter Wing stationed

at Hsinchu air base, and specialise in long-range air defence/air superiority missions. Within the 2nd TFW, one unit (41st TFS) is dedicated to conversion and training, whereas the two others (42nd and 48th TFS) are fully operational front-line squadrons. The 2nd TFW maintains an extremely high level of readiness, and constantly practises combat air patrols, escort missions and quick reaction alerts.

The ROCAF currently operates three types of modern fighters, each with a dedicated role. Whereas the F-16A/B Block 20 Fighting Falcon currently handles air-to-ground missions, the IDF (Indigenous Defence Fighter) specialises in the low-level, short-range air defence role. According to Taiwanese pilots, the Mirage 2000-5 carries much more fuel and has a much better range than the IDF. As such, it is ideal for long-range, high-altitude missions against multiple bogeys. It is the ROCAF's main air defence fighter, and a classified number of Mirages are held at six-minute readiness, ready to respond to any intrusion into Taiwanese airspace. In times of heightened tension, readiness can be brought down to two minutes,

The 2nd TFW has an alternative 499th TFW designation, introduced to hinder assessments of ROCAF strength.

with the pilots already strapped in their cockpits.

Soon after the first deliveries, a worrying problem appeared: in some conditions, mist formed on the windscreen, blocking forward vision. The ROCAF immediately asked Dassault to find a solution. French engineers and test pilots came up with a temporary answer to the problem, asking the Taiwanese pilots to turn on the heat to de-mist the windscreen. Although this was acceptable in the short term, a more effective solution had to be found, and Dassault decided to send the instrumented Mirage 2000 C01 to Hsinchu for an in-depth test programme spread over two months and about 30 flying hours. A permanent solution to the problem was found, and all aircraft were eventually modified.

In operational service, the Dash 5 is highly praised by ROCAF pilots for the efficiency of its fly-by-wire flight controls, which make the aircraft safe and easy to fly, even at very low speed where it excels. They say that the fighter is extremely agile, that its engine is powerful, and that the RDY radar and the countermeasures suite are excellent.

ROCAF Mirage 2000-5 pilots typically log about 180 flying hours per year, and flying at Hsinshu is spread over seven days per week, with no break during the weekend. Taiwanese Mirage pilots regularly practise dissimilar air-combat training with F-16s and IDFs. For combat mission debriefing, the Mirage 2000-5s can be fitted with the Serpam data restitution system which allows the pilots to accurately determine at squadron level who won the day, and how, and who lost the fight. ROCAF Mirages are not equipped with an inflight refuelling probe, and mission duration typically varies from 1 hour clean to 2 hours with two drop tanks.

For its Mirage 2000-5s, the ROCAF fields three weapons, allowing targets to be engaged from very long to very short distances: the MICA long-range missile, the short-range IR-guided Magic missile and the DEFA gun. The single-seat Mirage 2000-5EI is equipped with two DEFA 554 30-mm guns, and the two-seaters can carry a CC630 pod. Two wing pylons are available for 1700-litre (374-Imp gal) drop tanks, nearly doubling the operational range of the Mirage 2000-5, and a 1300-litre (286-Imp gal) external tank can also be fitted on the centreline station. The overall number of MICAs sold to the ROCAF is a well-held secret, but various sources, including – among others – *Jane's Air-Launched Weapons*, mention a total of 960 missiles. To supplement the RF-5 Tigers and the soon to be introduced reconnaissance pod for its F-16s, the ROCAF has ordered four Thales ASTAC pods for its Mirage 2000-5s.

2nd (449th) Tactical Fighter Wing
41st TFS	conversion and training unit	Hsinchu
42nd TFS	air defence unit	Hsinchu
48th TFS	air defence unit	Hsinchu

UNITED ARAB EMIRATES – UAE Air Force

The United Arab Emirates ordered 36 Mirage 2000EAD/RAD/DADs in May 1983. Based at Al Dhafra air base, the UAEAF Mirages are collectively known as SAD 8 (Standard Abu Dhabi 8). The aircraft were delivered in three variants: the single-seat 2000EAD which is fully multi-role, the single-seat 2000RAD which specialises in reconnaissance missions with the COR 2 and Harold reconnaissance pods, and the two-seat 2000DAD mainly used for conversion training. In all, 22 Mirage 2000EADs (731 to 752), eight Mirage 2000RADs (711 to 718) and six Mirage 2000DADs (701 to 706) were delivered by Dassault. These aircraft can carry the PGM-1A/2A/3A/1B/2B/3B Al Hakim missile, a precision guided munition also known as the PGM-500 when fitted with a 500-lb warhead or PGM-2000 with a 2,000-lb warhead. It is unclear which PGM variant is operational on the UAEAF Mirages, but recent Dassault pictures show a Mirage 2000 carrying an Al Hakim on the forward left fuselage station, suggesting this is a 500-lb version. UAEAF Mirage 2000s are all fitted with Elettronica ELT/158 radar warning receivers coupled with ELT/558 jammers.

The Mirage 2000RAD, the only dedicated reconnaissance variant of the Dassault delta fighter, is equipped with the COR 2 and Harold reconnaissance pods, which offer exceptional capabilities. Both can be fitted on the centreline station, but their side-to-side field of view is somewhat blocked when drop tanks are carried. The COR 2 is capable of both high- and low-level high-speed reconnaissance missions, and is equipped with a fan of four Omera 35 cameras with 150- to 600-mm (6- to 24-in) lenses suitable for a wide range of low- and high-level mission profiles, an Omera 40 horizon-to-horizon camera for low-altitude work, and a Super Cyclope sensor linked to an Enertec V1000ABF recorder for infrared imagery. With its 1700-mm (67-in) f1/5.6 focal length Omera 38 camera, the Harold pod is optimised for strategic reconnaissance. The imagery provided by the powerful Omera 38 camera is exceptionally good, and allows high-resolution images to be gathered from stand-off distances (1-m/3-ft resolution

at 100 km/62 miles in daylight). Three types of films are used: 75 m/246 ft (600 exposures), 105 m/344 ft (800 exposures) and 150 m/492 ft (1,300 exposures). To assist aiming, the Harold pod also comprises an Inspectronic CCD TV camera with a 30- to 300-mm (1.2- to 12-in) zoom lens.

In November 1998, it was officially announced that the UAEAF had signed a major contract to boost its operational effectiveness, becoming the fourth country to order second-generation Mirage 2000s. Called Bader 21, the new programme is the largest export contract won by Dassault, and it encompasses massive technology transfers, including the creation of a dedicated flight test centre in the UAE. Initially, it was planned that 30 new Mirage 2000-9s would be built in France, and that the surviving 33 SAD8 aircraft would be brought to the same standard. However, due to various reasons, this has now changed, and 32 new-build fighters (20 single-seaters and 12 two-seaters) will be delivered, plus 30 update kits. The first two SAD8 aircraft – a single-seater and a two-seater – will be upgraded in France to validate the process, while the remaining 28 fighters will be modernised in the UAE. As a result, more than 150 Dassault personnel will work in the UAE to help with that programme.

The Mirage 2000-9 is the most modern variant of the Mirage 2000 family. Its weapon system will be progressively developed as part of an incremental approach: the first aircraft will be of the SAD91 standard, which only offers air-to-air functionality, including the use of the new MICA IR

The original UAEAF buy included six 2000DAD two-seaters. This figure is to be swelled by 12 more new-build 2000-9s, which may have more of a combat role.

missiles. From 2004, the SAD92 standard offering the full spectrum of air-to-air and air-to-ground modes will be available. The first development Mirage 2000-9, a two-seater, first flew on 14 December 2000, while the second one, a single-seater, first took to the air on 25 January 2001. In all, four aircraft take part in the development programme. Moreover, four integration benches are dedicated to this development effort. It is worth noting that components of the new-build aircraft come from stored and uncompleted Mirage 2000s initially ordered by Jordan (the Jordanian contract had been cancelled in 1991). The first Mirage 2000-9 is slated to be delivered to the UAE in April 2003, and the aircraft will be delivered in batches of four every two or three months.

The Mirage 2000-9 is a highly modified variant of the Mirage 2000-5 Mk 2 which will only be operated by the

The UAE's Mirage 2000EADs have served in both desert and light grey camouflage. The new-build aircraft are being delivered in light grey.

UAEAF. It is equipped with a number of specific systems such as the Thales/Elettronica IMEWS (Integrated Modular Electronic Warfare System), the Thales Thomrad encrypted radio, the Damoclès pod (known as Shehab) and the FLIR pylon (Nahar), and a datalink compatible with the GATR (Ground/Air Transmit/Receive) system selected by the UAEAF. The appearance of new antennas under the forward fuselage and on the spine of the aircraft is linked to the adoption of the IMEWS. The improved air-conditioning/de-misting system developed for the Republic of China Air Force has also been adopted on the 2000-9.

The new variant will be armed with MBDA MICA EM and IR air-to-air missiles and with the MBDA Black Shaheen cruise missile, a derivative of the Scalp EG/Storm Shadow which was selected by the UAEAF, beating the PGM-4 Pegasus missile offered by GEC-Marconi. The UAEAF has also ordered six simulators from Sogitec. Mirage 2000-9 maintenance documentation is provided in both paper and electronic formats. With the advent of the 2000-9, there will not be any dedicated reconnaissance variant any more, as all 2000-9 aircraft will be able to carry the reconnaissance pods.

UAEAF – Western Air Command
I Shaheen Sqn	Al Dhafra	2000EAD
II Shaheen Sqn	Al Dhafra	2000RAD

Botswana Defence Force

Photographed by Chris Knott and Tim Spearman/API

To replace the elderly Strikemasters as the BDF's combat aircraft, 10 CF-5As and three CF-5Ds were acquired through Bristol Aerospace, which upgraded them prior to delivery. The aircraft were flown by An-124 transport direct into the base at Thebephatshwa ('colourful shield' in the local Setswana language), built in 1995 near the town of Molepolole. The first arrived in September 1996, and the initial batch was complete by October 1997. Two further CF-5Ds were acquired in 2000.

Botswana gained its independence from the UK peacefully in 1966, but it was not until 1977 that the Defence Force was formed to counter growing incursions from neighbouring Rhodesia/Zimbabwe. Today a small but competent Air Wing defends the nation and its valuable diamond/mineral deposits, while also playing a crucial role in the international fight against the rhino horn and ivory trade.

The BDF's first rotary-wing assets were two *AS 350B Ecureuils* which were delivered in 1985. Today the type remains an important type, with eight in use by Z23 Squadron. This group is seen at Francistown during an exercise to the base close to the Zimbabwe border.

Following the AS 350, the next helicopter type to be procured was the Bell 412SP (below), of which three were delivered in 1988. Subsequent deliveries raised Z21's complement to five, plus a single AB 412. The VIP Squadron, which flies from Gaborone's international airport, operates a single Bell 412EP (above) which was acquired in 2002.

The CF-5 fleet trains primarily for the attack mission, for which rockets and free-fall bombs are available. The two internal 20-mm cannon give a measure of air defence capability. Training for prospective F-5 pilots includes a brief spell in the United States flying the T-38 Talon, as the skills gap between the BDF's own PC-7 trainers and the CF-5 is considered too great.

Botswana Defence Force

Z10 is the main transport squadron, based at Thebephatshwa. On its roster are three C-130s, two CASA 212-300s (left) and two CN-235M-100s (above). The latter were acquired in 1987, superseding a pair of Britten-Norman BN-2A-111 Trislanders that served from September 1984 to 1991. The C.212s replaced two Skyvan 3Ms in 1993. Until Thebephatshwa opened in 1995 most of the Air Wing's transports operated from Notwane, which was subsequently closed and swallowed up by the expansion of the capital, Gaborone.

Above: For years the backbone of the BDF Air Wing has been the BN-2 Defender fleet. The Air Wing's first aircraftt, BN-2A-21 OA-1, was delivered in September 1977. A second aircraft was temporarily impounded in Nigeria after a forced landing during its delivery flight, and was later written off. A further five were delivered between 1978 and 1981, and six BN-2B-20s followed. Today the survivors operate from Francistown in northeast Botswana, where this group is seen. They are used for light transport and medevac duties, and are ideal for use in a large country with little transport infrastructure.

Above: The VIP Squadron operates a Beech King Air 200 and this single Gulfstream IV, which was delivered in 1991 as the successor to Botswana's second BAe 125. In August 1988 the previous 125 was attacked and forced to land by an Angolan MiG-23.

Left: In 1996 the US government offered surplus C-130Bs to various African nations, and Botswana picked up two, the first arriving in early 1997. Z10 Squadron later added a third Hercules.

Above: Ten Cessna O-2s were donated by the US government in 1994 for use by Z3 Squadron. Flying from Francistown, the five remaining in service are mainly used for anti-poaching operations.

Right: The BDF's first combat aircraft was the BAC 167 Strikemaster, of which nine were delivered in 1988 and four more in 1994. This example is retained as a gate-guard at Thebephatshwa.

Botswana Defence Force Air Wing

BDF serials consist of a two-letter type-specific designator (in parentheses below) and a sequential number. In some cases the two-letter type code has been reused: the 'OJ' of the CF-5 was previously used by the Strikemaster

Unit	Equipment	Base
Z3	5 x Cessna O-2 (OE)	Francistown
Z7	6 x Pilatus PC-7 (OD)	Thebephatshwa
Z10	3 x Lockheed C-130B (OM), 2 x Airtech CN-235 (OG), 2 x CASA 212-300 (OC)	Thebephatshwa
Z12	10 x Pilatus Britten-Norman BN-2A/BN-2B (OA)	Francistown
Z21	1 x Agusta-Bell AB 412, 5 x Bell 412SP (OH)	Thebephatshwa
Z23	8 x Eurocopter AS 350/AS 350BA (OF)	Thebephatshwa
Z28	9 x Canadair CF-5A, 5 x Canadair CF-5D (OJ)	Thebephatshwa
VIP Sqn	1 x Gulfstream IV (OK), 1 x Bell 412EP (OH) 1 x Beech King Air 200	Sir Seretse Khama IAP, Gaborone

Above: A few Aviatika 890U (background) and Streak Shadow microlights are used for rhino patrols from Thebephatshwa.

Right: The BDF's first training unit was established at Notwane in 1979 with a pair of Cessna 152s. Despite being replaced in the training role by Bulldogs, one of the Cessnas remained on flying strength until recently, and is now used for ground instruction by the BDF Technical Training School.

Below: Six Bulldogs were delivered for training in 1980, but in February 1990 were replaced by seven Pilatus PC-7s, of which these six remain in use with Z7 Squadron. The PC-7s fulfil most of the BDF's training requirements, although some special training needs have to be met overseas.

Former Yugoslavia
Part 2: Serbia and Montenegro, and Slovenia

Serbia and Montenegro (Yugoslavia)

Serbia's air force had its beginnings with two hot air balloons, bought in 1909 during a period of imminent war with the Austro-Hungarian Empire. The first air force unit was formed by order of the Ministry of the Army on 24 December 1912. An air force command was established at Nis, eastern Serbia, consisting of squads with 11 aircraft, a balloon unit, a plant for producing hydrogen, and facilities for pigeon post. Serbian pilots had their baptism of fire during the Balkan Wars of 1912 and 1913, against Turkey. A detachment of aircraft was sent as reinforcement to Montenegrin and Serbian forces during the siege of Skadar (presently Albania).

World War I started on 28 July 1914 when Austria-Hungary declared war on Serbia. During the three initial battles, which were won by the Serbian army, aircraft were used for reconnaissance. The joint German/Austro-Hungarian offensive beginning in October 1915 forced the Serbian army to retreat toward territories in the south Balkans. On the island of Corfu, the remains of the Serbian army were reorganised and sent to the Salonika front. There, under the auspices of the French army, the air force was restored.

Post-war, the dissolution of the Austro-Hungarian Empire enabled the creation of new Slavic states. The Kingdom of Serbs, Croats and Slovenes was created in December 1918 (from 1929, it became known as the Kingdom of Yugoslavia). Serbian squadrons formed the nucleus of the air force of the new state and were mainly equipped with French aircraft, such as the Breguet XIV, Dorand and the Nieuport family. Serbian aviators were then joined by Slovenes, who had fought against the Antante during World War I.

Prior to World War II, Yugoslavia tried to modernise its air force by acquiring technology from anyone who was prepared to sell it, and by building aircraft itself. The Hurricane, Blenheim and Dornier Do 17 were produced under licence, and IK-3 fighters were manufactured independently.

World War II

The attack by Axis forces on Yugoslavia began on 6 April 1941, countered by some 330 operational aircraft. Resistance lasted only two weeks. Some aircraft fled Yugoslavia and flew to the Middle East. The Yugoslav navy's Do 22H seaplanes and their crews became part of the UK Royal Navy, flying missions in the Mediterranean from the base at Aboukir, Egypt. Guerrilla resistance developed on Yugoslav soil, together with a civil war between many factions of which Communist forces, headed by Josip Broz Tito, dominated. His partisans had a modest air force, used in accordance with guerrilla tactics, equipped with aircraft that had been obtained by the defection of crews from the army of the independent state of Croatia and by seizure. The bias in London toward Tito's partisans was demonstrated in 1944 by the creation of two squadrons in north Africa equipped with Spitfire and Hurricane aircraft, which bore a combination of RAF roundels and red stars on their fuselages. After World War II, that symbol became the official emblem of the Yugoslav air force (Jugoslovensko ratno vazduhoplovstvo – JRV). During the war, Nos 351 and 352 Yugoslav Squadrons, as part of the RAF's Balkan Air Force, undertook support of partisans in western Yugoslavia.

In the eastern part of the country, two divisions of the partisan air force were formed with the support of the Soviet army and equipped with Yak-1 fighters and Il-2 attack aircraft. The first peacetime deployment of JRV units was to Yugoslavia's western borders in anticipation of conflict in defining post-war borders with Italy. Anglo-American forces did not allow Tito to spread to the west, resulting in border tensions and an incident in which Yugoslav Yak-3 fighters downed two American C-47 transport aircraft.

Post-war rebuilding

JRV units were organised into 40 squadrons. By the end of 1945, 677 aircraft were in use, of which 471 were combat aircraft: Pe-2FT, Il-2, Yak-1, Yak-3, Spitfire Mks Vc and IX, Hurricane Mk IV, and a small number of trophy aircraft of German and Italian origin. Development plans were produced that relied on Soviet and domestic aircraft, but Tito's ambitions for the creation of a regional force were halted by a conflict with Stalin in 1948. Yugoslavia found itself completely isolated from Eastern Bloc states which, together with the Soviet army concentrated their forces for armed intervention. The JRV withdrew the majority of its combat regiments deep inside the country, and for secrecy reasons changed the markings of all units, in expectation of a war which, in the event, never occurred.

Tito sought protection from the West, one result being that from the end of 1950 the JRV joined the US Military Assistance Program. New aircraft and helicopters for the modernisation of the air force were acquired from the USA, United Kingdom, Norway and France. The first series of shipments included aircraft from World War II, such as 150 F-47D Thunderbolts, 140 Mosquito Mk VIs and Mk 38s, 20 C-47s, and eight Avro Anson Mks I and V. In March 1953 the JRV received its first jet training aircraft, the T-33A, from the USAF. The period of friendly relations with the West ended in 1956 when the relationship between Tito and the new Soviet leader, Krushchev, normalised. Washington stopped MAP deliveries, but aircraft continued to arrive via commercial arrangements. The aircraft reached the RV i PVO (in 1959, the service became known as Air Force and Air Defence – Ratno vazduhoplovstvo i protivvaz-

Yugoslav military aircraft designations

In 1962 the Yugoslav Air Force introduced an internal system of marking aviation equipment, which is based on role abbreviations. The motive was the acquisition of Soviet MiG-21F-13 fighters, which by regulations within the purchasing contract had to be kept under special security conditions. The marking system was later used selectively. Exceptions were transport aircraft which were acquired in small numbers, and Alouette IIIs.

Role designations: *Helikopter* – helicopter, *Izvidzhach* – reconnaissance, *Jurishnik* – attack, *Lovac* – fighter, *Nastavni* – training, *Transporter* – transport aircraft, *Veza* – liaison.
Helicopter role designations: *Naoruzhani* – armed, *Osnovna namena* – utility, *Protivpodmornichki* – ASW, *Sanitetski* – medical.

Yugoslav designation	original designation	note
L-12	MiG-21F-13	
L-14	MiG-21PFM	
L-14i	MiG-21R	
L-15	MiG-21M/MF	
L-15M	MiG-21MF	recon variant
L-17	MiG-21bis	
L-17K	MiG-21bis-K	
L-18	MiG-29	
NL-12	MiG-21U	
NL-14	MiG-21US	
NL-16	MiG-21UM	
NL-16Sh	MiG-21UM	unarmed variant
NL-18	MiG-29UB	
J-20	Soko Kraguj	
J-21	Soko J-1 Jastreb	
IJ-21	Soko RJ-1 Jastreb	
NJ-21	Soko TJ-1 Jastreb	
J-22	Soko Orao 2	
NJ-22	Soko Orao 2D	two-seat J-22
IJ-22	Soko Orao 1	
INJ-22	Soko Orao 1	two-seater IJ-22
INJ-22M	Soko Orao 1	navy recon
N-60	Soko G-2A Galeb	
N-61F	Zlin 526F	
N-61M	Zlin 526M	
N-62	Soko G-4 Galeb	
N-62T	Soko G-4 Galeb	target tug
N-62Sh	Soko G-4 Galeb	unarmed variant
N-63	Utva Lasta	
T-70	Antonov An-26	
V-50	Utva-60H	floatplane
V-51	Utva-66	
V-52	Utva-66H	floatplane
NV-52	Utva-66H	training variant
V-53	Utva-75	
HT-40	Mil Mi-8T	
HT-40E	Mil Mi-8T	Elint variant
HT-41	WSK Mi-2	
HO-42	SA 341H Gazelle	
HSN-42	SA 341H Gazelle	SAR variant
HI-42	SA 341H Gazelle	Hera
HN-42M	SA 341H Gazelle	Gama, suffix M is abbreviation for Maljutka ATGM
HP-43	Ka-25BSh	
HP-44	Mil Mi-14PL	
HO-45	SA 342L Gazelle	
HN-45M	SA 342L Gazelle	Gama, suffix M is abbreviation for Maljutka ATGM
HP-46	Kamov Ka-28	

Above and left: Just four single-seat MiG-29s survived the NATO onslaught. The 127.Iae flies only sparingly due to the poor condition of the aircraft and the lack of funds for fuel.

Right: The single MiG-29UB (NL-18) which survived the war is a precious asset for the 127.Iae. Here it is armed with four B-8M 20-round rocket pods.

MiG-21bis fighters serve on both ground attack (above) and air defence (below) duties. 126.Iae MiGs are the only RV i PVO aircraft to mount an armed alert.

Above: The 353.iae at Ladjevci is the VaK's reconnaissance unit. As well as IJ-22 Oraos it has two MiG-21Rs (illustrated), the survivors of 12 delivered in 1968/70, and a single MiG-21M which carries a high-altitude recce pod.

dushna odbrana) at very favourable prices. The 1962 signing of a large contract for the acquisition of Soviet military technology brought an end to deliveries of US military equipment to Yugoslavia without special approval from the US Congress. By that time, through MAP and commercial acquisition, the JRV had received 231 F-84G Thunderjets, 121 F-86E Sabres, 130 F-86D Sabre 'Dogs', 95 T-33A/TV-2s, 22 RT-33As and 10 Westland S-51 Mk 1Bs.

Equipment from East and West

From the early 1960s until the break-up of the former Yugoslavia in civil war in 1991/92, Belgrade's foreign policy relied on balancing relationships with East and West. An effect of this was that the equipment of the RV i PVO reflected various influences. The following aircraft were acquired from the USSR: 260 MiG-21 and 16 MiG-29 fighter aircraft, two An-12B transport aircraft, 15 An-26s, six Yak-40s, and 24 Mi-4, 93 Mi-8T, four Mi-14PL, six Ka-25BSh and two Ka-28 helicopters. The domestic aviation industry equipped attack and reconnaissance squadrons with indigenous aircraft. The Soko factory in Mostar (Herzegovina) and Utva in Pancevo (Serbia) produced 130 G-2 Galeb training and light attack aircraft,

85 G-4 Galebs, 177 Jastreb attack and reconnaissance aircraft, and 43 Kraguj COIN aircraft. In co-operation with Romania, the Orao combat aircraft was produced and 115 were delivered to the RV i PVO.

The aviation industry relied on technological support from the UK, whose companies delivered jet engines and electronics. The first step in mastering helicopter production was the assembly of the S-55 under licence from

For training sorties the MiG-21UM (a 126.Iae example illustrated) is heavily used, as two pilots can receive the benefit of the flight at a time when fuel shortages curtail the RV i PVO's peacetime training schedule.

Westland in the 1960s. In October 1971 a contract for licensed production of the Gazelle was signed. The Soko factory assembled 163 helicopters in SA 341H and SA 342L variants for the RV i PVO and the police.

Aleksandar Radic

Ratno vazduhoplovstvo i protivvazdushna odbrana Vojske Srbije i Crne Gora – RV i PVO

The organisation of the RV i PVO when civil war broke out had been defined in 1986. The country was divided into three military regions representing areas of responsibility of three corps (Korpus). Every corps had units from all service branches. At the beginning of 1991 the corps that covered the northwestern region (5th Korpus RV i PVO, HQ Zagreb) and the southwestern region (3rd Korpus RV i PVO, HQ Nis) had identical flying units: one fighter aviation regiment (lovachki avijacijski puk – lap), one aviation brigade (avijacijska brigada – abr) for air support, two fighter-bomber aviation squadrons (lovachko-bombarderska avijacijska eskadrila – lbae), a reconnaissance aviation squadron (izvidzhachka avijacijska eskadrila – iae), one aviation brigade with transport aviation squadron (transportna avijacijska eskadrila – trae), one transport helicopter squadron (transportna helicopterska eskadrila – trhe), and two anti-armour helicopter squadrons (protivoklopna helicopterska eskadrila – pohe). The area of responsibility of the 1st Korpus RV i PVO, with its HQ in Belgrade, was the central region. It also had control over the maritime forces, because within its structure was the 97.abr for naval support, equipped with Jastreb attackers, G-4 Galebs, IJ-22 Orao for reconnaissance, CL-215 fire-fighting aircraft, Mi-8T transport helicopters, and Ka-25BSh, Ka-28 and Mi-14PL anti-submarine helicopters.

The Military Aviation Academy came under the control of the RV i PVO Command in peacetime. It consisted of three aviation and one helicopter regiments, which during war joined the corps structure according to areas of responsibility. The RV i PVO Command also used the 138th transport aviation brigade (transportna avijacijska brigada – trabr) from Batajnica, which had squadrons with fixed- and rotary-wing VIP aircraft, a utility helicopter squadron and Elint aircraft. In all, during the war the RV i PVO had seven iae, 12 lbae, four iae and four pohe, each with 12 to 16 aircraft at their disposal. Over 800 combat and auxiliary aircraft were on strength.

The end of the federal union found the RV i PVO at the beginning of an extensive modernisation plan. Under the codename Sloboda (Freedom), a domestic supersonic fighter jet was being developed that relied on Western technological support. Production of a medium transport helicopter had been planned. During the first months of open war, the Soko factory delivered the last J-22, NJ-22 Orao and G-4 Galebs, marking the end of the Yugoslav aviation industry. Some machines from the Soko factory were moved to the Utva factory in Serbia, and a number of machines that could not be moved were destroyed by explosives.

The civil war

Nationalist passions destroyed Yugoslavia in a brutal civil war. The RV i PVO was involved in clashes from the very first incident on 17 August 1990 when, faced with the threat of intervention by MiG-21s from Bihac air base, Croatian police had to cancel a helicopter raid against Serbian rebels in the region of Dalmatia. At the end of June 1991 the RV i PVO transported soldiers and federal policemen to Slovenia. Their mission was to assume control of border crossings and secure the supremacy of the federation. The Slovenes answered with fire, downing two helicopters and imprisoning many disembarked members of the federal armed forces. The RV i PVO reacted with attack missions, hitting TV and radio transmitters and positions of the Slovenian territorial defence. Subsequently, in accordance with a political agreement, federal forces left Slovenia and took all RV i PVO equipment with them.

The spiral of violence was spreading and during the summer of 1991 Croatian authorities and Serbian forces entered armed conflict in the historical region of Vojna Krajina. The RV i PVO initially performed low-level flights in a show-of-force. Counter-strikes on Croatian forces followed whenever aircraft were fired at from the ground. From August 1991 the federal armed forces commenced a war campaign against the Croats, in which the RV i PVO was engaged primarily in close air support missions and transport. Croatian forces blocking military barracks of the federal forces were attacked. The RV i PVO gradually abandoned airports positioned outside Serbian ethnic territories. Hostilities formally ended with a truce that came into effect on 3 January 1992. By then, Croatian forces had shot down 23 RV i PVO aircraft and helicopters: five MiG-21s, two Oraos, three G-4 Galebs, seven Jastrebs, three G-2 Galebs, one Mi-8T and two Gazelles. From March 1992 the armed conflict spread into Bosnia and Herzegovina. The RV i PVO attacked Croatian regular troops that penetrated across the river Sava, and paramilitary formations from Herzegovina. In April and May two MiG-21s, one G-4 Galeb and three Jastrebs were shot down.

Slobodan Milosevic, who assumed control over political and military power centres in Belgrade, decided to withdraw federal troops and station them on the territories of Serbia and Montenegro, and to form a new state called the Federal Republic of Yugoslavia (FRY). The armed forces, now called Vojska Jugoslavije (Yugoslav Army), or VJ, received orders to withdraw from Bosnia, Herzegovina and Croatia by 19 May 1992. The RV i PVO abandoned all bases west of the Drina river, and a portion of the equipment was given to local Serbian forces.

RV i PVO in 'small' Yugoslavia

Belgrade retained control over almost 85 per cent of the aircraft with which the RV i PVO had entered the civil war, and aircraft brought from other parts of the former federation overwhelmed the bases in the FRY. At one point, Batajnica near Belgrade accommodated over 250 aircraft alone. Military needs of the new state were much reduced and therefore a number of older aircraft, mainly Jastrebs, were withdrawn from operational use.

The old insignia of red stars were removed from aircraft, replaced by roundels on the fuselage and wings divided horizontally into three fields with the colours of the FRY flag: blue, white and red. The tail flag was reduced in size and placed centrally, where previously it had stretched across the whole vertical fin. Five-digit serials, usually placed above the flag, became smaller and thicker, making it harder to read the number and identify the aircraft. Under pre-war regulations, large tactical numbers representing the last three symbols of the serial were placed on the aircraft nose above the right wing and below the left wing, while helicopters had them on their fuselage. This practice spoiled the camouflage scheme and revealed unit movements to enemy intelligence, so at the beginning of the civil war the numbers were overpainted with the basic colour of the aircraft. The VJ regulating body adopted this solution and tactical numbers were no longer painted on aircraft.

From June to September 1992 the RV i PVO was reorganised. Instead of corps with units from all branches, two corps with specialised roles were formed. The Air Defence Corps (Korpus protivvazdushne odbrane – Ko PVO) unified all forces for defending territories from air attack. It included 12 rocket battalions organised in the 250th air defence rocket brigade (250. raketna brigada PVO – rbr PVO) and the 350th rocket regiment (350. raketni puk PVO – rp PVO), equipped with the S-125M Neva-M (SA-3 'Goa') system, plus the 126th radar surveillance brigade (126. brigada vazdushnog osmatranja i javljanja – br VOJIN), which unified the early radar warning system and fighter aviation guidance and co-ordinated all air defence systems. The Ko PVO took five Kub-M (SA-6 'Gainful') self-propelled rocket regiments (samohodni raketni puk PVO – srp PVO) from the land army.

When founded, the Aviation Corps (Vazduhoplovni korpus – VaK) had 520 aircraft, which was 135 per cent of the number required. The RV i PVO Command retained supervision over the logistics brigade, aviation experimental centre, 138.trabr with 35 aircraft, and a squadron of helicopters. Under control of the Yugoslav Army general staff was the squadron responsible for the selection of pilot candidates, equipped with Utva-75 piston aircraft. During initial extensive changes to the RV i PVO organisation in 1994, the VaK gave the Ko PVO two lap with four iae of MiG-29s and MiG-21s, and HQ RV i PVO took the 352.iae 'Oluj' (Storm) equipped with MiG-21Rs and MiG-21MFs. The VaK also took the branch's logistics from the RV i PVO Command.

According to the Dayton Peace Agreement, the West Balkan states that had fought each other were required to limit their war potential. From 1996 the Subregional Arms Control Agreement allowed the FRY 155 combat aircraft and 53 attack helicopters. Surplus aircraft were removed from the RV i PVO inventory by reclassifying them as non-combat types, destroying them, or giving them to the Aviation Museum at Belgrade airport for display. Fighter aviation was left with no MiG-21PFM, U, US or MF aircraft. Due to a lack of money for overhaul, a dozen MiG-21bis and MiG-21UM aircraft were stripped. The RV i PVO finally withdrew from use the outdated attack and reconnaissance IJ-21 Jastreb, but kept 17 G-2 Galebs for basic pilot training, after removing their machine-guns, pylons, targeting devices and weapons handling installations. Twenty-one G-4 Galebs and nine MiG-21UMs were similarly disarmed. In the attack helicopter category, the victims of the arms control agreement were the HI-42 Hera reconnaissance helicopters: the RV i PVO was left with only three examples. The majority of aircraft rendered surplus

From 1966 the Soko G-2 Galeb answered the RV i PVO's basic training needs. The entire fleet, however, was wiped out during NATO attacks on the main training base at Golubovci in Montenegro in 1999.

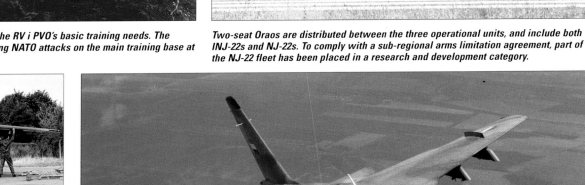

Two-seat Oraos are distributed between the three operational units, and include both INJ-22s and NJ-22s. To comply with a sub-regional arms limitation agreement, part of the NJ-22 fleet has been placed in a research and development category.

Backbone of the attack fleet is the J-22 Orao, flying from Ladjevci with the 241.lbae (above) and Batajnica with the 252.lbae (right). The IJ-22 is an earlier recce variant, lacking the afterburners of the J-22.

became exhibits at the Aviation Museum. Eight G-2 Galebs and three Jastrebs were sold to foreign buyers. Dozens of liaison Utva-66 aircraft and Gazelles were offered for sale.

In 1997, the organisation of the RV i PVO was adjusted to reflect the smaller number of aircraft. The 138.trabr was disbanded and formed the 677.trae under control of the RV i PVO HQ, and the 890th mixed helicopter squadron (890. meshovita helikopterska eskadrila – mhe) with its MiG-21Ms joining the surplus list. The 230.lbae was disbanded, its MiG-21Ms joining the surplus list. Eight MiG-21R/M/UMs from the 352.iae joined the 353.iae, which until then had been equipped solely with the Orao.

The end of the war in Bosnia and Herzegovina relieved the FRY of UN arms restrictions, but politicians did not take advantage of this. In 1996 Russia offered the delivery of 20 MiG-29s as compensation for pre-war

The single upgraded G-4M Super Galeb prototype flies with the VOC at Batajnica.

Primary training is accomplished using the Utva-75, assigned to the 251.lbae at Kovin and the 239.lbae at Golubovci.

debts, and the sale of the S-300PMU (SA-10 'Grumble') SAM system, at very favourable terms. However, Slobodan Milosevic was not interested in equipping the armed forces, and the RV i PVO lost its last opportunity to modernise before engaging in another war.

RV i PVO in the Kosovo campaign

The next war in the Balkans began in March 1998. Albanian rebels clashed with Serbian police forces in central Kosovo. The RV i PVO assisted anti-guerrilla operations with reconnaissance flights of 353.iae IJ-22 Oraos, which from low altitudes recorded movements of rebels and observed parts of the war-torn territory. The combat air force was not used at all –

it had even been ordered that MiG-21s from the 83.lap, based in Kosovo, were not to carry a combat load underwing during regular training. A detachment of two Mi-8 and two Gazelle helicopters was established at Slatina airport, near Pristina, tasked with maintaining communications and supplying troops concentrated along the border with Albania, in places the rebels used for procurement of arms and men in Kosovo.

NATO reacted to the Kosovo conflict by announcing the use of force against the FRY. In

Golubovci is home to the air force's training effort, but some of its aircraft also have secondary combat roles. The 242.lbae's G-4 Super Galebs support naval forces, as well as providing advanced/weapons training.

June 1998 an exercise called Determined Falcon took place, with 68 combat aircraft. Reinforcements arrived at bases in Italy. Both RV i PVO fighter regiments were brought to a state of readiness under the assumption that a possible NATO air strike would be directed against Kosovo. A detachment of MiG-29s transferred to the base at Nis, flying in the radar shadow of an An-26 in order to mask the relocation of the aircraft. The crisis came to a head on 4 October 1998 when the Supreme Council of Defence of the FRY issued an order to the armed forces to prepare for war. The RV i PVO mobilised reserve staff and returned reserve aircraft to operational status. Aircraft were dispersed around bases and the most valuable part of the fleet was accommodated in HASs (hardened aircraft shelters) and underground shelters. Missile units moved to their firing positions, warehouses were emptied and reserves dispersed.

However, conflict did not commence due to the Clark-Naumann Agreement, signed on 25 October 1998, under which the VJ agreed to withdraw from Kosovo a portion of those units that had been deployed there as reinforcement to the Pristina Corps. NATO acquired permission to patrol the terrain with its reconnaissance aircraft and UAVs. The text of the agreement stated that the VJ could keep 37 MiG-21s, from the 83.lap based in Slatina, in Kosovo.

Nevertheless, the peace agreement did not bring an end to clashes in Kosovo. On 15 February 1999, in the face of growing pressure from NATO, the RV i PVO once again ordered all units to prepare for war, and on 23 March it was announced that war was imminent. The following evening, NATO forces attacked the air defence system and airports of the RV i PVO, thus starting the war.

FRY plans had called for the use of the RV i PVO on a defensive basis. During the first days of the war and the initial attacks, the primary goal of branch commanders was to ensure the survival of materiel and personnel, and to prevent massive losses. After overcoming the crisis, the air defence system was gradually to join the battle. In the early stages of the war, only the MiG-29-equipped 127.lae received orders to attempt interception of NATO aircraft; this resulted in the loss of half the single-seat MiG-29s. After three days, the weakened fighter aviation restrained itself from participating in bigger actions. Later, on a few occasions, individual aircraft took off on interceptor missions, which led to the loss of yet another MiG-29 to superior NATO forces. The missile units were grouped in three belts around the largest cities in central Serbia, in Kosovo and the region of Vojvodina. Attack aviation received orders to attack Albanian rebels in Kosovo in a limited number of sorties, but in the event of a NATO land offensive they were to attack targets in neighbouring countries, primarily airports. The majority of helicopters were deployed to airstrips near the southern borders of Serbia, from where attack was expected.

RV i PVO authorities had no illusions about their forces' capabilities, but tried to make the most of their potential in a situation that could be described as checkmate by NATO. The infrastructure was exposed to systematic destruction: cruise missiles took out key targets, and smart weapons hit hangars, HASs, command posts, power stations, underground fuel reservoirs and warehouses. Carpet-bombing by USAF B-1B and B-52H bombers made runways unusable. According to official statistics, RV i PVO forces were attacked 512 times with 12,300 missiles. NATO specifically targeted the Slatina base, preventing the use of MiG-21s. An important target was the reserve base at Ponikve, in western Serbia near the border with Bosnia and Herzegovina, which was a starting point for assault units of the 98.lbap in actions in Kosovo and possible attacks against NATO's SFOR. All facilities necessary for fighter aviation, all hangars, and the Moma Stanojlovich depot in charge of overhauling RV i PVO aircraft, were thoroughly destroyed.

Air defence

The RV i PVO built up its air defence system over 40 years, combining Western and Eastern technology. Plans for its use had assumed that the system would probably be exposed to attacks from a superior opponent, so vital parts of the system were accommodated in underground facilities in granite mountains. The former Yugoslavia sold its knowledge of air defence systems to Iraq, Libya, Egypt, Syria and Angola, where the efficiency of the system was tested in real war conditions.

International isolation prevented the modernisation of this system, so the RV i PVO entered the conflict with NATO with outdated air defence equipment. Although the system was well thought out and very resistant to air attacks, it could not prevent relentless air strikes. NATO aircraft faced little danger from FRY air defences, relying on their superior technology of data-gathering, disruption of SAM radar, attacks mainly carried out from high altitudes, and the wide use of decoys. Ninety-six Neva-M and Kub-M SAMs launched by air defence succeeded in downing only two NATO aircraft. Both were attributed to the 3rd battalion of the 250.rbr PVO, which used Neva-M SAMs. That unit became the first in the world to destroy a stealth aircraft: on 26 March 1999, near the village of Budjanovci in the Srem region, a missile shot down an F-117A. On 1 May another F-117A was hit and heavily damaged, but managed to land at Spangdahlem base. A USAF F-16C from the 555th FS was shot down on 2 May above the mountain of Cer in western Serbia. During the war, the 3rd battalion was attacked 22 times, but vital parts of the system were never hit and it remained operational throughout.

The air defence systems were managed from an underground command centre of the Ko PVO, in the hill of Strazhevica in a southern suburb of Belgrade. Data from reconnaissance radar stations flowed into the command centre, which relied on a range of devices acquired from the USA, USSR and UK. For security reasons, crews were stationed outside radar cabins with a display connected to the main device by a cable more than 10 m (32.8 ft) long. Flight control service radar and meteorological radar were used for airspace reconnaissance. It transpired that the outdated Soviet P-12 and P-18 radars, which worked in decimetre and metre frequency ranges, could be used to detect stealth aircraft. Their employment was relatively safe during air attacks, as auto-guiding anti-radar missiles could not precisely locate the antenna: their primitive technology created an electromagnetic beam 1 km (0.62 miles) wide around the antenna, due to reflection from surrounding terrain, which completely confused sophisticated sensors. Nonetheless, persistent action by NATO's SEAD groups yielded results, and two-thirds of RV i PVO radars did not see the end of the war.

After the war

At the beginning of the NATO air campaign the RV i PVO had, on paper, 154 combat aircraft at its disposal, but could rely on only 40 each of operational fighter and attack aircraft. According to the arms register, which by the Vienna Document was handed over to the OSCE, the RV i PVO lost 50 combat aircraft during the conflict with NATO. However, many of the aircraft that survived were heavily damaged and lost forever to the RV i PVO, because it no longer had spare parts or depots and workshops for technical maintenance. The list of losses totalled 156 aircraft, and one-third of the missile units were destroyed. The infrastructure suffered great damage, with all hangars and three-quarters of the workshops being destroyed. Human losses, in such a high-technology war, were much lower than expected from pre-war estimates: 40 RV i PVO personnel were killed, mainly belonging to radar and SAM crews, and three pilots died on combat missions. Lieutenant General Ljubisha Velickovich, a member of the VJ general staff, was killed while visiting a station for guiding Kub-M SAMs when it was hit by an LGB. He had been commander of the RV i PVO during war preparations in October 1998.

Even after the end of the NATO air campaign, the FRY came under powerful pressure from the international community. The RV i PVO maintained a high level of combat readiness until 5 October 2000, when Slobodan Milosevic was ousted from power in a coup.

In the first months after the war, the RV i PVO restored a number of airfields to usable condition. Three months were needed for the resumption of flight training, although some units were out of action longer. Two squadrons of Oraos from the Ladjevci base were grounded until 11 May 2000, at which time the runway – which had been destroyed in April and May 1999 in a series of NATO air strikes – reopened. Due to a lack of money in the defence budget, the VJ proposed to the Serbian government a joint effort to restore the runways at Nis and Ponikve airfields and to undertake additional repairs at Ladjevci and Batajnica. Under the proposal, the restored bases would acquire equipment for navigation and for maintaining regular day and night flying as well as flying in any weather conditions, according to civilian standards. The airports would no longer exclusively be military, as they would also be used for civilian purposes such as regular airline traffic and cargo transport.

In the summer of 2002, civilian companies and air force engineering units started work on the restoration of the 3100-m (10,170-ft) runway at Ponikve, which is 920 m (3,018 ft) above sea level. NATO strikes had completely destroyed the runway and HASs. After it is restored, Ponikve will not be a permanent base for aviation units. The resident squadron was disbanded in 1997, leaving Ponikve free for air force manoeuvres. The runway at Nis has also been undergoing restoration since 2002, financed by a donation from the Norwegian

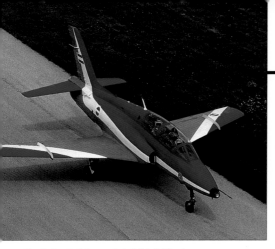

The Soko G-4 Super Galeb (right) is the most numerous fixed-wing type in RV i PVO service, mainly used for advanced training but with a limited combat role. It has four underwing pylons for the carriage of a variety of light stores. Seven were used by the 'Flying Stars' aerobatic team (above) until they were destroyed in a US Navy raid on Golubovci. The prototype G-4 (23601, below) is used today as a target-tug (N-62T) by the 252.lbae at Batajnica.

71504 is the only Yak-40 remaining in use with the 677.trae. An Elint-configured version has been retired.

The 677.trae operates two Dornier Do 28D-2s, configured for aerial survey and cartographic work.

Antonov products dominate the 677.trae inventory, which includes some An-2TDs detached to Nis (above). The most important type, however, is the An-26 (below), of which seven form the core of the fleet.

government. At Batajnica base, number one runway – 2500 m (8,202 ft) long and 45 m (148 ft) wide – was heavily damaged by some 30 bombs dropped from USAF strategic bombers, but is being restored. Two hangars have been constructed, for flight units and the technical workshop of the 177th air base.

Fighters

The RV i PVO entered the war with only 16 modern aircraft: 14 MiG-29 single-seaters and two MiG-29UB training two-seaters, which made up the 12.lae 'Vitezovi' (Knights) of the 204.lap at Batajnica. Sixty-seven outdated MiG-21bis and UM aircraft were distributed across three squadrons. The 126.lae 'Delta' was at Batajnica with the 204.lap, while the 123.lae 'Lavovi' (Lions) and 124.lae 'Gromovi' (Thunders) formed the 83.lap at Slatina.

The two-seaters were totally useless for fighter missions. The Ko PVO Command preserved the majority of MiG-29s at Batajnica, together with two-aircraft detachments at Nis

and Ponikve. One MiG-29 was at Golubovci. Part of the MiG-21-equipped 83.lap was deployed to Nis, Ponikve, Dubinje and Golubovci, although most aircraft were accommodated in underground halls in Golesh mountain, part of the Slatina base complex. The initial war deployment was meant to allow take-offs of pairs and small groups on interception missions, in any direction from which it was thought NATO aircraft might penetrate.

Faced with total air supremacy by NATO fighters, the RV i PVO performed 11 combat flights without a successful launch of an air-to-air missile. In air battles with AMRAAM missiles, four MiG-29s were destroyed, two pilots were killed and two ejected successfully. Two more pilots escaped by ejecting from a MiG-29, and the circumstances of the incident indicate they may have been victims of friendly fire. One MiG-29 was lost in an accident during a flight between two airports. One aircraft that was hit during the first night of the war, although damaged, made it to the base at Nis but was

later destroyed by a NATO Harrier GR.Mk 7. Three single-seaters and one two-seater were destroyed on the ground at Batajnica base.

Three MiG-21 squadrons did not engage in fighter missions, but were damaged in losses from air attacks. A total of 24 MiG-21bis aircraft was destroyed after attacks from various LGBs, cluster bombs and AGM-130s. The aircraft of the 83.lap were subjected to heavy attacks, but all aircraft that spent the war in underground galleries survived.

After a peace agreement was signed on 12 June 1999, 11 MiG-21bis and UM versions flew from Slatina to Batajnica. The loss of so many aircraft made it pointless to maintain the 83.lap in the organisational chart of the RV i PVO, so the unit was disbanded, its pilots and aircraft joining the 204.lap.

The 127.lae is in the Ko PVO structure and has only four MiG-29s and one MiG-29UB remaining. Aircraft acquired in 1987 have not been overhauled and are in poor technical condition, but are necessary for maintaining minimal flying qualifications. The inventory of the 126.lae includes 26 MiG-21bis and six MiG-21UM versions, significantly more aircraft than the 16 expected for general unit formation. However, the number of operational aircraft is frequently lower than the ideal squadron number. The 204.lap has expended much effort to restore its combat capabilities, and in October 2000 an exercise was staged where fighter aircraft, for the first time after the war, shot an R-60 (AA-8 'Aphid') air-to-air missiles at an illuminated target.

Attack and reconnaissance aircraft

When the war began, RV i PVO squadrons equipped with the J-22, NJ-22 Orao and G-4 Galeb light attack aircraft were tasked with combat actions around Kosovo against concentrations of Albanian rebel forces. Before the NATO air campaign, attack squadrons were forced to lie low because politicians wanted to avoid negative reaction from the international community. The Oraos from the 98.lbap flew into action from Ladjevci and Ponikve airports, and the G-4 Galebs of the 229.lbae 'Machevi' (Swords) from the airport at Nis. They avoided detection by NATO AWACS by flying at rooftop level and using mountains and river valleys as a shield from radar surveillance. Radio communication was banned, except in an emergency. To maintain mission secrecy, commanders of attack units received their orders from their commander and the chief of VaK HQ over the telephone after voice identification. Not one attack aircraft was intercepted in flight, although some missions were carried out just tens of kilometres from NATO strike groups. Attack squadrons made 31 combat flights, supporting the Pristina Corps in sweeping the terrain for rebels. Engagements of attack aviation had to be suspended due to NATO attacks on runways. Plans predicted a possible continuation of the war, so aircraft were saved for use in the event of a NATO land invasion. As the war progressed, many Oraos and G-4 Galebs were destroyed by LGBs in precise hits on shelters.

Today the 98.lbap forms the nucleus of the RV i PVO attack element. Its origin was the 98.abr, which prior to the civil war had two squadrons at Petrovec in Macedonia and one squadron at Ladjevci. In March 1992 all aircraft from the 98.abr ended up at Ladjevci, the present location of both the unit's command and the 241.lbae 'Tigrovi' (Tigers). The 252.lbae 'Kurjaci sa ushca' (Wolverine from mouth of river), which participated much more heavily in the civil war than other squadrons, is located at Batajnica. The two squadrons share aircraft according to pilot training needs. Due to large problems with maintenance and a lack of spare parts, only a few of the 17 J-22 and seven NJ-22 Oraos are operational. Three target-towing G-4 Galebs (N-62T) of the 252.lbae are used to train VJ anti-aircraft gun crews.

Attack aviation pilots gain additional flight time on a few G-4 Galebs. Twenty-eight armed G-4 Galebs were added to the 242.lbae 'Orlovi' (Eagles), 172.abr at Golubovci. The majority of the aircraft that lost their armament, according to the Subregional Arms Control Agreement, were destroyed in the NATO air campaign. The main task of the 172.abr is training student pilots, but all instructors must also be capable

of performing combat assignments.

RV i PVO reconnaissance assets were used in the Kosovo campaign to record training bases of Albanian rebels in that country. Officially, those flights never happened, but a TV shot showing two IJ-22 Oraos in a rooftop-level flight over the town of Tropoj is proof – despite the reporter's description of them as "Yugoslav MiGs". All RV i PVO reconnaissance aircraft are gathered in the 353.iae 'Sokolovi' (Hawks), which has a command and two flights at Ladjevci and one flight at Batajnica. A LORAP (long-range aerial photography) container carried by a modified MiG-21M is used for high-altitude reconnaissance. Two original Soviet MiG-21Rs are close to the end of their lives, having been manufactured in 1968. Eight IJ-22 Oraos are used for conventional photo-reconnaissance, and training is done on two INJ-22 Oraos.

Airlift capability

Transport aviation had already fulfilled its military mission before the first NATO air attack during Allied Force, a fleet of An-26s having transported equipment from units that had been moved to other airports. On the morning of 25 March 1999, all functional aircraft from the 677.trae 'Labudovi' (Swans) flew at low altitude to civilian airports in Belgrade and Echka near Zrenjanin, where they stayed until the end of the conflict. On 26 March one An-26, by special order, flew between Batajnica and Golubovci, for which the crew received a medal. During the conflict with NATO, the 677.trae made 19 flights totalling 5 hours and 40 minutes. Three An-26s and three An-2TDs were destroyed on the ground.

After the war, all remaining RV i PVO transport aircraft were concentrated in the 677.trae at Batajnica. There are seven An-26s, all but one acquired in 1976 and therefore all currently close to the end of their service lives. The training of paratroops from the 63rd parachute brigade is conducted at Nis airport with two An-2TDs and one An-26, which take-off from a grass runway. Two Do 28D-2s from the 677.trae are used for photo-survey work for the VJ's Military-Geographic Institute.

The RV i PVO has a very modest fleet for VIP transport, currently consisting of only one Yak-40. Two aircraft of the same type are not in flying order, one of which is equipped for Elint.

Helicopters

During the NATO war in 1999, three squadrons from the 119.hp were used for medevac from Kosovo to the Military Medical Academy in Belgrade. They also transported ammunition and food to checkpoints on the border with Albania, in mountains above 1500 m (4,920 ft). The Mi-8-equipped 787.rhe 'Medvedi' (Bears), 897.mhe 'Strshljeni' (Hornets) and 890.mhe 'Pegazi' (Pegasus) were organised into SAR groups with members of the 63rd parachute brigade, an elite special unit of the VJ that was subordinated to the RV i PVO during the conflict. In the path of a potential penetration by NATO armoured forces in the south of Serbia, the 712.pohe 'Shkorpioni' (Scorpions) and 714.pohe 'Senke' (Shadows) with Gama tank-hunting helicopters were hidden and waiting. Utility Gazelles from those squadrons were available for battlefield transport in detachments at Kosovo.

Mi-8s and Gazelles made 179 flights in the face of NATO air superiority, transporting 94 wounded, 113 passengers and 5 tonnes (4.92 Imp tons) of freight. Sixteen helicopters were destroyed in air raids, all of which had been in non-working order and unable to fly to secret airstrips. The RV i PVO saved only two Ka-28s, from the 784th anti-submarine squadron 'Ajkule' (Sharks). Three non-operational Mi-14PLs and two Ka-25BShs were transformed into piles of molten metal on the grassy apron at Golubovci.

Today the RV i PVO has 110 Mi-8 and Gazelle helicopters at its disposal. However, the service life of the Mi-8 is nearing its end and, of 33, only one-third are in usable shape, making the acquisition of new aircraft a necessity. Of the Gazelles, 44 are of the Gama attack variant and three are Hera reconnaissance versions, used for the correction of artillery fire. Lack of funds has forced the air force to sell off the utility Gazelles to private users in the FRY and abroad. In 2002, the 30 HO-42 and HO-45 Gazelles were reduced by five, which were put up for sale. In 2000, two Ka-28s were withdrawn from use due to lack of money for overhaul, and were handed over to the Aviation Museum at Belgrade airport.

Reorganisation

War experiences revealed that the organisation of the VJ did not meet realistic needs. Those responsible for its reorganisation believed that a massive army burdened with outdated equipment should be transformed into a smaller but more compact and more efficient armed force. The search for a new model began in 1999.

After a long delay, a plan was adopted in December 2001 at the meeting of the Supreme Council of Defence. The land force retained its dominant position, to the detriment of the air force and the navy. The concept of a joint general staff was adopted, with the armed forces operational command directly subordinated. On 27 February 2002 the VJ disbanded command HQs of strategic groups, i.e., three armies of the land army, the RV i PVO Command and the Navy Command. In the new organisation, six land army corps, two air force corps (VaK and Ko PVO), and the navy are directly subordinated to the chief of staff of the general staff.

All structures previously under direct control of the disbanded RV i PVO command HQ were dispersed between two corps. The building in Zemun houses the VaK (Aviation Corps) Command, which received 12 aviation and helicopter squadrons and four air bases (vazduhoplovna baza – vb), brigade-sized units that perform maintenance of aircraft from both corps, and oversee the use of all peacetime and reserve airfields. Every air base has a technical-aviation battalion, logistics battalion and combat battalion for defence and protection of equipment and infrastructure. Base defences are organised into battalions or regiments for protection against low-altitude attack and are equipped with Strela-1M (SA-9 'Gaskin'), Strela-2M (SA-7 'Grail'), Igla-1 and Igla (SA-16 'Gimlet' and SA-18 'Grouse') MANPADS, 40-mm Bofors L/70 anti-aircraft guns, and 20-mm M-55A4 triple-barrelled mobile systems.

The headquarters of the 177.vb is at Batajnica airport, 20 km (12.4 miles) north of

Above: Soko built Gazelles in both the SA 341H (H-42) and SA 342L (H-45) variants. This quartet of SA 342Ls is from the 897.mhe, which occasionally performs as the 'Flying Hornets' team.

Right: Local production was preceded by 21 SA 341Hs from Aérospatiale, supplied for training. Such utility aircraft have since been retired and put up for sale.

Locally developed Gazelle versions still in use are the HI-42 Hera (above), HN-42M Gama (below) and HN-45M Gama (left). The Hera ('Helikopter-Radar') has surveillance radar under the rear fuselage and is used for artillery fire correction. The Gama ('Gazelle-Maljutka') is armed with the 9K11 Maljutka (AT-3 'Sagger') anti-armour missile, and is available in either SA 341H or SA 342L forms.

Belgrade, which is the central base of the VSiCG air force. At Kovin airport, east of Belgrade, the 177.vb has a logistics company that supports the utilisation of Utva-75 piston aircraft from the 251.lbae 'Pume' (Pumas). Ladjevci airport, near Kraljevo in western Serbia, is the home of the 285.vb. The 161.vb is located at the airport of Nis in eastern Serbia. There is only one permanently occupied airport – Golubovci, which is 15 km (9.3 miles) from the Montenegrin capital city of Podgorica. The 423.vb holds the Golubovci area of responsibility. The air base at Ponikve airport was disbanded in 1997 and is presently occupied by a logistics unit of the 65th battalion for support and logistics (65. bataljon za obezbedzhenje avijacije – boa). Directly reporting to the VaK-a are also single engineer and signalling battalions.

The Ko PVO (Air Defence Corps) is in charge of protecting the state territory, important buildings and groups of armed forces. All parts of the air defence system are connected into a single structure by an automatic Swedish system, the AS-84, which is installed in the main command centre inside the hill of Strazhevica and at sector command centres. At Batajnica two MiG-21bis from the 126.lae, each armed with four R-60 (AA-8 'Aphid') air-to-air missiles, are on constant interception alert duty. Six S-125M Neva-M (SA-3 'Goa') SAM battalions from the 250.rbr PVO protect the wider area of the Serbian capital city of Belgrade. Four battalions from the 450.rp PVO are dispersed in western and central Serbia.

The Neva-M system is intended for use from stationary positions, but the war with NATO forced Ko PVO crews to practise quickly

moving equipment to other locations. A mobile component of the air defence system are four regiments with Kub-M (SA-6 'Gainful') SAMs. In peacetime, these regiments are dispersed to cover the state's entire territory.

The northern region of Vojvodina houses the 240.srp PVO, with a compound in Novi Sad. The western and eastern part of central Serbia are covered by the 310.srp PVO from Kragujevac and the 230.srp PVO from Nis. The Danilovgrad garrison has the 60.srp PVO in charge of defending the territory of Montenegro. Radar is the basic equipment of the 126.br VOJIN.

The Ko PVO has its own engineer and signal battalion, and an Elint centre. The Air Force and Air Defence Sector of the VSiCG general staff has control over the aviation testing centre (Vazduhoplovni opitni centar – VOC), which is

at Batajnica and is responsible for testing aviation equipment. The break-down of the domestic military industry meant the VOC nearly ended up without work. A series G-4 Galeb was remodelled into a modernised variant armed with R-60 (AA-8 'Aphid') AAMs and improved electronics, and made its maiden flight two days before the start of the NATO air campaign. The G-4M was slightly damaged by bomb fragments. Repairs took some time due to a lack of both spare parts and interest in the project. Flight testing resumed only in March 2002.

The future

The political agreement between the states of Serbia and Montenegro, ratified in January 2003, transformed the federation into a loose state union. The term Yugoslavia has been

Left: *The three Mil Mi-14PLs last flew from Golubovci in April 1998. A year later they were all destroyed by LGBs dropped by NATO aircraft, along with two Ka-25BShs.*

Above: *Mi-8Ts serve with units at Batajnica, Nis and Golubovci. The latter are flown by the 897.mhe, which combines rotary-wing training with transport duties*

dropped, so the name of the VJ (Yugoslav Army) has been changed to the Army of Serbia and Montenegro (VSiCG). The armed forces have remained united, but will be reformed and establish closer relations with NATO members and the Partnership for Peace programme; until the end of 2000, countries participating in those organisations had been considered to be potential opponents.

The first step in reform is the reduction of the number of peacetime personnel, to meet NATO criteria. The role of the air force will be defined through analyses and discussions within the army, but circumstances do not favour a more prominent role for the air force. The land forces have demonstrated an understanding only of the Ko PVO.

As if by rule of inertia, common to all armed forces, after the war with NATO the VJ started preparing itself for a war that had already ended. A modernisation 'wish list' included multi-purpose combat aircraft with strong fighter capabilities, long-range and high-mobility missile systems for air defence, an automatic system to integrate air defence, and advanced surveillance radars. Aviation generals who had advanced in their careers through fighter aviation and rocket units supported these plans.

The air force has since changed its thinking about its future needs. Earlier plans assumed the acquisition of products from Russian industry, but times have changed. Many leading aviators are convinced that the only realistic approach to modernisation is an orientation toward the West. In December 2002 the commander of VaK, General Vladimir Starcevich, published a modernisation plan that reflected the realistic financial state of the country and its armed forces. The air force would rely on one type of multi-purpose combat aircraft, of Western production. The country

does not have the means to purchase new equipment, but General Starcevich believes that there is a way out of the air force's crisis. He talks of sending a group of 10 promising pilots for training to a NATO member country. They would form a nucleus for the retraining of other pilots, which is necessary to form a squadron organised along NATO standards with 20 to 24 aircraft. The aircraft would probably be acquired through a lease contract: initially, four two-seaters for conversion would be necessary, and the squadron would not be completely equipped until 2010. The financial situation would be the decisive factor in the selection of an aircraft, and the one that would most closely meet the needs of Serbia and Montenegro is the F-16A/B. Only after 2010 could a long-term solution be examined, together with potential contracts for multi-purpose combat aircraft.

The 677.trae would be equipped with new transport aircraft and a new transport helicopter for one of the squadrons that currently uses the Mi-8. The modernisation of vertical transport is necessary for the adoption of the Partnership for Peace programme. One of the problems is the choice of helicopters with a wider variety of combat roles than the Gama, which is dedicated to the anti-armour role.

Four Mi-8Ts were extensively modified with Western equipment to become HT-40Es. They perform an Elint role, although not all are in use due to the poor operational state of the Mi-8 fleet.

Ratno vazduhoplovstvo i protivvazducna odbrana Vojske Srbije i Crne Gore

Direct reporting unit, General staff armed forces

VOC	G-4, Mi-8, Utva-75, Gazelle	Batajnica

Vazduhoplovni Korpus (Air Force corps), HQ Belgrade

677.trae, HQ Batajnica		
	An-26, An-2, Do 28D-2, Yak-40	Batajnica, Nis
890.mhe, HQ Batajnica		
	Mi-8, Gazelle	Kovin
353.iae, HQ Ladjevci		
	IJ-22/INJ-22 Orao, MiG-21R/M	Ladjevci, Batajnica
172.abr, HQ Golubovci		
239.lbae	Utva-75	Golubovci
242.lbae	G-4	Golubovci
251.lbae	Utva-75	Kovin
897.mhe	Mi-8, Gazelle	Golubovci
98.lbap, HQ Ladjevci		
241.lbae	J-22/NJ-22 Orao, G-4	Ladjevci
252.lbae	J-22/NJ-22 Orao, G-4	Batajnica
119.hp, HQ Nis		
712.pohe	Gazelle	Nis
714.pohe	Gazelle	Ladjevci
787.trhe	Mi-8	Nis

Korpus Protivvazducne Odbrane (PVO Corps), HQ Belgrade

204.lap, HQ Batajnica		
126.lae	MiG-21bis/UM	Batajnica
127.lae	MiG-29, MiG-29UB	Batajnica

Glossary of unit designations
abr: *avijacijska brigada, aviation brigade*
br VOJIN: *brigada vazdushnog osmatranja, javljanja i navodzhenja, radar surveillance brigade*
iae: *izvidzhachka avijacijska eskadrila, reconnaissance aviation squadron*
lae: *lovacka avijacijska eskadrila, fighter aviation squadron*
lap: *lovacki avijacijski puk, fighter aviation regiment*
lbae: *lovacko-bombarderska avijacijska eskadrila, fighter-bomber aviation squadron*
lbap: *lovacko-bombarderski avijacijski puk, fighter-bomber aviation regiment*
mhe: *meshovita helikopterska eskadrila, mixed helicopter squadron*
pohe: *protivoklopna helikopterska eskadrila, anti-armour helicopter squadron*
rbr PVO: *raketna brigada PVO, air defence rocket brigade*
rp PVO: *raketni puk PVO, air defence rocket regiment*
srp PVO: *samohodni raketni puk PVO, air defence self-propelled rocket regiment*
trae: *transportna avijacijska eskadrila, transport aviation squadron*
the: *transportna helikopterska eskadrila, transport helicopter squadron*
vb: *vazduhoplovna baza, air base*
VOC: *vazduhoplovni opitni centar, aviation test centre*

The air force urgently needs training aircraft, as all the G-2 Galebs used for basic training were destroyed in NATO raids on Golubovci. New pilots are being trained according to a four-year programme. During the first two years of their education at the Belgrade Military Academy, student pilots learn general and aviation subjects. In their third year of study they undergo basic training on Utva-75 piston aircraft with the 251.lbae at Kovin. In their fourth year, students spend their greatest number of hours in the air with the 172.abr. Jet pilots must overcome the great difference in performance between the Utva-75 and the G-4 Galeb, which is used for advanced and combat training. The air force cannot complete training with the 172.abr, so young pilots become qualified in operational units.

The system of training desperately needs another aircraft type, one that would be more economical for basic training. The first proto-

The Mil Mi-8T (HT-40) fleet is long overdue for replacement, but the survivors continue to form the basis of the army's mobility capability. Some are used for medevac missions, including two painted in this all-white scheme. During the war, Mi-8s were used on resupply missions, for combat search and rescue missions, and for transporting elite ground units.

type of Utva's Lasta piston aircraft took place as long ago as 2 September 1985, but the type never reached series production. NATO jets destroyed five experimental examples in the factory. The air force has not abandoned the

project and has announced the development of a modified Lasta-3 or Lasta-95 variant – dependent, of course, on securing financing for the new prototype's production and testing.

Aleksandar Radic

Vazduhoplovna jedinica MUP-a Srbije (Serbian Interior Ministry air wing)

Serbia's police helicopter unit was formed on 1 January 1967. To start with it used a single Agusta-Bell AB 47J-2A for traffic control. In the early 1970s the fleet was enlarged with the purchase of three AB 206A JetRanger Is and in 1976 a single JetRanger II. Three SA 341G Gazelles were acquired for the needs of the Serbian police and Belgrade's city police

Terrorism, which became a global occurrence in the mid-1970s, spurred the Yugoslav police plans for stronger air support. After 1978, helicopter units of all federal units were strengthened and pilot training began in the Yugoslav Air Force's Aviation Military Academy (VVA JRV i PVO). From the USA six Bell 206B JetRanger IIIs and three Bell 206L-1 LongRanger IIs were acquired. AB 212s arrived from Italy for transport of special forces, which were organised to combat terrorists, dangerous criminals and rebels. Three examples were placed at the disposal of the federal special brigade and one was given to the Serbian police.

In the 1980s the use of police helicopters in Yugoslavia was decentralised: Federal, Serbian

and Belgrade police helicopters were based at Belgrade airport, while police formations of the Vojvodina and Kosovo autonomous provinces, and regional police in Kraljevo, Smederevo, Kragujevac and Zajecar had their own helicopters. Other federal states of former Yugoslavia had helicopter forces in their capitals. The Serbian police centralised its fleet at the end of the 1980s for economic maintenance and all were placed under the command of the 135. helikopterska eskadrilache (helicopter squadron) at Belgrade airport. SA 341Gs and earlier variants of JetRanger I and IIs were withdrawn from use. The last new helicopters to enter service were two new JetRanger IIIs bought in 1991, at the beginning of civil war.

When riots erupted in Kosovo in 1989 and 1990, helicopters were used to control and break up mass demonstrations by ethnic Albanians. Four AB 212s from federal and Serbian helicopter units transported anti-terrorist units to remote villages in which were rebel strongholds. The Yugoslav Air Force supported police with the loan of four Mi-8s, which were painted in a blue-white colour scheme.

A detachment of Serbian police helicopters was continually present in Kosovo as support to special forces. On the beginning of open Albanian armed action in 1998, helicopters participated in the invasion of rebel strongholds in the area of Drenica, together with the helicopter squadron of the Serbian security services (RDB), which had existed since 1992. At the start it consisted of a small helicopter section with a single Bell 206B JetRanger III and one Bell 206L-1 LongRanger II. Fleet expansion came through the take-over of AB 212 and Gazelle helicopters from disbanded federal

police units. From one confiscated armament shipment which arrived at the port of Bar from Lebanon, the RDB took over an SA 342L armed with HOT ATGMs and GIAT 621 guns. In combat activities in Bosnia and Herzegovina, and Croatia, the helicopters were used on the secret operations of an RDB squad known as 'red berets'. On their missions they flew without any markings, or with Serbian flags only, and only the best-informed observers could recognise that these helicopters did not belong to local Serbian forces.

In 1997, RDB combat units were reorganised into the brigade-level Unit for Special Operations (Jedinica za Specijalne Operacije – JSO), which included helicopter units in its composition. In the short period between 1996 and 1998 when the FRY was exempted from the UN arms embargo, the JSO acquired two Mi-24Vs and two Mi-17s from Russia. From the United States arrived a single Sikorsky S-76B, which became the personal helicopter of regime leader Slobodan Milosevic.

In combat actions in Kosovo the RDB carried out hundreds of transport and medevac flights. During the fight for Donji Prekaz village on 1 March a Mi-24 was damaged. It performed an emergency landing and was later repaired. Mi-24Vs attacked Albanian rebel training camps on several occasions. In July 1998, when Serbian citizens and around hundred policemen were surrounded at Kijevo in western Kosovo, the siege was lifted by the intervention

Before the two helicopter fleets were brought together, both the Serbian police (above) and JSO (below) used Agusta-Bell AB 212s. JSO aircraft were employed to transport anti-terrorist teams in Kosovo.

The two Mi-24V 'Hinds' acquired by the RDB/JSO were heavily involved in fighting in Kosovo prior to NATO intervention. They now form the 'sharp end' of the VJ MUP's combined police/security service force.

The JSO acquired two Mi-17s, one in a passenger configuration (left) for use on SAR, insertion and general transport duties, and a fully armed aircraft (above). The latter has six weapon pylons, automatic 30-mm grenade launcher and a self-protection system.

This ex-JSO Gazelle operated in a civilian-style scheme rather than the normal camouflage.

This ex-Iraqi SA 342L is the aircraft confiscated at Bar. It now flies with the VJ MUP after service with the JSO.

The Bell 206B JetRanger III is used by the VJ MUP for training and support missions.

of JSO members who had been inserted deep into territory controlled by rebels. It was the beginning of a wider operation during which rebel forces were suppressed in the area of Pec. During the NATO air campaign police and JSO helicopters flew liaison and medevac missions. No helicopter was destroyed in air raids, but a cruise missile demolished part of the JSO's hangar at Belgrade airport.

The JSO maintained the secrecy of its actions until 5 October 2000, when it supported the democratic parties in the overthrow of the Milosevic regime. In October 2001 the helicopter unit was publicly presented for the first time during a joint exercise with military forces at the Nikinci range. An Mi-24V and an

SA 342M supported the attack of an anti-terrorist team with three Humvees, while an AB 212 transported special unit members to simulated terrorist hideouts. Serbia's Interior Ministry (MUP) was reorganised in 2002, and the JSO was detached from the national security service and subordinated directly to the Minister. The helicopters, together with the police squadron, were combined in a joint aviation unit (Vazduhoplovna jedinica MUP). Rationalisation of equipment is being planned in order to make maintenance more efficient. The first step was the sale of a single SA 365N and single JetRanger III in 2001 and 2002.

In a final twist to the tale, the JSO was disbanded on 25 March 2003 in the wake of the

assassination of Prime Minister Zoran Djindjich, who had been killed by the JSO's deputy commander on 12 March. By this time the helicopter units had already been divorced from the JSO, although part of the VJ MUP was assigned to JSO support. This has now ended, and all helicopters are assigned to anti-terrorist and other police duties.

Aleksandar Radic

Vazduhoplovna jedinica MUP-a Srbije

Base Belgrade airport   27 helicopters: Mi-24V, Mi-17, AB 212, SA 365N Dauphin II, SA 341H/SA 342L/M Gazelle, Sikorsky S-76B, Bell 206B JetRanger III and Bell 206L-1 LongRanger II

Vazduhoplovna jedinica Vlade Crne Gora (Montenegrin government aviation unit)

Golubovci airport, near Podgorica, is the base of the state aviation unit which is responsible for VIP transport and for standard police assignments. Montenegro's government Learjet 35A was sold in March 2000. It was replaced with a US-registered Learjet 45, and the Montenegrin authorities used two Cessna Citation Xs which

were registered as the property of Fort Aviation and Brook Aviation from Wilmington, Delaware, USA. Following an organised crime battle in 2001 for control of the cigarette market in the Balkans, it was discovered that the Citations were owned by Stanko Subotic, tobacco and mafia boss. Montenegrin authori-

ties suspected of having connections with organised crime quit using the 'exposed' aircraft and now use only their Learjet 45 and single Cessna 421 Golden Eagle, officially owned by the Geodetic Institute of Montenegro, whose role is aerial survey.

At the time when Slobodan Milosevic was

Above: Seen at Golubovci is one of the two Cessna Citation Xs used by the Montenegrin government until corruption allegations ended their use by officials.

Right: This SA 341H is one of three Gazelles which serves with the Montenegrin police helicopter support unit.

Vazduhoplovna jedinica Vlade Crne Gore (Montenegrin government aviation unit)	
Base Golubovci-Podgorica	1 x Learjet 45, 1 x AB 412, 1 x AB 212, 1 x AB 206A Jet Ranger I, 3 x SA 341H Gazelle

threatening Montenegro with intervention at the end of the 1990s, the Montenegrin police developed special units organised, equipped and trained in military style. These special forces have two AB 412EP and AB 212 helicopters at their disposal, with VIP transport and medevac as secondary roles. Three SA 341H Gazelle helicopters taken over from the JRV i PVO in the early 1990s fly on utility missions.

An AB 206A, acquired in 1972 and which was the first Montenegrin police helicopter, is no longer in use because it was uneconomical to overhaul it and it remains in a hangar at Golubovci. Due mainly to maintenance problems, only the AB 412EP is used.

Attempts to build a new hangar for the police helicopters in December 1999 almost turned into open armed conflict between Montenegrin police and federal armed forces.

The commander of the 423. air base challenged the right to build on land owned by the Army, and threatened the use of force. The police reacted by sending in reinforcements. On the military side of the airfield the resident unit was put on readiness and four G-4s and two Gamas were armed, but conflict was avoided. After the fall of Slobodan Milosevic, relations between Belgrade and Podgorica returned to normal.

Aleksandar Radic

Serbian/Montenegrin government aviation units (VIP and calibration)

Originally, the government's aviation service reflected the federal organisation of the country as a whole. The federation had its own VIP fleet, while each of the constituent states had one or two aircraft based in the state capitals. Each weekend these aircraft would fly home the politicians from their week's work in Belgrade. The exception was Serbia, which relied on Federal Aviation Service aircraft as its politicians needed them only for trips abroad.

A special unit of the RV i PVO, the 675.trae at Batajnica air base, was formed to support international journeys by the Yugoslav leader, and was responsible for VVIP transport. From 1992, the 675.trae was engaged on government transport duties. The unit's single Falcon 50 and

two Learjet 25Bs received civil registrations, which were carried in addition to the military serial number to ease overseas journeys. In a 1995 rationalisation all three were handed over to the federal government's Aviation Service.

The Aviation Service of Serbia and Montenegro currently operates two Falcon 50s, YU-BNA and YU-BPZ, which are used to Euro-Mediterranean destinations and on routes within the country. For long-range flights federal authorities lease aircraft from JAT, the national airline. It is planned to re-equip the fleet, and four Learjet 25Bs have been sold.

The Federal Directorate for Air Traffic Control (SUKL – Savezna Uprava za Kontrolu Letenja) is under the control of a joint Serbia-

Montenegro administration. SUKL operates two Yakovlev Yak-40s – YU-AKT and YU-AKV – which are used for calibration of radars and guidance systems. In the past, SUKL acted under the strong influence of the army and therefore aircraft wore military serial numbers. Pilots from the JRV's transport squadrons were allocated to work for SUKL, which is now in the process of separation from military influence.

Aleksandar Radic

Government aviation units **Base Belgrade airport**	
Avio-servis Saveta ministara Srbije i Crne Gore	2 x Falcon 50
Savezna Uprava za Kontrolu Letenja	2 x Yakovlev Yak-40

Ministerial VIP transport for the Serbia and Montenegro government is handled by these two Falcon 50s which fly from Belgrade's international airport.

Also based at Belgrade are the SUKL's two Yak-40s. As the fuselage titles suggest, the aircraft are equipped for calibration of navigation and landing aids.

Kosovo Force, NATO (KFOR)

The entry of NATO peacekeeping troops into Kosovo in June 1999 was supported by significant aviation assets. British Royal Air Force Boeing Chinook HC.Mk 2 and Westland Puma HC.Mk 1 support helicopters, backed by US Army Boeing AH-64A Apache attack helicopters, played a key role in the seizure of Kosovo's capital, Pristina, from withdrawing Serb troops.

Over the summer of 1999, NATO's Kosovo Force (KFOR) moved to secure the borders of the province and provide security for the local population until UN police could be deployed in strength. Ethnic killings were rife as old scores were settled, while NATO troops also had to clear minefields and other debris of war. KFOR established five brigades areas, led by British, French, Italian, German and US headquarters, to run its operations. Each had an integral air component attached to them.

The composition of the air elements in KFOR has evolved over the years to meet changing circumstances and requirements. Up to early 2002, KFOR required each of its brigades to hold forces in reserve to counter a conventional Yugoslav military attack. The French, Italian and US brigades included strong attack helicopter forces, equipped with Eurocopter

Gazelles, Agusta A 129 Mangustas and AH-64A Apaches, as part of this reserve force but with the reduction of the threat, these have significantly reduced. Russian troops assigned to KFOR also had their own attack helicopter support in the shape of a squadron of Mi-24s.

Command of KFOR rotates every six months or so, with the country providing the commander also being responsible for deploying a command and control aviation element. The British KFOR commander had a flight of Westland Lynx AH.Mk 9s, the Germans used Bell UH-1Ds, the Norwegians Bell 412s and the Spaniards UH-1Hs to fly them around the province.

Within the brigades, support helicopters of various types – including British Pumas and Gazelles, German Sikorsky CH-53Gs and UH-1Ds, Ukrainian and Russian Mil Mi-8s, Italian Agusta-Bell AB 212s and French Pumas – are used to move troops and supplies around, as well as evacuate military and civilian casualties. Not surprisingly, the largest aviation contingent deployed in Kosovo is provided by the US Army, which has maintained a strong aviation task force at its large Camp Bondsteel base in the south of the province. Until spring 2002 this task force boasted two companies of

Apache attack helicopters, a scout troop of Bell OH-58D Kiowa Warriors, a company of Sikorsky UH-60 transport helicopters and a company of Chinook heavy-lift machines. A United Arab Emirates AH-64 squadron was also based at Bondsteel under US command until late 2001. More recently, the composition of task forces has changed to two OH-58 and two UH-60 companies, backed by Chinooks, to better support KFOR's increasing internal security focus.

The activities of ethnic Albanian rebel groups along the demilitarised zone along the Kosovo-Serbia boundary in 2000 and cross-border support for other rebels in Macedonia during 2001, led to KFOR having to beef up its surveillance capabilities. KFOR made extensive use of unmanned aerial vehicles (UAVs) to monitor these groups, including French and US IAI/TRW Hunters, British BAE Systems Phoenix, German EMT Ingenieurgesellschaft LUNA X-2000s and US General Atomics RQ-1 Predators.

KFOR has made extensive use of aviation assets in all aspects of its operations and this looks set to continue as long as NATO troops are needed to keep the peace in the troubled province.

Tim Ripley

Slovenia

1. OpBM VLZ

In June 1991 Slovenia declared its independence from the federation of Yugoslavia. The first signs of aggression from the federal forces came on 21 June, when the authorities in Belgrade halted the departure of an Adria Airways Airbus A320. Soon after the warning, Mil Mi-8s landed in the military part of Ljubljana-Brnik airport, carrying special forces from Nis. This group included pilots, who flew the Slovenian Territorial Force's Soko J-20 Kraguj aircraft out of Brnik to the military air base at Cerklje in the southern part of Slovenia.

The situation intensified following the formal declaration of independence on 25 June. At 14:00 on 26 June the federal authorities closed Brnik airport and the whole airspace over Slovenia. Brnik became the main target for the aircraft of the Jugoslovensko Ratno Vazduhoplovstvo (JRV – Yugoslav Air Force). An attack using Soko J-22 Oraos armed with Durandal bombs was planned, only to be cancelled due to bad weather. On the first day of fighting the JRV only used helicopters, and one of them was damaged when it attacked the Karavanke tunnel near the border with Austria. JRV crews underestimated the determination and morale of the Slovenian defenders and they flew their helicopters low. That proved crucial for one Mi-8, which was shot down with an SA-7 Strela missile near the town of Ig. Low flying over the capital, Ljubljana, also proved disastrous for a Gazelle, which was also shot down.

On 28 June a Gazelle crew defected from the JRV to the Territorial Defence Forces (TDF – Teritorijalna Obramba/TO). They defected from the barracks in Maribor and escaped first to Rogla mountain and then to Golte mountain, where they hid the Gazelle. The helicopter was hidden until 14 November, when it was sent abroad for overhaul.

On 26 June two Soko G-4 Super Galebs attacked Brnik airport, making two passes firing unguided rockets and cannon. The attack damaged a Dash 7, a DC-9-33, an A320, the hangar of Adria Airways and cars parked in front of the terminal building. On the same day, Territorial Defence Forces shelled Cerklje military airfield, but missed the parked aircraft. However, this determined attack surprised the JRV, forcing it to remove its 15 J-22 Orao strike aircraft, 20 Mi-8s, six Gazelles and one Falcon.

On 29 June the JRV was in action again, attacking the border town of Sentilj near Maribor. In the process some MiG-29s overflew the Austrian border although Austrian Drakens were unable to react. The most important radio and TV transmitters in Slovenia were also attacked but, despite some damage, broadcasts continued. On 30 June two G-4s attacked a barricade of trucks near Medvedjek on the main Ljubljana-Zagreb road with cluster bombs, killing some truckdrivers. Afterwards, a pair of Jastrebs attacked TDF positions in the Krakov wood near the town of Novo Mesto. Two MiG-21bis's attacked Dravograd, a border town with Austria, where members of the TDF were blockading the military barracks. In following days JRV aircraft routinely overflew Slovenia,

but did not fire any weapons. On 18 July the federal authorities announced the withdrawal of their military forces from Slovenia.

After the withdrawal of federal forces from Slovenia and the formal declaration of independence, a logical consequence was the establishment of new national armed forces, and an air element was quickly formed – the 15th Aviation Brigade of the Slovenian Army (15. Brigada Vojnega Letalstva Slovenske Vojske). Before 1991 the TDF existed as a regional reservist force, but after the brief fighting it became the framework for the new Slovenian army. The defection of the SA 341 Gazelle and its crew was the foundation stone for the new 15th Brigade. Following the end of the fighting in Slovenia, in July 1991, the air force unit of the TDF had three Bell 412s, an Agusta A 109 and the Soko SA 341. Although the helicopters wore civilian schemes, they all carried the TDF crest. The first tasks were to provide SAR, air ambulance and firefighting cover, and VIP transport.

Official formation of the 15th Air Force Brigade of the Slovenian Army occurred on 9 June 1992. At that time it consisted of 21 pilots and engineers and five helicopters, and training was the priority task. Thanks to a healthy sport flying scene and many ex-JRV Slovene pilots there was a ready pool of skilled recruits and combat-trained personnel. It is estimated that in 1991 the JRV had at least 110 Slovenian pilots, two of them assigned to the 'Flying Stars' aerobatic team.

At that time, all of the former Yugoslavia was under a UN arms embargo and Slovenia was unable to acquire any new equipment. However, after Slovenia joined NATO's Partnership for Peace programme in March 1994, three Zlin Z-242L trainers and a single Let L-410UVP-E Turbolet aircraft were bought. In June of the same year, a contract was made using with Bell Helicopter Textron for three Bell 206B-3 JetRanger helicopters for rotary-wing training. In March 1995 the 15th Air Force Brigade accepted three Pilatus PC-9s, which had earlier been under test by the US Army, after overhaul at Stans-Buochs.

On 5 November 1997 the Slovenian forces officially became part of the Bosnian Stabilisation Force (SFOR) and 15th Aviation Brigade helicopters began peacekeeping operations in Bosnia. Daily flights were made using three Bell 412s and the L-410UVP-E. Helicopters flew around 100 hours a month and the L-410UVP-E 40 hours a month. Two pilots are designated as permanent liaison officers to the SFOR co-ordination centre at Zagreb, Croatia, while selected crews are on a four-hour standby seven days a week. All flights are undertaken from Brnik.

In the meantime the manpower of the 15th Air Force Brigade grew steadily, a new building was constructed and the hangar was enlarged. Nine new PC-9 Mk IIs joined the 15th brigade in 1997, and one of them was subsequently modernised to PC-9M Swift standard by the Israeli Radom company and Slovenian engineers. Upgrades were made to the flight instrumentation, communications and weapon systems, while a head-up display was fitted. Also including Alkan, El-Op and FN Herstal, the Swift team developed a system which enables the PC-9 to use a variety of modern weapons. In combat configuration, the upgraded PC-9M

Swift is capable of carrying up to 1040 kg (2,100 lb) of armament on six underwing pylons. The first example was displayed at the 1997 Paris Air Show.

In the late 1990s the 15th Air Force Brigade consisted of two flying squadrons – one equipped with the fixed-wing aircraft and the other with the helicopters, and a technical and maintenance squadron. All aircraft and helicopters were maintained and overhauled at Brnik. There was also a logistical section and the headquarters. Each squadron was divided into four flights: training flight, fighter training flight, transport flight and fighter flight. A ninth independent flight comprised only three aircraft: the L-410UVP-E and two PC-6s, the latter being delivered on 13 May 1998.

Initial Slovenian wishes to set up an efficient air defence system prompted a search for around 16 multi-role jet fighters, with used F-16s then being seen as the most likely candidate. A requirement was also identified for new transport aircraft and attack helicopters. Negotiations between the Slovenian defence minister and the Pentagon over the possible purchase of 12 Bell AH-1W Super Cobras equipped with anti-tank missiles took place in the summer of 1997. The Agusta A 129 Mangusta/International was also considered. However, NATO's refusal to allow Slovenian membership had negative consequences for all these new equipment programmes.

In 2000/2001 Slovenian military aviation was substantially reorganised. Some parts of its operation were transferred from the military side of Brnik International airport to the renovated military airbase at Cerklje, in the south of Slovenia. The introduction of the ASOC surveillance system, purchased from Lockheed Martin, was an important step forward toward the introduction of a Roland 2 air defence system.

For new pilots, selected candidates first have to perform seven months of compulsory military training before beginning a private pilot's licence and studying a degree in mechanical engineering (aeronautical section). After obtaining a PPL, the students undergo air training at Cerklje air base. There are three Bell 206B-3 helicopters, one equipped for IFR operations, used for rotary-wing training, and two four-seater Zlin Z-143Ls and eight twin-seat Zlin Z-242Ls to provide basic and navigation training. Once the students obtain their wings, they undergo a probation period. For career officers this lasts nine months, for others six months. Operational pilots fly about 100 hours a year, but actually spend about 200 hours more than this in the air as co-pilots or crew members.

Current organisation

Military aviation and all air defence responsibilities are now under the control of the 1st Air Force and Air Defence Force Command (1. Operativno Poveljstvo Vojaskega Letalstva in Zracne Obrambe – 1. OpPM VLZ) located at Kranj. Its main task is maintaining the integrity of Slovenian airspace, air training, air support for the Slovenian army, search and rescue duties, and international co-operation. Subordinate to the 1. OpPM VLZ are the 15th Air Force Brigade, the 9th Air Defence Force Brigade, the 16th Air Surveillance Battalion and the 107th Logistic Base.

The 15th Air Force Brigade, which represents the main Slovenian air unit, has six components

Slovenia's largest aircraft is the L-410UVP-E (above), used for transport and patrol. It wears SFOR titles for work over Bosnia, as do the Bell 412s (below). The Bell was chosen as the air arm's standard transport helicopter from the outset, and the fleet stands at eight.

Above: The sharp edge of Slovenia's air arm are the nine PC-9M Swifts. This aircraft carries seven- and 19-round rocket pods and an El-Op/Radom laser rangefinder.

The training effort is concentrated at Cerklje na Dolenjsekm, home to the Bell 206B-3s (above) and the Zlin lightplanes. A batch of eight two-seat Z-242s (below) was bought for basic training, together with two four-seat Zlin Z-143s.

(squadrons) of which the helicopter element is still located on the military side of Brnik airport. The fixed-wing and training components are now located at the renovated Cerklje ob Krki airfield. The training group has three Bell 206B-3 helicopters, eight Zlin Z-242L and two Zlin Z-143L basic trainers. The helicopter squadron has eight Bell 412HP/SP/EP helicopters while the fixed-wing squadron flies 12 Pilatus PC-9/PC-9Ms, two Pilatus PC-6s and one L-410UVPE. Part of the fixed-wing squadron is a combat air unit with nine upgraded PC-9M Swifts, which are now combat-ready. The Bell 412s are now also capable of using unguided rockets and machine-guns.

Looking to the immediate future, the first and most important task is to equip the airfield infrastructure with modern navigation and landing systems for use by military aircraft and helicopters. In 2001 the Slovenian government signed a contract with Eurocopter to purchase two (plus one optional) AS 532 Cougar helicopters. The first Cougars will arrive in the spring of 2003, but six pilots and a number of technicians are already in France to begin train-

ing. All six pilots will achieve also an instructor rating to allow them to train more crews once back in Slovenia.

In November 2002 Slovenia was invited by NATO to begin discussions on joining the treaty organisation. This may alter the air force's make-up considerably. In the worst scenario currently proposed by some politicians, the PC-9Ms would be sold and only the helicopters would be kept. A more likely scenario involves the use of the PC-9Ms for training foreign pilots, whose air forces have the same type of aircraft in their inventory.

A genuine need exists for a transport aircraft for which the CASA C-295-300 and Alenia/Lockheed C-27 Spartan are prime contenders. However, such an acquisition has been threatened since the government decided to buy a new VIP aircraft (Falcon 900EX) from the defence budget. This purchase (more than US$ 54 million) caused considerable controversy. Meanwhile, any purchase of NATO-compatible modern combat aircraft and helicopters has been postponed until at least 2010.

Marko Malec

Two Turbo-Porters are in use for general transport and liaison tasks. They are sometimes used as jump platforms for parachute training.

1. Operativno Poveljstvo Vojaskega Letalstva in Zracne Obrambe

9th Air Defence Brigade (9. Raketna Brigada Zracne Obrambe)
15th Air Force Brigade (15. Brigada Vojaskega Letalstva)
Air Combat Sqn	Cerklje na Dolenjskem	9/3 x PC-9M/A
Multirole Helicopter Sqn	Brnik	5/2/1 x Bell 412EP/HP/SP
Air Transport Sqn	Brnik	1 x Let L-410UVP, 2 x PC-6
Training Sqn	Cerklje na Dolenjskem	2/8 x Zlin 143L/242L, 3 x Bell 206B-3
Parachute Training Sqn	Cerklje na Dolenjskem	

16th Air Surveillance Battalion (16. Bataljon za Nadzor Zracnega Prostora)
107th Logistic Base (107. Logisticna Baza)

Current inventory
Pilatus PC-9	3	training
Pilatus/Radom PC-9/M Swift	9	training/strike
Pilatus PC-6 Turbo Porter	2	transport/liaison
Let L-410UVPE	1	transport
Zlin Z-143L	2	training
Zlin Z-242L	8	training
Bell 412EP/HP/SP	5/2/1	patrol/SAR/transport
Bell 206B-3	3	training
Eurocopter AS 532AL	4	(on order, delivery 2003/04)

PfP Helos

Photographed by Carlo Brummer and Marnix Sap

NATO's Partnership for Peace (PfP) scheme involves many European nations – mostly from the East – who share the aim of joining the treaty organisation. The annual Co-operative Key exercise provides a chance for their forces to operate together, and alongside those from existing NATO members. The exercise also brings together a fascinating collection of rotary-wing equipment.

Macedonia – Mi-17 *(left and above)*, **Mi-8MT** *(below)*

In 1994, Macedonia acquired four Mi-17 'Hip' helicopters from the Ukraine. Because of the arms embargo they were delivered with civil registrations. The Security Council excluded Macedonia in 1996 from the UN arms embargo and, shortly after, all four Mi-17s were painted in camouflage schemes and received military serials. When the crisis between Albanian terrorists and Macedonian Government forces broke out in March 2001, Macedonia received four Mi-8MTs. Having taken part in CK'00, Macedonia returned to the Co-operative Key exercise cycle at Saint-Dizier and, besides one of the originally delivered 'Hips', one of the former Ukrainian combat helicopters was also present, complete with self-defence flare package attached to the tail boom.

Croatia – Mi-8MTV *(left)*

A first timer in the 2002 Co-operative Key exercise at Saint-Dizier in France was the participation of Croatia with an Mi-8MTV. Following positive results, Croatia plans to participate in the CK'03 exercise to be held in Bulgaria with an additional An-26, and perhaps even some of its MiG-21s.

Slovenia – Bell 412 *(right*

Regular CK participants are the Bell 412s from the 15 Brigade Vojaskega Letalstva/Helikopterska Eskadrila at Ljubljana-Brnik, which are used for transport, border patrol and SAR. During CK'02 at Saint-Dizier 10 NATO and 12 PfP countries brought together 29 fighters, 12 transport aircraft and 20 helicopters. New exercise elements were the establishment of a field air operations centre (AOC) and the introduction of night (air) operations for the more experienced participants. The assessment programme was eliminated and CIMIC operations were downsized.

Hungary – Mi-17 (above)
Like many of the CK 'Hip' helicopters, this Hungarian Mi-17 from the 4 Szállító Helikopter Század/87 HE 'Bakony' at Vesprem-Szentkirályszabdja – still wearing its markings from SFOR – performed in the PSO role. A typical CK'00 peace support mission began with the insertion of special operation teams to secure a landing zone after aerial reconnaissance missions had located a refugee concentration. Meanwhile, close air support was given by aircraft like the Bulgarian Su-25, Romanian Lancer-A or Greek and Turkish F-16s. French, Romanian or Greek C-130 Hercules dropped a multinational paratroop force made up of Austrians, French, Moldovans and USMC soldiers to properly secure the area. Medical teams with supplies were flown in to tend to the wounded and prepare them for evacuation. Once a secure perimeter was established, Romanian medevac IAR-330s, accompanied by Hungarian and Slovakian Mi-17s in air transport configurations, picked up the refugees and flew them to a secure landing strip, where a Bulgarian An-26, Moldovan An-72 or Slovenian L-410 took them over and airlifted them out of the crisis area.

Latvia – Mi-8MTV (above)
The main aim of the Co-operative Key exercise cycle is to practise and refine interoperability in air and land operations in support of Peace Support Operations (PSO), to which many Partnership for Peace (PfP) member countries contribute air assets, as well as troops and observers. During CK'02 at Saint-Dizier the Latvijas Gaisaspeki (Latvian Air Force) sent a single Let L-410 and a brand-new Mil Mi-8MTV-1. This hoist-equipped helicopter is operated by 1 Aviacijas Eskadrila at Lielvarde airbase, and mainly used for SAR missions along the Latvian coast.

Lithuania – Mi-8MTV (right)
During CK'02 another Baltic participant came in the form of a new Lithuanian Mi-8MTV1 coming from Antroji Aviacijos Baze (airbase no. 2), better known as Pajoustis, where it is operated by the Sraigtasparniu Eskadrile (helicopter squadron).

Romania – IAR-330 SOCAT *(above)*

The Israeli company Elbit Systems Ltd was selected as foreign system supplier in 1994, and in September 1995, IAR S.A. signed the contract for the upgrade of 24 IAR-330L Pumas with the Sistem Optic de Cautare si lupta AntiTanc (Anti-tank Optronic Search and Combat System) or SOCAT system. At the heart of the upgrade are a new integrated mission management system, a new 'glass' cockpit and digital avionics, advanced sensors, a Modular Integrated Display and Sight Helmet (MIDASH) for day/night observation and target acquisition, an integrated navigation, armament and attack system plus a completely new weapon system. For CK'02 two SOCATs operated mainly in the CSAR escort role.

Romania – IAR-330L *(above and right)*

The Cooperative Key series began in Romania in October 1996, hosted by the 90th Otopeni Air Transportation base with a total of 12 nations and eight helicopters and transport aircraft taking part. Among them were the Romanian IAR-330L Pumas. In 1974 a license-agreement was reached with Aérospatiale to build the SA 330L Puma in Romania, and since then 104 IAR-330Ls have been produced under license by IAR Brasov for delivery to the Romanian Air Force, which uses them for transport and SAR. During the second CK exercise hosted by Romania – CK'00 held at Mihail Kogalniceanu airbase – the participant numbers rose to 650 personnel and 50 aircraft from 15 NATO and PfP countries. Here the new element of assessment was introduced for the first time.

Moldova – Mi-8MTV-1 *(left)*

CK 1998 was held in July of that year at three different locations in Turkey and involved six NATO nations and eight PfP nations contributing some 46 aircraft in six types of missions. New elements, such as a multi-national combined air operations centre (MNCAOC), air defence and 'No-fly' zone enforcement missions, were introduced. During this exercise Moldova sent observers but no aircraft. During CK'00 a Moldovan Antonov An-72 'Coaler' participated, and during CK'02 at Saint-Dizier a single Mi-8MTV-1 made its appearance, coming from the Mixed Aviation Group at Markulesthy.

Bulgaria – Mi-24 (above – Mi-24V foreground, Mi-24D background)

Based on experiences from Kosovo, combat search and rescue (CSAR) was introduced in 2001 in Bulgaria as a new CK exercise element, when a downed pilot behind enemy lines had to be rescued. In a pre-planned CSAR operation, a helicopter formation comprising two Austrian Bell 212s packed with US and Austrian special ground forces was guided into hostile territory by the AOC and AWACS, protected by Bulgarian 'Hinds' which shadowed the formation, looking for possible threats. The rescue package also included A-10s circling around the formation at altitudes of between 600 and 1,500 ft to provide air cover. The downed pilot identified himself by using a simple hand-held mirror. With the pilot located, the pair of Agusta-Bell 212s landed. Immediately after the go-ahead was given, US and Austrian Special Forces disembarked to set up a perimeter around the helos and escorted the downed pilot into the cabin. During this phase of the operations the Bulgarian 'Hinds' kept their distance in order not to give away the position of the downed pilot and to distract possible hostile forces.

Slovakia – Mi-17 (below)

Co-operative Key '97 was held in Slovakia at both Sliac and Piestany airbases and saw the participation of some 20 combat and transport aircraft, 12 helicopters and 600+ personnel in search and rescue (SAR), medical evacuation (MEDEVAC), close air support (CAS) and peace support operations (PSO) missions. The Slovakian helicopter participation included one of its Mil Mi-17s from 34 Letecká Základna Presov, located in the eastern part of Slovakia. The 34 Vrtulníkové Krídlo (34VrK), was established in late 1995 and has four squadrons: 1 and 2 Bitevná Vrtulníková Letka with Mi-24 gunships, and 3 and 4 Vrtulníková Letka with Mi-2s and Mi-17s.

Bulgaria – Mi-17 (above)

Bulgaria offered to host CK'01 in September 2001 and used its completely renovated Graf Ignatievo and Krumovo airbases for hosting the largest ever CK exercise, with 24 nations contributing more than 32 fighter aircraft, 11 transport aircraft, 20 helicopters and 1,600 soldiers. This exercise saw the official introduction of assessment, combat search and rescue (CSAR), the use of photo-reconnaissance aircraft and civil-military co-operation (CIMIC). As well as Mi-24 'Hinds', the Bulgarians have participated in several CK exercises with Mi-8 and -17 'Hips' from the 24 VAB at Plovdiv-Krumovo, used for medevac and air transportation tasks.

Austria – AB 212 (above)

A regular participant in Co-operative Key is the Austrian air force, with Agusta-Bell AB 212s from Hubschrauber Fliegerregiment 1 based at Langenlebarn. Until the 2002 exercise the Austrians sent several Twin Hueys to operate in the medevac and CSAR roles. In 2002 the Österreichische Luftstreitkräfte was gearing up for the process of equipping with the Sikorsky S-70 Black Hawk. Several experienced personnel were on a course in the US, learning to fly the new helicopter. The first crew began training in July 2002 at Sikorsky's plant at Stratford, Connecticut. Consequently, the Austrians did not participate in CK'02.

Italy – AB 212 (below)

The Italian air force employed its hoist-equipped Agusta-Bell AB 212AMs for the first time during CK'01 in Bulgaria, where they operated alongside the Austrian Twin Hueys. During CK'01, a typical scenario involved the medical evacuation of wounded civilians on the ground to a nearby field hospital. Before the ground troops had established their positions, Close Air Support (CAS) was called in to assist in suppressing and eliminating enemy forces. USAF A-10 Thunderbolts and Bulgarian Su-25K 'Frogfoots' provided the necessary CAS during the exercise, while Bulgarian Mi-24 'Hind' gunships flew over the targets in support of the transport or medevac helicopters, ensuring their safe arrival. As soon as the area was secured, a medical facility was established to give initial treatment to the injured. Medevac helicopters, like the Hungarian Mi-17s and the AB 212AMs from the Grazzanise-based 9° Stormo 'Francesco Baracca' and its 609ª Squadriglia Collegamenti e Soccorso, took the wounded and flew them out to the field hospital at Krumovo.

During CK'02 in France, refined but still unclassified CSAR tactics and procedures were practised and employed, as were more regular Search and Rescue (SAR) operations. For these purposes a single Aeronautica Militare AB 212AM was present at Saint-Dizier, this time belonging to the 651ª Squadriglia Collegamenti e Soccorso of the 51° Stormo.

Switzerland – AS 332M (above)

The Force Offering by the Troop Contributing Nations (TCNs) at Co-operative Key 2002 at Saint-Dizier, France, amounted to 29 fighters, 12 transport aircraft, 20 helicopters and one French AWACS, nearly reaching the maximum force permitted by the size of the air base. The Schweizer Luftwaffe sent in a brand new TH98 Cougar, piloted by Captain Ines Widner, one of several female helicopter pilots in the Swiss Air Force. It only had 60 hours on the clock when it arrived for the exercise. The Swiss had first participated in Co-operative Key exercise during CK'01 at Graf Ignatievo, providing a TH89 Super Puma. (Note that the numerical suffix corresponds to the fiscal year in which the aircraft were ordered).

The culmination of the CK'02 combat rescue missions was CSAR Task Force 6 flown on 3 October. It was a classic downed fighter aircraft scenario with two survivors (one Swedish and one American), and was led by two French Pumas. Two Romanian SOCAT gunships served as helicopter escorts, while a Swiss Cougar served as a mobile surgical centre. Mission Lead and Escort was provided by four USAF A-10s, supported by a USAF AC-130H Gunship which provided the initial on-scene commander. A French E-3F AWACS served as Airborne Mission Coordinator.

Boeing B-52 variants: Part 2 B-52D to 'B-52J'

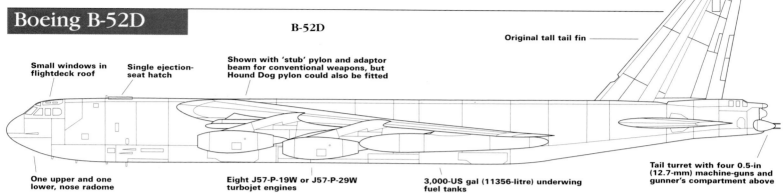

Boeing B-52D

B-52D

Original tall tail fin

Small windows in flightdeck roof

Single ejection-seat hatch

Shown with 'stub' pylon and adaptor beam for conventional weapons, but Hound Dog pylon could also be fitted

One upper and one lower, nose radome

Eight J57-P-19W or J57-P-29W turbojet engines

3,000-US gal (11356-litre) underwing fuel tanks

Tail turret with four 0.5-in (12.7-mm) machine-guns and gunner's compartment above

Whereas the RB-/B-52B and B-52C were built in small numbers (50 and 35, respectively), the B-52D (Model 464-201-7) was the first large-scale Stratofortress production model, and was the first B-52 variant to be manufactured by two different plants. Some 101 B-52Ds were built in the main Boeing plant in Seattle, and the remaining 69 examples were built at Boeing's Wichita, Kansas, facility (bought from Stearman in 1934 and extensively used for B-29 and B-47 production).

Locating a B-52 line at Wichita kept the plant operating at its optimum capacity as B-47 production there wound down, and also freed up valuable space at Seattle for the burgeoning jet airliner production programme.

An initial batch of 50 B-52D bombers was ordered under contract AF33(600)-28223, on 31 August 1954. These aircraft (55-0068/0117) were built at Seattle. Another 27 aircraft (55-0049/0067 and 55-0673/0680) were built at Wichita to fulfil letter contract AF33(600)-26235 of 29 November 1954. Contract AF33(600)-31267 (of 26 October 1955) ordered 51 more B-52Ds (65-0580/0630) from Seattle, and contract AF33(600)-31155, signed on 31 January 1956, ordered another 42 B-52Ds (56-0657/0698).

The new Wichita line accounted for 69 of the 170 B-52Ds. Wichita would build more examples of later models than did Seattle, until, with the B-52G and B-52H, it became the sole production source.

A Wichita-built aircraft was the first D model to fly, making its maiden flight on 14 May 1956. The first Seattle-built B-52D made its first flight on 28 September 1956. The first examples were delivered to the 42nd Bomb Wing during late 1956.

As originally built, the B-52D was virtually identical to the last B-52Cs, powered by the same J57-P-19W or J57-P-29W engines and having the same MD-9 fire control system for the

Photographed on a Big Belly test mission over the Eglin range on 7 March 1966, B-52D-80-BO 56-0629, was a Seattle-built aircraft. It was damaged by a SAM over Vietnam on 26 December 1972. Such was the quality of Boeing's work, however, that the machine landed safely at U-Tapao, complete with 14 punctures in its skin.

tail turret. The only major difference was that the B-52D was built without provision for the reconnaissance capsules used by the RB-52B and B-52C, although, ironically, a handful of B-52Ds were used for the weather reconnaissance role after being appropriately modified in service.

From 1959, the B-52Ds were modified for the low-level role with improvements to their navigation and bombing systems, and the addition of a terrain clearance radar, Doppler and low altitude radar altimeters. The structural modifications required for each B-52C and D for low-level operations proved almost twice as expensive as those needed for any other B-52 variant. The three-phase structural modification programme began with the strengthening of the fuselage bulkhead and aileron bay, and reinforcement of the booster pump panels and wing root splice plates as aircraft reached 2,000 flying hours. The upper wing splices inboard of the inner engine nacelles, the lower wing panels supporting the nacelles, the upper wing surface fuel probe access doors, and the lower portion of the fuselage bulkhead were replaced under Phase II of the programme, at 2,500 flying hours. Phase III consisted of an IRAN (Inspect and Repair As Necessary) programme specifically targeting wing cracks.

Plans to incorporate a new AN/ALQ-27 ECM system were cancelled because it was considered to be too complex and costly. Instead, a four-phase ECM improvement

programme was instituted. Phase I aimed to provide the minimum ECM equipment that was necessary to counter the Soviet radar and SAM threat. Phase II was included in the Big Four or 'Mod 1000' programme (the same Quick Reaction Capability package installed in new-build B-52Hs) that was retrofitted to earlier B-52 variants between November 1959 and September 1963. Under Phase III, the B-52D finally received ECM equipment comparable to the AN/ALQ-27 system.

Before the Vietnam War broke out, it

would have been difficult to predict the importance the D model would assume. Despite being built in large numbers, it had been superseded on the production line by the superior B-52E and F, which formed the tip of SAC's spear. But Vietnam changed all that.

The first Stratofortresses committed in Southeast Asia were the 'hot-rod' B-52Fs, but when it became clear that the US involvement would be both protracted and huge, the availability of much larger numbers of D models led to the older model gaining a new lease

Above: The first ever B-52D, 55-0068, a product of Boeing's Seattle plant, is shown here later on in life, with short-tail B-52Gs in the background.

Left: By contrast, this aircraft was part of the last batch of 18 B-52Ds ordered from Wichita. Although they wore later serials than the Seattle aircraft, the Wichita B-52Ds were actually ordered first.

Photographed in December 1958, these B-52Ds were standing alert at Loring AFB, Maine. Although they have protective covers over their tail turrets, the fact that their outrigger wheels are touching the ground suggests that they are fuelled ready for launch at short notice.

of life. SAC decided that the B-52D fleet would be modified for conventional bombing duties.

Under the so-called 'Big Belly' modification, B-52Ds were upgraded to carry much larger conventional warloads. The B-52Fs initially deployed to Southeast Asia had carried up to 27 500-lb (227-kg) or 750-lb (340-kg) bombs internally, with 12 more under each inner wing. 'Big Belly' raised the internal capacity to 42 750-lb bombs or an astonishing 84 500-lb bombs, by using new high-density clips of bombs. There was no increase in the internal dimensions of the bomb bay, no use of bulged bomb bay doors, and no change to external capacity. A new 'preload system' was developed under which clips of bombs could be pre-prepared for rapid loading. 'Big Belly' B-52Ds could also carry mines, and retained their ability to carry up to four free-fall nuclear weapons.

When being modified under the 'Big Belly' programme, the B-52Ds gained a new colour scheme (to T.O.1-1-4 specifications) featuring black (FS17038 or 27038) undersides and tailfin and a set pattern of Dark Green (FS34079), SAC Bomber Green (FS34159) and SAC Bomber Tan (FS24201 or 34201) on the upper surfaces of the fuselage, wings and tailplanes.

During 1967-1969, under the Rivet Rambler or Phase V ECM programme, B-52Ds assigned to conventional warfare missions in Southeast Asia were given further electronic warfare updates which transformed them into the best protected B-52s in service.

The Phase V suite included one AN/ALR-18 automated receiving set, one AN/ALR-20 panoramic receiver set, and one AN/APR-25 radar homing and warning system. Countermeasures included four AN/ALT-6B or AN/ALT-22 continuous wave jamming transmitters, two AN/ALT-32H and one AN/ALT-32L high- and low-band jamming sets, and two AN/ALT-16 barrage-jamming systems. Expendables included eight AN/ALE-24 chaff dispensers (a total of 1,125 bundles of 'window') and six AN/ALE-20 flare dispensers (with 96 IR decoy flares).

During the Vietnam War, the B-52D was preferred by most aircrew: it generally had a better ECM suite than

A precision attack capability was given to some B-52Ds through the use of the GBU-15 TV-guided glide bomb. The bomb required a datalink pod for guidance.

most B-52Gs and a much larger bombload, and the manned tail turret gave the crew a better ability to monitor SAM launches. Eleven B-52D wings undertook Arc Light deployments to Southeast Asia, some as many as three times and the type flew Linebacker missions later in the war.

Even after its participation in the Vietnam War, the B-52Ds continued to be modified to extend their service lives and to give improved capabilities. Though they retained their tactical-looking camouflage, the D models returned to the strategic role, and some even gained Hound Dog capability.

Once expected to retire in 1971, many of the B-52Ds had amassed nearly double their anticipated 5,000-hour lives, but the decision was taken to retain these useful bombers. Accordingly, between 1972 and 1977, the B-52D fleet underwent the $US219.4 million Pacer Plank upgrade, under which wings and fuselages were strengthened. After restricting the B-52's flight envelope, SAC selected the 80 best D models (of 128 remaining) for an extensive wings-off rebuild that incorporated much new internal structure, new leading and trailing edges, new wing skinning, and 31 rebuilt fuel tanks. The aircraft were rewired, and any damage was expensively repaired. Some 37 of the remaining Ds were retired to 'active' and inviolate storage under Project Crested Dove in 1978.

From 1977, some B-52Ds were modified to carry GBU-15 glide bombs under a $US5 million programme. The aircraft could carry two GBU-15s in tandem under each wing, or one GBU-15 and the necessary datalink pod. Between 1978 and 1982, the B-52Ds also underwent the $US149.1 million Mod F18411B (the B-52D OAS Supportability Improvement (DBNS) programme), under which the original offensive avionics system was replaced by new digital equipment.

Pacer Plank gave the B-52D the theoretical ability to serve until 2000. However, the type was destined for a much shorter career, and the last operational B-52D sortie was flown by the 20th BMS (Bomb Maintenance Squadron) at Carswell AFB on 1 October 1983.

B-52Ds equipped the 7th Bombardment Wing (Heavy) at Carswell AFB, Texas, the 22nd BW(H) at March AFB, California, the 28th BW(H) at Ellsworth AFB, South Dakota, the 42nd BW(H) at Loring AFB, Maine, the 43rd Strategic Wing at Andersen AFB, Guam, the 70th BW(H) at Clinton-Sherman AFB, Oklahoma, the 91st BW(H) at Glasgow AFB, Montana, the 92nd BW(H) at Fairchild AFB, Washington, the 93rd BW(H) at Castle AFB, California, the 96th SAW at Dyess AFB, Texas (this unit was redesignated as the 69th BW(H) in March 1972), the 99th BW at Westover AFB, Massachusetts, the 306th BW(H) at McCoy AFB, Florida, the 307th SW at U-Tapao AB, Thailand, the 340th BW(H) at Bergstrom AFB, Texas, the 376th SW at Kadena AB, Okinawa, the 454th BW(H) at Columbus AFB, Mississippi, the 461st

55-0049 was the first Wichita-built B-52, and the first B-52D to fly. Its 2-hour 20-minute maiden flight on 14 May 1956 was captained by Rod Randall.

Strategic Wing (Heavy) at Amarillo AFB, Texas, the 462nd SAW at Larson AFB, Washington, the 484th BW(H) at Turner AFB, Georgia, the 494th BW(H) at Sheppard AFB, Texas, the 509th BW(H) at Pease AFB, New Hampshire, the 4047th SW at McCoy AFB, Florida, the 4128th SW at Amarillo AFB, Texas, the 4130th SW at Bergstrom AFB, Texas, the 4138th SW at Turner AFB, Georgia, the 4141st SW at Glasgow AFB, Montana, the 4170th SW at Larson AFB, Washington, the 4245th SW at Sheppard AFB, Texas, the 4252nd SW at Kadena AB, Okinawa, and the 4258th SW at U-Tapao AB, Thailand.

Block	Serial	Quantity
B-52D-55-BO	55-0068/0088	21
B-52D-60-BO	55-0089/0104	16
B-52D-65-BO	55-0105/0117	13
B-52D-70-BO	56-0580/0590	11
B-52D-75-BO	56-0591/0610	20
B-52D-80-BO	56-0611/0630	20
Sub-total (Seattle): 101		
B-52D-1-BW	55-0049/0051	3
B-52D-5-BW	55-0052/0054	3
B-52D-10-BW	55-0055/0060	6
B-52D-15-BW	55-0061/0064	4
B-52D-20-BW	55-0065/0067	3
B-52D-20-BW	55-0673/0675	3
B-52D-25-BW	55-0676/0680	5
B-52D-30-BW	56-0657/0668	12
B-52D-35-BW	56-0669/0680	12
B-52D-40-BW	56-0681/0698	18
Sub-total (Wichita): 69		
Total:		**170**

Boeing B-52E

Externally identical to the B-52D, the E model (Model 464-259) was built by both Seattle (42 aircraft) and Wichita (58 aircraft). It was the cheapest B-52, having a pre-inflation, amortised unit cost of $US4.1 million. Wichita received the first and fourth letter contracts for the B-52E, these being AF33(600)-31155 (signed on 10 August 1955) for 14 aircraft, and AF33(600)-32864 (signed on 2 July 1956) for 44 B-52Es. Seattle contracts comprised AF33(600)-31267 (signed on 26 October 1955), originally a B-52D contract but also including 26 B-52Es, and AF33(600)-32863 (signed on 2 July 1956) for 16 B-52Es.

Though hard to tell apart from a B-52D, the E model introduced a number of new systems. It was the first variant built from the start for low-level missions, although B-52Cs and B-52Ds later received equipment that allowed them to perform similar tasks. The B-52E was also the first version to test the new GAM-77 (later AGM-28) Hound Dog missile – fitted with the first thermonuclear warhead (hydrogen

bomb) deployed by SAC – and it was the first version to receive the GAM-72 (later ADM-20) Quail decoy.

The switch to the low-level role demanded the integration of what was supposed to be a more accurate and reliable navigation system, so the highly

automated AN/ASQ-38 was fitted. The AN/ASB-4 bomb navigation system replaced the AN/ASB-15, AN/APN-89A Doppler replaced AN/APN-108, and an MD-1 automatic astrocompass and AN/AJA-1 or AJN-8 true heading computer system were included. The navigator-bombardier's station was refined, giving improved crew comfort.

A Seattle-built B-52E made the variant's first flight, on 3 October 1957, followed into the air by the first Wichita B-52E on 17 October.

Unfortunately, the B-52E's AN/ASQ-38 navigation system proved unreliable, inaccurate and difficult to maintain, requiring a major modification effort that was code named Jolly Well. This was also applied to the B-52F-H, and was finally completed in 1964.

Unlike the B-52D which preceded it and the B-52F which came afterward, the B-52E was never used for conventional bombing, was not used in Vietnam, and remained in the strategic role for the whole of its

Using its new radar and navigation systems, the B-52E used terrain masking to cover its run in to the target. Attacks were planned to be pressed home at altitudes around 200 ft (61 m).

There was little to outwardly distinguish a B-52E from a B-52D, the differences being avionics-based and therefore mostly internal. The B-52E was a specialised low-level penetrator.

career. The variant served with the 6th BW at Walker AFB, New Mexico, the 11th BW at Altus AFB, Oklahoma, the 17th BW at Wright-Patterson AFB, Ohio, the 70th BW at Clinton-Sherman AFB, Oklahoma, the 93rd BW at Castle AFB, California, the 4043rd SW at Wright-Patterson AFB, Ohio, and the 4123rd SW at Clinton-Sherman AFB, Oklahoma.

Some time-expired B-52Es were retired in 1967, but the bulk served until committed to the 'Boneyard' for long-term storage in 1969 and 1970.

Block	Serial	Quantity
B-52E-45-BW	56-0699/0712	14
B-52E-50-BW	57-0095/0109	15
B-52E-55-BW	57-0110/0130	21
B-52E-60-BW	57-0131/0138	8
	Sub-total Wichita:	**58**
B-52E-85-BO	56-0631/0649	19
B-52E-90-BO	56-0650/0656	7
B-52E-90-BO	57-0014/0022	9
B-52E-95-BO	57-0023/0029	7
	Sub-total (Seattle):	**42**
	Total:	**100**

Boeing NB-52E and engine testbeds

The second Seattle-built B-52E (56-0631) was never delivered to the USAF, and was instead assigned to test and trials duties. It initially retained its standard 'bomber' designation, and was used for trialling prototype landing gears, engines and other major subsystems. The aircraft later became an NB-52E after undergoing more permanent modifications.

The aircraft became the first in a series of disparate research platforms used for the development and assessment of Active Control Technology, flying two quite distinct research programmes in two different configurations.

The NB-52E was initially used for the Load Alleviation and Mode Stabilization (LAMS) study. Several operational B-52s had already been lost while flying at low level due to structural and fatigue failures. These had necessitated expensive and often heavy structural reinforcements and modifications, and it was clear that a better approach might be to use improved control technology to reduce the buffeting encountered by the structure, and thereby to reduce fatigue damage.

On the LAMS testbed, wind gusts were detected and measured by sensors in a long nose boom and by a

battery of gyros and accelerometers distributed along the length of the fuselage. If these detected any abrupt change in vertical, horizontal, pitch, roll or yaw acceleration, a signal was transmitted to the onboard flight control computers, which automatically actuated the flight controls to cancel out the effects of the turbulence. On the LAMS aircraft in its initial configuration, this system, collectively known as the Ride Control System (RCS), controlled the outboard spoilers (operating symmetrically around a 15° 'datum'), the ailerons (which were actuated symmetrically and differentially as required) and the elevators. The system

performed well, reducing wing fatigue by about 50 per cent, and producing similar reductions in fatigue damage in the centre fuselage.

A Direct Lift Control study was also conducted as part of LAMS, using spoilers, symmetrical aileron and elevators to uncouple pitch from translational aircraft movement up and down, and later to decouple pitch and roll. This facilitated precise manoeuvring during, for example, inflight refuelling. The LAMS programme, which ran between 1966 and 1968, had some direct relevance to the operational B-52 fleet, whose flight control systems were modified in response in order to reduce fatigue during low-altitude flight.

The NB-52E test aircraft was further

modified and returned to airworthy status for a new Control Configured Vehicle (CCV) study that grew out of LAMS. The aim was to provide automatic, electronic 'active flutter suppression' in order to lessen structural loads on the aircraft by using an onboard automatic control system that would immediately actuate the appropriate control surface in response to sensed structural motion, in order to control wing torsion and bending loads and to suppress flutter. To this end, the NB-52E gained an array of new control surfaces, all with new electronic actuators replacing the traditional mechanical and hydraulic linkages. The aircraft was fitted with small swept canard winglets on each side of the nose and another vertical surface below the nose, plus outboard ailerons and inboard flaperons.

The system functioned very well, and in mid-1973 the NB-52E flew 10 kt (11.5 mph; 18.5 km/h) faster than its open-loop flutter velocity – the speed at which flutter normally would have caused the aircraft to break up. Despite this success, active flutter suppression

GE's 57-0119 is seen at Fairbanks in early 1972 during cold-start trials of the CF6-50 turbofan. The engine was started in temperatures as low as -17°F (-27.2°C).

has remained largely experimental, and has still not achieved operational status on any large aircraft.

Two more B-52Es were modified for use as engine testbeds, sometimes referred to as JB-52Es and sometimes as NB-52Es. Certainly, neither was returned to operational status, and the NB-52E designation therefore seems most appropriate, though the USAF Museum records both as having been JB-52Es.

One of these aircraft, 56-0636, was modified to test the Pratt & Whitney JT9D turbofan engine designed for the Boeing 747, for which a JT9D in a representative nacelle replaced the inboard pair of J57 engines on the starboard side. Thus modified, the NB-52E flew in mid-June 1968.

Another aircraft, 57-0119, was assigned to General Electric and used to test the company's XTF99 (later TF39) engine, a high-bypass-ratio turbofan designed for the Lockheed C-5A Galaxy transport. The 40,000-lb st (177.89-kN) thrust TF39 was mounted on the right inboard engine pylon in place of the two J57s normally installed, and produced more thrust than the four J57 turbojets carried under each wing of a standard production B-52E. The aircraft made its first flight with the XTF99 installed on 30 June 1968.

The two NB-52E engine testbeds looked almost identical, though two quite separate airframes were used.

Apart from their serial numbers, they differed in that the JT9D testbed aircraft retained a SAC band around its upper nose.

Various design studies were undertaken to investigate the feasibility of using new-generation, very large, high-bypass turbofan engines on the B-52. A B-52 powered by XTF99s could theoretically have been a twin, but most fan studies envisaged the new B-52 in a four-engined configuration.

Boeing B-52F

The B-52F (Model 464-260) was the last of the 'tall-tail' B-52 variants, and was also the last Stratofortress version to be built at both Wichita and Seattle.

Letter contract AF33(600)-32863 signed on 2 July 1956 included all 44 Seattle-built B-52Fs (as well as 16 B-52Es). AF33(600)-38264 signed on the same day covered the Wichita-built B-52Fs (plus 44 B-52Es).

The B-52F differed from the B-52E principally in its powerplant, which consisted of eight J57-P-43W, J57-P-43WA, or J57-P-43WB engines, each rated at 13,750 lb st (61.15 kN) with water injection. The B-52E and earlier variants had used fuselage-mounted air-driven turbines and alternators, and these had sometimes caused catastrophic fires due to their proximity to the main fuselage fuel tanks. The B-52F's pairs of J57-P-43W engines were fitted with 'hard-drive' alternators to supply electrical power to the aircraft. These new alternators were attached to the left-hand side of each podded pair of engines, necessitating the addition of a bulged housing and ram air intake on the bottom left-hand 'corner' of each engine pod. A new fairing was added between the two engine intakes, and small ram air intakes were set into the lower lip of each intake to provide cooling air for the engine oil and constant-speed drive units.

The first Seattle-built B-52F made its maiden flight on 6 May 1958, followed on 14 May by the first Wichita-built example. Fiscal restrictions imposed by the Defense Department in late 1957 limited authorised overtime work at Boeing, and delayed deliveries slightly. The 93rd Bomb Wing began receiving its first F models in June 1958, and all had been handed over to the USAF by the end of February 1959.

Before delivery, most of the B-52Fs were fitted with Quickclip safety straps around their Hard Shell stainless steel strap clamps, which had been found to have faulty latch pins. The Hard Shell clamps themselves had been developed to replace stress corrosion-prone aluminium Blue Band straps, which had been added to B-52s in 1957 in an attempt to prevent fuel leaks caused by the breaking of the original Marman clamps, the flexible fuel couplings that interconnected fuel lines between tanks. The remaining B-52Fs and other early versions were modified retrospectively with the new Quickclip straps.

Like its predecessors, the B-52F could carry some 27 500-lb (227-kg) or 750-lb (340-kg) bombs internally, for the conventional role. This hardly scratched the surface of the aircraft's potential

Arc Light saw B-52Fs, including this pair from the 454th BW attacking a target in South Vietnam in 1965, based at Andersen AFB, Guam. The raids began on 18 June 1965, the aircraft involved having had their conventional weapons capabilities upgraded.

payload, and in June 1964 the Air Staff approved the modification of 28 B-52Fs under Project South Bay, adding another 24 750-lb bombs on two external pylons installed underneath each wing inboard of the inner engine pods. The B-52F, with its multiple alternators and excellent performance, was the obvious choice for deployed operations (especially in the tropics) and South Bay merely increased its usefulness. The

pylons were modifications of the pylons originally designed to carry the Hound Dog cruise missile. The South Bay upgrade doubled the B-52F's conventional warload, increasing the total bombload to 51 750-lb bombs.

When the war in Southeast Asia expanded, 46 more B-52Fs received similar modifications under Project Sun Bath. The South Bay and Sun Bath B-52Fs were replaced by Big Belly

M117 750-lb bombs fall from a 2nd BW B-52F over South Vietnam in October 1965. The aircraft is equipped with wing racks for 24 bombs, added during either the South Bay or Sun Bath programmes. Bomb release was sequenced to deconflict the weapons as they fell.

B-52Ds from April 1966, and reverted mainly to strategic duties. Some were later modified as Hound Dog missile carriers, and they received a SIOP camouflage finish.

Operators of the B-52F included the 2nd Bombardment Wing at Barksdale AFB, Louisiana, the 7th Bombardment Wing at Carswell AFB, Texas, the 93rd Bombardment Wing at Castle AFB, California, the 320th Bombardment Wing at Mather AFB, California, the 454th BW at Columbus AFB, Mississippi, the 4134th SW at Mather AFB, California, the 4228th SW at Columbus AFB, Mississippi, and the 4238th SW at Barksdale AFB, Louisiana. The last B-52F (57-0171) in service was retired from the 2nd BW at Barksdale on 7 December 1978.

Block	Serial	Quantity
B-52F-65-BW	57-0139/0154	16
B-52F-70-BW	57-0155/0183	29
	Sub-total (Wichita):	**45**
B-52F-100-BO	57-0030/0037	8
B-52F-105-BO	57-0038/0052	15
B-52F-110-BO	57-0053/0073	21
	Sub-total (Seattle):	**44**
	Total:	**89**

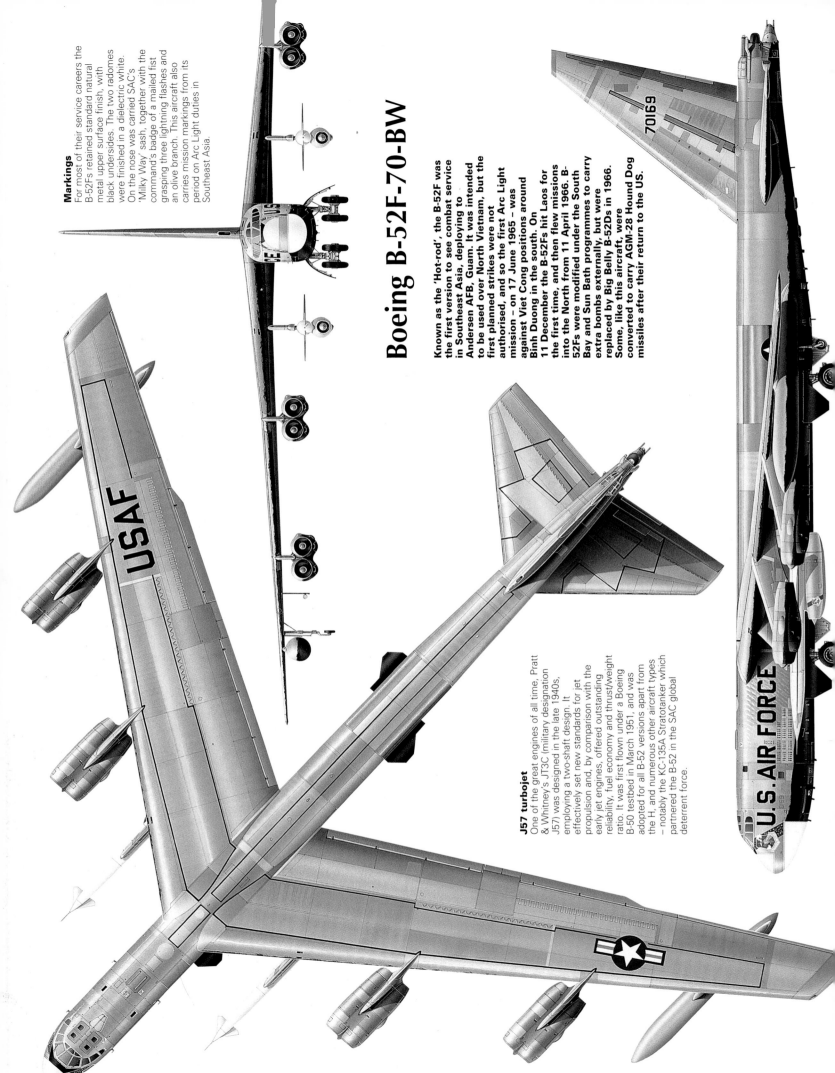

Markings
For most of their service careers the B-52Fs retained standard natural metal upper surface finish, with black undersides. The two radomes were finished in a dielectric white. On the nose was carried SAC's 'Milky Way' sash, together with the command's badge of a mailed fist grasping three lightning flashes and an olive branch. This aircraft also carries mission markings from its period on Arc Light duties in Southeast Asia.

Boeing B-52F-70-BW

Known as the 'Hot-rod', the B-52F was the first version to see combat service in Southeast Asia, deploying to Andersen AFB, Guam. It was intended to be used over North Vietnam, but the first planned strikes were not authorised, and so the first Arc Light mission – on 17 June 1965 – was against Viet Cong positions around Binh Duong in the south. On 11 December the B-52Fs hit Laos for the first time, and then flew missions into the North from 11 April 1966. B-52Fs were modified under the South Bay and Sun Bath programmes to carry extra bombs externally, but were replaced by Big Belly B-52Ds in 1966. Some, like this aircraft, were converted to carry AGM-28 Hound Dog missiles after their return to the US.

J57 turbojet
One of the great engines of all time, Pratt & Whitney's JT3C (military designation J57) was designed in the late 1940s, employing a two-shaft design. It effectively set new standards for jet propulsion and, by comparison with the early jet engines, offered outstanding reliability, fuel economy and thrust/weight ratio. It was first flown under a Boeing B-50 testbed in March 1951, and was adopted for all B-52 versions apart from the H, and numerous other aircraft types – notably the KC-135A Stratotanker which partnered the B-52 in the SAC global deterrent force.

70169

USAF

U.S. AIR FORCE

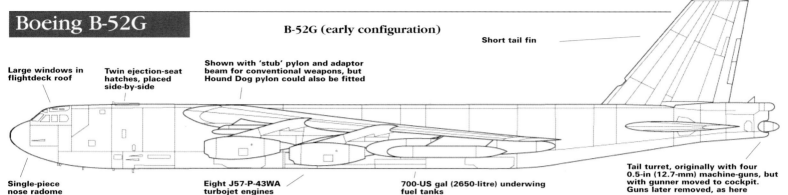

Boeing B-52G

B-52G (early configuration)

Short tail fin

Large windows in flightdeck roof

Twin ejection-seat hatches, placed side-by-side

Shown with 'stub' pylon and adaptor beam for conventional weapons, but Hound Dog pylon could also be fitted

Single-piece nose radome

Eight J57-P-43WA turbojet engines

700-US gal (2650-litre) underwing fuel tanks

Tail turret, originally with four 0.5-in (12.7-mm) machine-guns, but with gunner moved to cockpit. Guns later removed, as here

The Model 464-253 designation was initially applied to a loosely defined 'Super B-52' which was offered as an alternative to the Convair B-58. It was then proposed as an interim solution when it seemed as though the B-58's service entry might be delayed for a very long time, and while a decision was taken on Boeing's definitive WS-110A – a highly advanced bomber powered by turbojet engines that used boron-based 'zip fuel'. The redesigned B-52 initially featured an entirely new wing, and was to be powered by eight non-afterburning Pratt & Whitney J75 engines (reheated versions of which powered the F-105 and F-106). Each of these engines produced 5,000 lb (22.24 kN) more thrust than the standard B-52's J57s, permitting much higher take-off weights.

Boeing informally presented its new design to the Air Research and Development Command in March 1956. The company received $US1.2 million of funding and formally began design work in June 1956, only three months after the RB-52B had entered USAF service. The Air Staff committed $US8.8 million more shortly afterward.

The scope of the changes to the 464-253 were scaled back before the B-52G designation was applied to the aircraft, which thus became a much more modest improvement of the existing B-52, retaining the Pratt & Whitney J57-P-43WA turbojets used by

SAC might in the late 1950s: the fourth B-52G refuels from one of the original short-finned KC-135As.

the B-52F, each rated at 11,200 lb st (49.81 kN) dry and 13,750 lb st (61.15 kN) with water injection. Water capacity was increased (to some 1,200 US gal/4542 litres), giving an increase in injection duration, and engine-driven pumps were added for the water injection system and the hydraulics, replacing the original pneumatic packs.

In order to improve performance further, Boeing set itself extremely stringent weight reduction targets, aiming to shave 15,000 lb (6804 kg) off the empty weight of the B-52F. The company achieved this target – actually saving 15,421 lb (6995 kg) – which helped Boeing raise the maximum take-off weight by a staggering 19 US tons (17.2 tonnes), to 488,000 lb (221357 kg).

The wing was extensively redesigned so that the aileron system was deleted (leaving only the spoilers for roll control) and integral tanks replaced the original rubber 'bladders'. The new wing used lighter-weight alloys, and incorporated new integral stiffeners and a new trailing edge structure. The flap drive motors were replaced by lighter units (which increased retraction time from 40 to 60 seconds). The new integral fuel tanks increased internal fuel capacity by 8,719 US gal (33005 litres), allowing the huge 3,000-US gal (11356-litre) tanks of the B-52C/D/E/F to be replaced by much smaller, lighter 700-US gal (2650-litre) ones. The prime purpose of these tanks was to reduce the upward bending moment on the wing tips, and to act as

anti-flutter weights. Overall fuel capacity was still 330 US gal (1249 litres) more than that of the B-52F – at 47,975-48,030 US gal (181605-181813 litres), rather than 41,553 US gal (157295 litres) – and the fuel system was considerably simplified, although it was fitted with an all-new fuel management system.

The vertical tailfin was reduced in height by 91 in (2.31 m), from 48 ft 3 in (14.71 m) to 40 ft 7 in (12.37 m). The tail gunner was relocated to the main cockpit, facing aft alongside the EWO, allowing his pressurised cabin to be deleted and making it possible to redesign and relocate the brake parachute stowage above the tailcone. The fire control radar for the tail guns was changed from an MD-9 set to an Avco-Crosley AN/ASG-15. This featured separate radar dishes for search and track, like earlier B-52 tail gun fire control systems, but was also fitted

B-52G 57-6468, the first of the species, posed for photographs on a sortie over Wichita in November 1958. The B-52G emerged as a shadow of the far more ambitious configuration that Boeing had originally schemed for it.

with a television camera (although this was later replaced by ALQ-117 countermeasures gear).

The elimination of the ailerons, in conjunction with the smaller fin and rudder, caused some handling problems, including a marked tendency to Dutch roll (which could not adequately be countered by the existing yaw damper) and to buffet and mildly pitch up in turns. Until modifications were incorporated, the G model B-52 was fatiguing to fly, and especially difficult to fly accurately during inflight-refuelling contacts.

Designed from the outset as a missile carrier, the B-52G initially carried Hound Dog. Like the B-52E, the B-52G/Hound Dog combination was expected to penetrate to its target at low level. Here a crew practises a scramble.

The B-52G's lighter structure and higher fuel capacity soon gave rise to serious structural problems. The weakness of the redesigned wing led to the loss of a B-52G near Goldsboro, North Carolina, on 24 January 1961, and this accident resulted in the imposition of flight restrictions on the variant while a wing-strengthening programme was developed.

The chosen fix was known simply as ECP1050 and provided for an entirely new wing box beam, using thicker aluminium and steel taper locks in place of the original titanium locks. Wing skins gained extra clamps, brackets and panel stiffeners, and a new coating for the integral tanks reduced leaks. The final 18 B-52Hs were built with the redesigned wing, but all earlier B-52Gs and B-52Hs received the modifications during routine IRAN depot-level maintenance. The programme was completed in September 1964.

The rewinging of the G and H models was conducted in parallel with the more general High Stress structural reinforcement programme for the entire B-52 fleet.

Great emphasis was placed on improving crew comfort in the G model, incorporating redesigned seats and a new heating and ventilation system that included electrically heated windows to prevent icing and misting. More systems were increasingly automated, thereby reducing crew workload, and the instrument panel was revised, lowered and tilted. The pilots' rudder pedals were redesigned and the footwell floors were lowered, improving comfort and allowing the pilots to rest their heels on the floor when taxiing, which helped to reduce over-controlling on the ground. The pilots' ejection hatches were enlarged, and the other ejection hatches were fitted with aerodynamic hatch lifters.

Externally, the nose of the B-52G was subtly changed in outline: a one-piece radome replaced the separate upper and lower radomes used on earlier variants.

The first B-52A flew with a shortened tailfin (actually a cut-down XB-52 tailfin) during 1957, thereby serving as a partial aerodynamic prototype for the G model (and using the XB-52G designation, according to some sources). The first true B-52G flew initially on 31 August 1958.

The B-52G was destined to be the most numerous Stratofortress sub-variant, 193 being built at the Wichita plant between October 1958 and February 1961. It was thus the first B-52 variant to be built exclusively at the Wichita factory.

The 193 B-52Gs were ordered under three letter contracts. The first 53 used FY57 funding in AF33(600)-35992, finalised on 15 May 1958. The biggest B-52G order came with AF33(600)-36470, covering 101 examples; although finalised on the same day, this contract used FY58 funds. The last B-52G contract, AF33(600), finalised on 28 April 1959,

A B-52G cruises with its precious Hound Dog cargo. The engines of the two Hound Dog missiles were routinely used to augment the eight J57s for heavyweight take-offs.

covered the final 39 aircraft and was paid for using FY59 funds.

The B-52G entered service with the 5th Bomb Wing at Travis AFB, California, on 13 February 1959, and with the 42nd Bomb Wing in May. Although some earlier B-52 models were destined to receive the new Hound Dog cruise missile, the B-52G was the first version designed 'from scratch' as a missile carrier, and from the 55th aircraft (58-0159), the type introduced inboard underwing pylons to accommodate the new missiles. Hound Dog capability was added to all earlier B-52Gs by 1962.

As the demands of the Vietnam War overstretched the B-52D fleet, some B-52Gs assumed a conventional attack mission, though they lacked the Big Belly increased-capacity modifications of the D models. Some of the B-52Gs assigned to conventional missions were given the same Phase V Rivet Rambler electronic warfare upgrades as the B-52D fleet. This EW suite included one AN/ALR-18, one AN/ALR-20, one AN/APR-25, four AN/ALT-6B or AN/ALT-22, two AN/ALT-32H, one AN/ALT-32L, and two AN/ALT-16 barrage-jamming systems, together with a comprehensive set of countermeasures dispensers. These are all more fully described under the B-52D entry. Some of the Rivet Rambler-modified B-52Gs

were also fitted with pylons between the inboard and outboard nacelles, used to carry AN/ALE-25 forward-firing chaff dispenser rocket pods containing 20 Tracor AN/ADR-8 2.5-in folding-fin chaff rockets. The chaff rockets could be fired manually by the EWO, or automatically by the AN/ASG-21 fire control system.

The Phase V EW suite fitted to some B-52Gs gave these aircraft some protection against enemy SAMs, but only half of the B-52Gs based on Guam for the Linebacker II raids of 1972 had received the Rivet Rambler updates. Some of the Linebacker II B-52Gs were fitted with AN/ALQ-119(V) ECM pods in place of the AN/ALE-25. Despite these modifications, the lightly-built B-52G proved extremely vulnerable to hostile SAMs and six were shot down. Only one B-52G was able to survive being damaged by a SAM, although several B-52Ds hit by SAMs landed safely. During the later stages of Linebacker II, B-52Gs were sometimes diverted in flight to less dangerous targets.

In 1970, the Air Force decided to arm the B-52Gs with a new stand-off weapon, the Boeing AGM-69A short-range attack missile (SRAM). This was to be carried in place of the ageing Hound Dog, though its relatively small size and low weight allowed aircraft to carry a much larger number of missiles,

up to 20 in total: 12 externally on the underwing pylons and eight more on a rotary launcher in the rear section of the bomb bay. SRAM modifications began on 15 October 1971, with the delivery of the first G model to the Oklahoma City AMA, and the first B-52Gs with SRAM re-entered service with the 42nd Bomb Wing during March 1972. The San Antonio AMA made the same modification to B-52Hs under the same engineering change number, (ECP2126), which saw a total of about 270 B-52Gs and Hs becoming SRAM carriers. External carriage of the SRAM proved troublesome and was abandoned during the early 1980s. Warhead safety, reliability and supportability issues led to the withdrawal of AGM-69s from the operational inventory in 1990.

Between 1972 and 1976, all surviving B-52Gs gained the AN/ASQ-151 electro-optical viewing system (EVS). The system consisted of two separate sensors contained in two fairings below the nose, immediately behind the radome, and was designed to give the B-52 crew an enhanced view of the outside world when flying at low level at night. When not in use, the EVS sensors could be rotated to face aft for protection against damage from foreign objects. The sensor windows included built-in 'washers'. A steerable

Although it is shown here as it appeared during compatibility trials with the AGM-69A SRAM, 58-0204 also carries the nose art and a tail modification that were added as part of the Rivet Ace Phase VI ECM defensive avionics test programme.

Westinghouse AN/AVQ-22 low-light-level television camera was housed in the port fairing, with a Hughes AN/AAQ-6 forward-looking infra-red (FLIR) sensor to starboard. Pictures from these sensors could be displayed on new CRT screens which were installed at the pilot, co-pilot, and both navigator stations. A variety of information could be overlaid on the TV or FLIR picture, including alphanumerics representing sensor position (invaluable if the sensor was not boresighted), indicated airspeed, time to weapons release, a radar altimeter height readout, and heading error, and a graphic overlay of the terrain avoidance profile trace or an artificial horizon overlay.

Almost simultaneously with the installation of EVS, surviving B-52Gs gained a Phase VI ECM defensive avionics systems (ECP2519) upgrade

under the Rivet Ace programme. Some aircraft had the new EW equipment fitted at the same time as they gained EVS, or immediately afterward, and it was very unusual to see aircraft fitted with EVS alone.

Although the programme was launched in December 1971, development was protracted and the modifications took several years to incorporate across the B-52G fleet, escalating in price from $US362.5 million to $US1.5 billion. The Rivet Ace testbed (58-0204) trialled a variety of ECM systems and flew with a number of different antenna configurations. Unlike production Phase VI aircraft, the prototype carried ALQ-153 tail warning radar in a pod mounted on the tip of the port stabilator, instead of inside the fin, covered by deep blisters on each side. With ALQ-172 (later adopted as the Phase VI+ fit), the aircraft had unusual antennas below the rudder and 'cheek fairings' farther aft and higher, close to the cockpit side windows. The equipment re-used the standard ALQ-117 antenna fairings on the

'production' Phase VI+ conversions.

Threat warning systems used in the Phase VI EW suite included an AN/ALR-20A countermeasures receiver, an AN/ALR-46(V) digital radar warning receiver set, and the AN/ALQ-153 tail warning radar set. The Phase VI upgrade added an AN/ALQ-117 active countermeasures set, with a pair of antennas inside prominent teardrop-shaped cheek fairings on each side of the nose above the EVS turrets, and additional antennas in the extended tailcone; the latter was stretched 40 in (1.02 m) to accommodate the extra electronic equipment. Phase VI also added the AN/ALQ-122 false target generator system (sometimes known as smart noise operation equipment), AN/ALT-28 noise jammers (with a prominent antenna fairing projecting upward from the top of the nose radome), AN/ALT-32H and AN/ALT-32L high- and low-band jamming sets, and an AN/ALT-16A barrage-jamming system.

Expendables included 12 AN/ALE-20 flare dispensers (containing a total of 192 flares) and eight AN/ALE-24 chaff

dispensers (with 1,125 chaff bundles). The latter were housed in the wing trailing edge, just outboard of the inner engine pod in the vacant space between the two sets of flaps. On previous variants this space had occupied the aileron.

During the 1970s, the B-52G and B-52H bombers formed the backbone of the USAF's strategic nuclear deterrent, while the B-52D continued primarily to serve in the tactical and conventional roles. With the steady improvement of Soviet nuclear weapons, and the abandonment of the airborne alert operation, greater emphasis was placed on rapid-reaction take-offs by nuclear-armed bombers. Under the $US35 million Project Quick Start, launched in 1974, cartridge starters were fitted to all unmodified B-52G and B-52H engines (since 1963-1964 each aircraft had already received such starters on two engines). The programme was completed in July 1976.

The addition of EVS and Phase VI Rivet Ace EW systems to the B-52G and B-52H fleets was followed closely by improvements to the aircraft's avionics systems.

By the mid-1970s, the B-52G's AN/ASQ-38 bombing/navigation system was becoming obsolescent and increasingly unreliable. From 1980, 168 B-52Gs and 96 B-52Hs received a new digital (MIL STD 1553A-based) AN/ASQ-176 offensive avionics system (OAS). This was optimised for low-level use and was hardened against electromagnetic pulse. The system included dual AN/ASN-136 inertial navigation systems, AN/APN-218 Doppler, an AN/ASN-134 attitude heading reference system, and an AN/APN-224 radar altimeter, together

For Vietnam War duty the B-52G fleet received a three-tone 'SEA' camouflage. The undersides were white, instead of the black applied to the B-52Ds.

with new digital missile interface units. Major modifications were made to the primary attack radar. A B-52G flew with the new system on 3 September 1980, quickly demonstrating massively improved reliability. Weapons integration work culminated in an AGM-69A launch in June 1981.

Operators of the B-52G included the 2nd BW at Barksdale AFB, Louisiana, the 5th BW at Travis AFB, California, the 17th BW at Beale AFB, California, the 19th BW at Robins AFB, Georgia, the 28th BW at Ellsworth AFB, South Dakota, the 39th BW at Eglin AFB,

Florida, the 42nd BW at Loring AFB, Maine, the 43rd SW at Andersen AFB, Guam, the 68th BW at Seymour Johnson AFB, North Carolina, the 72nd BW at Ramey AFB, Puerto Rico, the 92nd SAW at Fairchild AFB, Washington, the 93rd BW at Castle AFB, California, the 97th BW at Blytheville AFB, Arkansas, the 319th BW at Grand Forks AFB, North Dakota, the 320th BW at Mather AFB, California, the 366th Wing headquartered at Mountain Home AFB, Idaho but based at Castle AFB, California, the 379th BW at Wurtsmith AFB, Michigan, the 380th SAW at

Plattsburgh AFB, New York, the 397th BW at Dow AFB, Maine, the 416th BW at Griffiss AFB, New York, the 456th SAW/BW at Beale AFB, California, the 465th BW at Robins AFB, Georgia, the 4038th SW at Fairchild AFB, Washington, the 4039th SW at Griffiss AFB, New York, the 4126th SW at Beale AFB, California, the 4135th SW at Eglin AFB, Florida, the 4137th SW at Robins AFB, Georgia, the 4241st SW at Seymour Johnson AFB, North Carolina, the 4300th BW(P) at Diego Garcia AB, Indian Ocean, and the 1708th BW(P) at Prince Abdulla AB, Saudi Arabia.

Block	Serial	Quantity
B-52G-75-BW	57-6468/6475	8
B-52G-80-BW	57-6476/6485	10
B-52G-85-BW	57-6486/6499	14
B-52G-90-BW	57-6500/6520	21
B-52G-95-BW	58-0158/0187	30
B-52G-100-BW	58-0188/0211	24
B-52G-105-BW	58-0212/0232	21
B-52G-110-BW	58-0233/0246	14
B-52G-115-BW	58-0247/0258	12
B-52G-120-BW	59-2564/2575	12
B-52G-125-BW	59-2576/2587	12
B-52G-130-BW	59-2588/2602	15
Total:		**193**

Boeing B-52G CMI

The AGM-69 was effectively replaced as the B-52's primary strategic stand-off weapon by the Boeing AGM-86B Air-Launched Cruise Missile (ALCM), after a competitive evaluation against the General Dynamics AGM-109 Tomahawk ALCM. The original AGM-86A version of the ALCM had been developed for the Rockwell B-1A, and had been sized to fit inside the B-1's short bomb bays. When the B-1A was cancelled, the ALCM was redesigned for the B-52, becoming the longer-bodied AGM-86B.

Powered by a 600-lb st (2.67-kN) Williams F107-WR-100 turbofan, and having folding wings with a full span of 12 ft (3.67 m), the 3,200-lb (1452-kg) cruise missile combined a very long range (about 1,500 miles/2414 km) with a powerful, selectable yield (150-170 kT) W80-1 nuclear warhead. The ALCM was relatively slow and relied on very low-level flight to penetrate hostile defences. It used a combination of inertial guidance and terrain contour matching for navigation and terrain avoidance, with a highly accurate radar altimeter and downloaded mapping data.

The B-52G could carry six AGM-86Bs on each of its two underwing pylons,

and the new missile became operational with the 416th Bomb Wing at Griffiss AFB, New York, in December 1982.

A total of 1,715 ALCMs was built, equipping 98 modified B-52Gs and all 96 surviving B-52Hs. Under the provisions of the unratified SALT II treaty, cruise missile-carrying aircraft had to be identifiable easily as such by reconnaissance satellites. The turbofan-engined B-52Hs were considered to be sufficiently distinctive, but a means had to be found to differentiate between cruise- and non cruise-equipped B-52Gs. The chosen solution was to add a distinctive, rounded non-functional wing root fairing, or 'strakelet', to the AGM-86B-equipped B-52Gs. The modification had to be structurally integral with the aircraft so that the change could not be removed easily, or moved from one aircraft to another.

While stringently adhering to this provision of the SALT II treaty, the USAF disregarded the Treaty's limit of 130-cruise missile-carrying aircraft. It received its 131st ALCM-capable B-52 Stratofortress on 28 November 1986, and eventually accepted a total of 194 such aircraft.

From 1985, some B-52Gs received the same Norden AN/APQ-156 strategic

radar that was simultaneously fitted to all surviving B-52Hs, in place of the original ASQ-176. The $US700 million programme enhanced the B-52's autonomous targeting capability, and included the provision of new controls, displays and software.

By the end of 1988, the original 193-aircraft B-52G fleet had been reduced to 166 aircraft. Ninety-eight were converted to carry the AGM-86B ALCM, leaving 68 conventionally-armed ICSMS B-52Gs. Although conventionally-armed aircraft formed the backbone of the B-52 force used in Operation Desert Storm, many of them were actually CMI (Cruise Missile Integration) aircraft; seven of the 2nd Bomb Wing's B-52Gs opened the war by firing a salvo of conventionally-armed Boeing AGM-86C cruise missiles.

Part of the B-52G force had first received the conventionally-armed AGM-86C conversion of the AGM-86B ALCM during the late 1980s. This new weapon was known as Senior Surprise (the F-117A was Senior Trend) and was developed under a black programme. The missile had a 992-lb (450-kg) high-explosive blast-fragmentation warhead in place of the usual nuclear warhead, and a GPS-based navigation system replaced the terrain profile matching/ inertial navigation system used by the

nuclear version. The conventional AGM-86C had a slightly shorter range (1,200 mile/1931 km) than the nuclear version, due to its heavier warhead.

The existence of the AGM-86C was kept secret until 1992, when it was finally revealed that 35 had been used against Iraqi targets during the opening phase of Operation Desert Storm.

Operators of the CMI B-52G and the ALCM included the 2nd BW (one squadron only, the 596th BS), the 93rd BW (which used small numbers for training), the 97th BW, the 379th BW, and the 416th BW.

Ninety-eight aircraft received the CMI modifications, including 41 which were still in service when Air Combat Command formed in 1992. These were 57-6471, 57-6472, 57-6480, 57-6490, 57-6492, 57-6495, 57-6498, 57-6503, 57-6508, 57-6511, 57-6515, 58-0160, 58-0164, 58-0165, 58-0166, 58-0170, 58-0173, 58-0176, 58-0179, 58-0181, 58-0182, 58-0193, 58-0211, 58-0222, 58-0227, 58-0229, 58-0231, 58-0236, 58-0239, 58-0244, 58-0245, 59-2566, 59-2567, 59-2568, 59-2580, 59-2581, 59-2583, 59-2590, 59-2591, 59-2594, and 59-2602. These last survivors served with the 2nd Bomb Wing and the 379th Bomb Wing. The last aircraft to be retired was 58-0166, on 21 December 1992.

Above and right: General Dynamics' AGM-109 was not selected as the USAF's new cruise missile, but was tested extensively on the B-52G. Worthy of note on the aircraft above are the open bomb bay doors and crew entry hatch. The EVS sensors have their heads reversed in the stowed position.

Boeing B-52G ICSMS

Those B-52Gs retained in the conventional role when the bulk of the fleet were converted to carry cruise missiles were fitted with an integrated conventional stores management system (ICSMS). These conventionally-armed B-52Gs featured a shorter 'stub' pylon than that fitted to ALCM carriers. This pylon was compatible with the original I-beam used on the B-52D and on other versions using converted Hound Dog pylons, and with a new heavy stores adaptor beam (HSAB) optimised for large and heavy

weapons. When used to carry smaller weapons such as the M117 and Mk 82, the I-beam had a larger capacity than the HSAB, carrying 12 bombs each, rather than eight. The HSAB permitted heavier weapons to be carried, however, including five 2,000-lb (907-kg) Mk 84 bombs on each pylon, or six AGM-84 Harpoon anti-ship missiles.

Some 30 B-52Gs were fitted with a Harpoon aircraft command launch control set (HACLCS) at the navigator's station, enabling the carriage of up to

12 missiles on the underwing HSABs. The 320th BW's 441st BS at Mather AFB acted as the test and evaluation wing for the Harpoon. Harpoon-armed B-52Gs were delivered to the 42nd BMW's 69th BMS at Loring AFB in 1984, and then to the 43rd Wing at Andersen AFB, Guam. All surviving ICSMS-modified B-52Gs were eventually fitted with the HACLCS.

Retirement of the B-52G began in 1989, but the variant received a temporary lease of life after the 1991 Gulf War, and some 83 were still on charge in 1992, when SAC gave way to Air Combat Command. Roughly half –

42 – were conventionally-armed, comprising 57-6473, 57-6476, 57-6488, 57-6497, 57-6520, 58-0163, 58-0192, 58-0206, 58-0195, 58-0197, 58-0202, 58-0203, 58-0210, 58-0212, 58-0213, 58-0214, 58-0216, 58-0218, 58-0221, 58-0226, 58-0230, 58-0233, 58-0235, 58-0240, 58-0242, 58-0248, 58-0250, 58-0253, 58-0255, 58-0257, 58-0258, 59-2565, 59-2569, 59-2570, 59-2572, 59-2573, 59-2585, 59-2586, 59-2588, 59-2595, 59-2598, and 59-2599. They served with the 42nd BW, the 93rd BW, the 366th Wing and the 412th Test Wing.

Eight of these aircraft were equipped

As well as launching cruise missiles during the 1991 Gulf War, the B-52G also returned to Vietnam-style 'iron' bombing, mostly using the 750-lb (340-kg) M117 bomb, as here. The aircraft flew missions out of Diego Garcia, Saudi Arabia, Spain and the UK.

Underlying the USAF's willingness to retire the B-52Gs were increasing concerns about the water-injected J57 engines. One B-52 pilot explained, "The idea of dumping tons of water into a fire is as absurd as it sounds. If the pumps don't work or you lose the water augmentation, conditions can become critical. What can also happen is that you can put out the fire in the engine. If you put out two engines in the outboards, you not only lose the engines, but you give yourself a nightmare of a directional control problem. It's a problem that is serious. It gets your attention."

The last B-52G to be retired was the 93rd BW's 58-0240, which flew to the 'Boneyard' on 3 May 1994.

to carry the AGM-142 Have Nap, these being 57-6520, 58-0202, 58-0203, 58-0212, 58-0242, 59-2569, 59-2570, and 59-2598, all of which served with the 366th Wing's 34th Bomb Squadron.

The AGM-142A Raptor is a derivative of the Rafael Popeye precision-guided air-to-surface missile jointly developed by Israel's Rafael and Lockheed Martin under the Have Nap programme. The weapon is designed for use against high-value ground and sea targets using inertial mid-course autonomous guidance with TV or imaging infra red (IR) terminal guidance, controlled by the B-52G's radar navigator using a joystick and TV display. Four sub-variants of the missile are in use, with different combinations of warhead and guidance system. The AGM-142A has a 1,000-lb (454-kg) blast fragmentation warhead and TV guidance, the AGM-142B combines a blast fragmentation warhead with IIR, the AGM-142C has a penetrator with TV guidance, and the AGM-142D has a penetrator warhead and IIR guidance. The B-52G can carry

four AGM-142 Raptors on each of its underwing pylons (or three missiles and a datalink pod).

The tail turrets of all surviving B-52s were removed between 1991 and 1994, allowing gunners to be taken off the crew complement. Tail guns were no longer effective against fighters

armed with off-boresight all-aspect missiles, and represented a maintenance headache. The disarmed 'BUFFs' were known as 'Bobbited BUFFs' to their crews. A perforated pressure equalisation panel, sometimes known as the 'cheesegrater', replaced the guns.

Although maritime operations were an important part of the B-52G's brief, the aircraft's part in over-water missions has been little publicised. While Harpoon was the weapon of choice for anti-ship operations, mines could also be deployed, as this aircraft shows.

Boeing JB-52G

About five of the many B-52Gs used for test and trials duties received the JB-52G designation, indicating their temporary experimental test status. They included 57-6470 used for GAM-87 Skybolt tests at Wichita and Eglin, and 57-6473 used for the service

evaluation of the new integral tanks' coating, and as the GAM-87A Skybolt aerodynamic test aircraft.

57-6477 and 58-0159 were used for GAM-87 tests at Eglin, 58-0182 was used as an experimental test aircraft for the AN/APN-150 radio altimeter, and for ECM and defensive avionics testing including trials of the AN/ALE-25 chaff

rocket pods. The aircraft was used by the 3246th Test Wing, AFSC at Eglin, and later by the AFTTC at Edwards, where it eventually operated in an unusual all-white colour scheme that saw it nicknamed 'Snowbird'. At Edwards it was used for trials associated with SCAD (subsonic cruise armed decoy) and ALCM-A.

57-6470 was retired to AMARC (the Aerospace Maintenance and Regeneration Center at Davis-Monthan AFB, also known as the 'Boneyard') on 23 October 1990. 57-6743 went to AMARC on 25 February 1993, 57-6477 on 21 September 1990, 58-0159 on 10 October 1991, and 58-0182 on 10 June 1992.

58-0182 was involved in early tests with the Boeing ALCM and is seen above launching an AGM-86A test round. The all-white paint scheme (left), which led to the nickname 'Snowbird', was later spoiled by Day-Glo underwing tanks (above).

Boeing B-52H

The B-52H (Model 464-261) was the last production version of the Stratofortress, and is the only variant remaining in front-line service. The H model was manufactured only at Wichita, under separate B-52H contracts. Letter contract AF33(600)-38778 was signed on 6 May 1960 and covered 62 aircraft, paid for using FY60 funding. Letter contract AF33(600)-41961 was signed on 28 June 1960, but was not finalised until late 1962. The 40 aircraft ordered were paid for using FY61 funding.

The B-52H was a derivative of the earlier G model, and its design was initiated one month before the first B-52G was delivered to the USAF. The aircraft was, in essence, originally planned to be little more than a turbofan-powered version of the B-52G, and it shared the same basic airframe and had the same 22-ft 11-in (6.99-m) short vertical tail fin.

As the B-52H design was refined and developed, the USAF switched to low-level penetration tactics. The aircraft was not designed for the new role, and the first 18 aircraft were not built with the low-level equipment and capabilities that characterised later B-52Hs, though these were soon retrofitted.

The B-52H did have a new tail turret, with a single 20-mm M61A1 cannon and 1,242 rounds of ammunition in place of four 0.50-in (12.7-mm) machine-guns. The new weapon had a maximum rate of fire of 4,000 rounds per minute. The AN/ASQ-15 fire control system was replaced by an Emerson AN/ASG-21 set.

The B-52H also had a revised spoiler system that allowed the outboard sections to be extended by about 10°, enabling the pilot to make small lateral corrections without changing pitch; this was invaluable when manoeuvring during air-to-air refuelling.

The most obvious and most important difference between the B-52H and earlier versions lay in its powerplant. The H model introduced

The B-52H was intended to be the USAF's carrier for the Douglas Skybolt missile, carrying four under the wings. Skybolt was cancelled, leaving the B-52H with free-fall weapons and AGM-28s.

Pratt & Whitney TF33-P-3 turbofan engines, which were military adaptations of the JT3D turbofan developed for the airliner market from the J57 used by other B-52 versions.

As a turbofan, the TF33 was much quieter than the J57, especially at full power, and did not produce the same plume of thick black smoke. Quite apart from their environmental advantages, these features offered real operational benefits. The B-52H had a less noisy cabin than previous versions, bringing a corresponding reduction in crew fatigue and facilitating crew co-operation. The black smoke left by the B-52G could, under some circumstances, make an enemy gunner or fighter pilot's job much easier.

With a maximum thrust rating of 17,100 lb st (76.05 kN), the new TF33 offered 30 per cent greater thrust than the B-52G's J57s, and did so without recourse to water injection. This gave the B-52H superb take-off performance, with a roll about 500 ft (152 m) shorter than that of the B-52G, and conferred a useful safety margin during heavyweight take-offs.

The TF33 had very fast throttle response compared to the J57, which was useful in the event of a missed approach or go-around, but which caused some problems. If the eight throttles were slammed forward too quickly, the aircraft could pitch up more quickly than the pilot could correct using the B-52's modest elevator authority. The rapid increase in power was aggravated by a rapid rearward movement of the fuel in the wing tanks, moving the centre of gravity farther aft. Adjustable mechanical thrust gates were soon added to the throttle quadrant, allowing the pilot to pre-set the maximum thrust required.

The TF33 also offered much lower fuel burn, providing a 15+ per cent increase in range. The B-52G had a combat radius of 3,550 nm (4,148 miles; 6575 km) while carrying a 10,000-lb (4536-kg) bombload, but the B-52H had a 4,176-nm (4,806 miles; 7734-km) radius. The use of water injection by earlier versions had required the maintenance of large stocks of distilled water at B-52 operating bases, and this had sometimes made deployments more difficult.

Above: Although externally this aircraft looks like a B-52H, it is in fact the B-52G (57-6471) that was used to test the TF33 turbofans for the new model.

Below: 60-0006 was part of the first batch of 13 B-52H-135-BW aircraft. Note that even brand new B-52s had signs of skin 'rippling' on their fuselages.

Installation of the TF33 led to a distinctive change in the shape of the B-52's engine nacelles. The new engine's large-diameter fan required a bigger intake and a nacelle with a wider, deeper front. This was stepped down after the bypass air outlet, making it easy to differentiate between the B-52G and the B-52H.

The first B-52 to fly with TF33 engines was a converted B-52G (57-6471), and the first true B-52H flew on 20 July 1960. Deliveries of the first B-52Hs for the USAF began on 9 May 1961, to the 379th Bombardment Wing at Wurtsmith AFB. The last B-52H was delivered to the 4136th SW at Minot

AFB on 26 October 1962.

The B-52H would also have been distinguished by its armament, since it had originally been intended to carry four Douglas GAM-87 Skybolt air-launched ballistic missiles, two each on an inverted Y-section pylon underneath each inboard wing hardpoint. Each Skybolt carried a Mk 7 re-entry vehicle containing a W59 nuclear warhead. Initial deployment of the Skybolt originally had been scheduled for 1964 (and the UK government ordered 100 more to equip its V-bombers). However, the missile was cancelled by President Kennedy in December 1962 for political and economic reasons, forcing the new

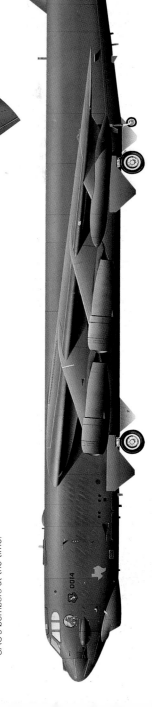

TF33 turbofan

Pratt & Whitney produced the excellent JT3D turbofan (military designation TF33) in very short time by modifying the low-pressure section of the proven JT3C (J57) which had powered earlier B-52 variants. Because the B-52 engines were mounted in pairs, the fan duct could not wrap around the entire engine, as it did on single-mounted instllations. Consequently, the fan duct was arranged in a "banana nozzle' arrangement on one side of each engine. The TF33-P-3 of the B-52H was 11 ft 4 in (3.45 m) long, had a diameter of 4 ft 5 in (1,346 m) and weighed 3,900 lb (1769 kg).

Boeing B-52H-140-BW

7th Bomb Wing, Carswell AFB, Texas

At the end of the 1980s, just before the end of the Cold War, Strategic Air Command had 11 B-52 wings operating a mix of B-52Gs and Hs. The force was split between two numbered Air Forces – the 8th responsible for the bombers in the eastern half of the US and the 15th responsible for units in the west. The 7th Bomb Wing was controlled by the 8th AF, and was one of the many units to lose its B-52s during the dramatic cutbacks of the 1990s, deactivating in 1993 with its aircraft passing to the 2nd BW at Barksdale AFB, Louisiana.

Markings

In the late 1980s the B-52 fleet adopted an all-over Gunship Gray scheme. It was standard for units to carry the Strategic Air Command badge on the starboard side of the nose, with the wing badge to port. The 7th BW's badge consisted of a bald eagle carrying a bomb and the legend 'Mors ab alto' (death from above). Most units applied a less formal marking on the tail, often in the form of a fin-stripe. The 7th used a 'low-viz' version of the Texas state flag, complete with longhorn steer's head, and the state's silhouette decorated the front fuselage. Individual aircraft nose art was common among SAC's bombers at the time.

B-52H (1980s configuration)

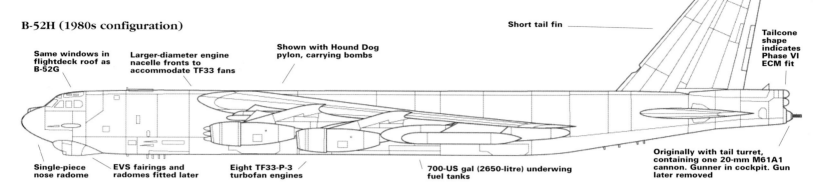

Same windows in flightdeck roof as B-52G

Larger-diameter engine nacelle fronts to accommodate TF33 fans

Shown with Hound Dog pylon, carrying bombs

Short tail fin

Tailcone shape indicates Phase VI ECM fit

Single-piece nose radome

EVS fairings and radomes fitted later

Eight TF33-P-3 turbofan engines

700-US gal (2650-litre) underwing fuel tanks

Originally with tail turret, containing one 20-mm M61A1 cannon. Gunner in cockpit. Gun later removed

Both the B-52G and the H wore this three-tone camouflage over gloss white finish, late on in their SAC service. A hatch in the top of the B-52's tailcone, just aft of the rudder trailing edge, streams the aircraft's brake 'chute.

B-52H to rely on the same combination of thermonuclear gravity bombs and AGM-28 Hound Dog missiles as earlier B-52 variants. The situation was even more serious for Britain: its V-bombers lost their strategic role, which was assumed by Polaris-armed submarines. The B-52H also had a conventional capability equivalent to that of the pre-Big Belly B-52D, but this remained latent until the end of the Cold War.

The first 18 B-52Hs were delivered with the same avionics as the B-52G, though all subsequent B-52Hs were fitted with an improved avionics suite optimised for the low-level role. This was subsequently retrofitted to surviving examples of the B-52D, E, F, and G (together with structural modifications and Hound Dog capability on the older versions) under the Big Four or Mod 1000 upgrade, which differed according to the variant to which it was applied.

An advanced capability radar (ACR) with terrain-following capability was fitted to all but the first 18 B-52H bombers (which later received it as a retrofit). This had anti-jamming and low-level radar mapping capabilities. It presented a three-dimensional picture on three 5-in (12.7-cm) dual-mode CRT displays on the pilot's, co-pilot's and navigator's instrument panels. The pilot could select a plan model which gave a map-like 'god's eye view', or a profile model which showed the terrain ahead as a cross-sectional view. The new integrated flight instrumentation system included a flight director, and the MA-2 autopilot was improved. These features were retrofitted to most earlier B-52s in operational service through the Big Four programme, which was also performed in the early 1960s.

The new avionics system required greater electrical power, and the H (and retrofitted earlier versions) gained new 120-kVA alternators, a static transformer-rectifier, and NiCad batteries.

Low-level terrain avoidance capabilities were further improved with the Jolly Well programme, under which the modification and replacement of major components of the AN/ASQ-38 bombing/navigation system fitted to the B-52E, F, G and H models were made. The Jolly Well programme was completed in 1964, after 480 aircraft had been modified.

Like all Stratofortress variants, the B-52H was continuously upgraded in service and received many of the same modifications as the closely related B-52G. Such modifications are not described in detail, since they are covered in the B-52G entry.

The B-52H suffered a succession of teething troubles, and in 1961, shortly before the major rewinging programme began, stress corrosion cracking was discovered at the fuselage/wing joint of two Homestead-based B-52Hs. The Straight Pin modification was designed by Boeing to fix the problem by replacing the corrosion-prone taper locks and mechanically cleaning cracked holes. The modification programme was interrupted by the Cuban missile crisis, but was completed by the end of 1962.

Another B-52H-specific modification programme was similarly disrupted by the Cuban missile crisis. This was Project Hot Fan, instituted to improve TF33 reliability generally, and to address specific known problems that included uneven throttle alignment, throttle creep, slow and difficult engine starting, flame-outs, excessive oil consumption, turbine blade failures, and inlet case cracking. The programme (at the Oklahoma City AMA) was completely halted by the Cuban crisis, when every available B-52 stood alert, but was resumed in January 1963 and completed by the end of 1964.

The B-52H was in most respects little more than a re-engined G model Stratofortress, using the same extensively redesigned lighter-alloy wing with integral tanks and lightweight flap drives, and having no ailerons and smaller underwing tanks. The variant therefore required exactly the same ECP1050 wing strengthening programme, which was applied to all but the final 18 B-52Hs – they were the

only Stratofortresses to be built with the final redesigned wing.

The B-52 force maintained a Chromedome (later Steel Trap) airborne alert commitment until the loss of an aircraft (and its live hydrogen bombs) at Thule, Greenland in January 1968. Early hopes that the airborne alert force would entail having one-quarter (and later one-eighth) of the force airborne at any one time were soon dispelled, except during brief contingencies. This placed greater emphasis on the aircraft standing ground alert, and in particular on achieving rapid-reaction take-offs, in the face of improved enemy missile technology which steadily reduced the warning time available.

In an attempt to reduce reaction times, a pair of cartridge starters was installed in two engines of every B-52 in order to start all eight engines more quickly and thereby allow the aircraft to get airborne in less time. Another benefit of using cartridge starters was that it promised to free the Stratofortress from its dependence on electrical power carts and other ground support equipment. The programme succeeded in reducing the reaction time by about two minutes.

As enemy defences improved and their reach increased, it became clear that relying on the stand-off range of the Hound Dog would not guarantee the destruction of enemy targets, and that some B-52s would have to be used for defence suppression, blasting their way through enemy defences. The chosen solution was to use multiple short-range low-yield nuclear missiles (thereby complicating the defenders' problems, and obviating the need for great precision or accuracy). The AGM-69 SRAM was developed to meet the requirement, for carriage by the B-52 and FB-111A. The weapon had a warhead broadly

equivalent to that carried by a Minuteman ICBM, with a similar degree of accuracy, and had a tiny radar cross-section and hypersonic speed. The missile was usually fired in a semi-ballistic trajectory, powered by a two-stage (boost/sustain) Thiokol XSR-75-LP solid fuel rocket and guided by a Singer Kearfott guidance system. Range from medium level was about 100 miles (161 km), or 30 miles (48 km) when fired from low level. The B-52 could carry up to 20 SRAMs, six under each inboard wing and eight more on a rotary launcher in the bomb bay. The firing interval was a mere five seconds.

Both the B-52G and B-52H were modified with the AN-ASQ-151 advanced electro-optical viewing system (EVS) between 1972 and 1976, and both types received the Phase VI ECM defensive avionics systems (ECP2519) upgrade almost simultaneously. These two upgrades changed the appearance of the B-52H, which gained EVS sensor turrets undernose and ECM antenna fairings on the sides of the fuselage and tailfin, together with an extended tailcone. The 12 new AN/ALE-20 flare dispensers were located on the lower surfaces of the horizontal stabilisers (six per side), and eight AN/ALE-24 chaff dispensers were scabbed onto the underside of each wing forward of and between the flap sections (four dispensers per side).

The addition of two cartridge starters to each B-52 improved reaction times, but by the early 1970s it was clear that quicker times were possible, and (following the end of airborne alert in 1968) necessary. Under the Quick Start programme in 1974, all surviving B-52Gs and B-52Hs received cartridge starters on all eight engines, allowing the simultaneous, instantaneous ignition of all eight engines.

From 1980, 96 surviving B-52Hs

SRAM replaced Hound Dog as the B-52's counter to enemy air defence systems. This image shows the revised shape of the TF33 engine nacelle to good effect, as well as the vast flap area of the Stratofortress family.

Right into the mid-1990s the B-52H retained a nuclear-only tasking, while the B-52G assumed conventional duties. Free-fall weapons gave way to ALCMs in the 1980s as the main weapon.

and 168 B-52Gs received a new digital (MIL STD 1553A-based) AN/ASQ-176 offensive avionics system (OAS). The upgrade programme was completed by the end of 1986. At much the same time, in a programme beginning in 1982, all surviving B-52Hs (and 98 B-52Gs) were made compatible with the Boeing AGM-86 ALCM. This gave the USAF a total of 194 cruise-carrying B-52s, exceeding the limit set under the unratified SALT II agreement which stipulated that no more that 130 ALCM bombers should be deployed by the USA. The delivery of the 131st ALCM-capable aircraft, which put the USA in technical violation of this treaty, took place on 28 November 1986. This significant aircraft (B-52H 60-0055) was appropriately named *Salt Shaker*.

The 96 surviving B-52Hs modified under these programmes were 60-001, 60-0002, 60-0003, 60-0004, 60-0005, 60-0007, 60-0008, 60-0009, 60-0010, 60-0011, 60-0012, 60-0013, 60-0014, 60-0015, 60-0016, 60-0017, 60-0018, 60-0019, 60-0020, 60-0021, 60-0022, 60-0023, 60-0024, 60-0025, 60-0026, 60-0028, 60-0029, 60-0030, 60-0031, 60-0032, 60-0033, 60-0034, 60-0035, 60-0036, 60-0037, 60-0038, 60-0040, 60-0041, 60-0042, 60-0043, 60-0044, 60-0045, 60-0046, 60-0047, 60-0048, 60-0049, 60-0050, 60-0051, 60-0052, 60-0053, 60-0054, 60-0055, 60-0056, 60-0057, 60-0058, 60-0059, 60-0060, 60-0061, 60-0062, 61-0001, 61-0002, 61-0003, 61-0004, 61-0005, 61-0006, 61-0007, 61-0008, 61-0009, 61-0010, 61-0011, 61-0012, 61-0013, 61-0014, 61-0015, 61-0016, 61-0017, 61-0018, 61-0019, 61-0020, 61-0021, 61-0022, 61-0023, 61-0024, 61-0025, 61-0026, 61-0027, 61-0028, 61-0029, 61-0031, 61-0032, 61-0034, 61-0035, 61-0036, 61-0038, 61-0039, and 61-0040.

Like the B-52G, the B-52H could carry 12 AGM-86B missiles underwing, with six weapons on each of the long ALCM pylons.

From 1985, the original AN/ASQ-176 of the B-52H was replaced by a new Norden AN/APQ-166 synthetic aperture strategic radar. This retained a modified version of the existing antenna, but featured a new antenna electronics unit, an improved radar processor, and new controls, displays and software. The strategic radar was also fitted to some B-52Gs. This $US700-million programme enhanced the B-52's autonomous targeting capability.

A Phase VI+ ECM defensive avionics systems upgrade was applied to the B-52G and B-52H from 1986 and 1988, respectively. Separate configurations were developed for each version under Project Pave Mint. The B-52G used the new AN/ALQ-172(V)1 in conjunction with the existing antennas from the ALQ-117 system which was being replaced; the B-52H used the new AN/ALQ-172(V)2 with a new electronically steerable phased-array antenna farm.

From 1988, some 82 B-52Hs were modified at Tinker AFB and Kelly AFB with fittings, electrical and hydraulic lines to allow a common strategic rotary launcher (CSRL) in the bomb bay. The 500-lb (227-kg) rotary launcher was adapted to yokes within the bomb bay. Eight AGM-86B missiles can be carried on the new CSRL (which was not fitted to any CMI B-52Gs), bringing the potential maximum load to 20

cruise missiles, with six on each underwing pylon. Alternatively, the CSRL can carry four 70- to 350-kT B28 free-fall nuclear bombs, eight 10- to 500-kT B61-7s, or eight 1- to 2-mT B83-0s or B83-1s. Mixed weapon loads cannot be carried on the CSRL due to software limitations.

The CSRL programme began in 1988 and the first CSRL-equipped B-52Hs were delivered later that year.

Some 82 of the 96 surviving OAS/CMI/Phase VI+ modified B-52Hs were provided with CSRL capability, comprising 60-001, 60-0003, 60-0004, 60-0007, 60-0008, 60-0009, 60-0010, 60-0011, 60-0013, 60-0014, 60-0015, 60-0016, 60-0017, 60-0018, 60-0019, 60-0020, 60-0022, 60-0023, 60-0025, 60-0026, 60-0028, 60-0029, 60-0030, 60-0031, 60-0032, 60-0033, 60-0034, 60-0035, 60-0036, 60-0037, 60-0038, 60-0040, 60-0041, 60-0042, 60-0043, 60-0044, 60-0045, 60-0046, 60-0047, 60-0049, 60-0050, 60-0051, 60-0052, 60-0053, 60-0054, 60-0055, 60-0056, 60-0057, 60-0058, 60-0059, 60-0060, 60-0061, 60-0062, 61-0001, 61-0002, 61-0003, 61-0004, 61-0006, 61-0007, 61-0008, 61-0009, 61-0010, 61-0011, 61-0013, 61-0016, 61-0017, 61-0019, 61-0020, 61-0021, 61-0022, 61-0023, 61-0024, 61-0026, 61-0027, 61-0028, 61-0029, 61-0031, 61-0032, 61-0036, 61-0038, 61-0039, and 61-0040.

Two of these (61-0026 and -0040) have subsequently been lost. Some 14 surviving aircraft were not fitted with CSRL, comprising 60-0002, 60-0005, 60-0012, 60-0021, 60-0024, 60-0048, 61-0005, 61-0012, 61-0014, 61-0015, 61-0018, 61-0025, 61-0034, and 61-0035. For the purposes of SALT II, all B-52Hs are counted as '20-shooters' since the non-CSRL equipped aircraft cannot be distinguished externally. By not counting the 28 aircraft which the USAF regards as attrition replacements, it is just within the new 66-aircraft limit set on cruise missile-carrying heavy bombers.

The common strategic rotary launcher cannot be used to carry the B-52H's most potent and newest strategic weapon, the General Dynamics/McDonnell Douglas AGM-129 advanced cruise missile (ACM). This was developed during the Reagan administration as a stealthy replacement for the AGM-86B air-launched cruise missile, and plans were drawn up to equip the B-52 force with 1,461 missiles.

The 2,750-lb (1247-kg) AGM-129A was powered by a Williams F112-WR-110 turbofan engine, and armed with a 200-kT yield W80-1 nuclear warhead, like that fitted to the AGM-86B. The advanced cruise missile had a range of more than 1800 nm (2,072 miles; 3334 km), and used GPS inertial guidance with a laser radar for terrain contour matching during the terminal phase. Details of the AGM-129B remain unclear: some reports suggested that it featured 'structural and software changes' and a 'different nuclear warhead' and was used for a 'classified cruise missile mission'; other sources suggest that the B model had a conventional capability.

Deliveries of the AGM-129 began in June 1990. The programme was cancelled by George Bush Sr in 1992 after only 640 AGM-129s had been built (or 500, according to some sources, including 29 test rounds). All B-52Hs in the active inventory have cruise missile integration (CMI) and advanced cruise missile integration (ACMI), whether or not they are fitted with the common strategic rotary launcher. Twelve of the weapons can be carried, six on each underwing pylon. The ACM can only be carried externally, because it is too large to be carried on the internal rotary launcher.

Deployment details relating to the AGM-129A were classified, and although it is known that the B-52Hs of the 410th BW at K. I. Sawyer AFB were modified to receive the ACM in the late 1980s, it is not known whether these missiles were ever actually fitted. The 7th BW at Carswell was earmarked to be the next unit to receive the ACM, but was deactivated in 1992. The 5th Bomb Wing at Minot (already an operator of the H model) received AGM-129s in 1993, as it gained suitably-equipped B-52Hs from the deactivating 410th Bomb Wing, which finally gave up its last B-52Hs in November 1994. The 2nd Bomb Wing at Barksdale took over the 7th Bomb Wing's H models from 1992, retiring its own B-52Gs at the same time. It is not known whether the 92nd BW at Fairchild and the 416th BW at Griffiss AFBs received AGM-129s before they deactivated in 1994 and 1995, respectively.

The tail turrets of all surviving B-52Hs were removed between 1991 and 1994, just as they were from the B-52G. The gunner's ejection seat (located alongside that of the EWO) was retained, and can now be occupied by an instructor, examiner, or supernumerary crew. The M61A1 Vulcan 20-mm cannon was replaced by the pressure-equalisation 'cheesegrater'. The wiring and instruments necessary for using the gun were all retained so that it could theoretically be reinstalled, though it is not known whether this equipment is still being maintained and supported – and there are no longer any trained gunners to operate it.

Operators of the B-52H include the 2nd BW at Barksdale AFB, Louisiana, the 5th BW at Minot AFB, North Dakota, the 17th BW at Wright-Patterson AFB, Ohio, the 28th BW at Ellsworth AFB, South Dakota, the 92nd BW at Fairchild AFB, Washington, the 93rd BW at Castle AFB, California, the 96th BW at Dyess AFB, Texas, the 319th BW at Grand Forks AFB, North Dakota, the 379th BW at Wurtsmith AFB, Michigan, the 410th BW at K. I. Sawyer AFB, Michigan, the 416th BW at Griffiss AFB, New York, the 449th BW at Kincheloe AFB, Michigan, the 450th BW at Minot AFB, North Dakota, the 4042nd SW at K. I. Sawyer AFB, Michigan, the 4133rd SW at Grand Forks AFB, North Dakota, the 4136th SW at Minot AFB, North Dakota, the 4200th Test Wing at Beale AFB, California, the 4239th SW at Kincheloe AFB, Michigan, and the Air Force Reserve's 917th Wing at Barksdale AFB, Louisiana.

The B-52 force was significantly reduced following the Gulf War upon the withdrawal of the B-52G (which still equipped four wings, including the 2nd BW at Barksdale). At the same time, four of the five remaining first-line B-52H wings were inactivated. The 5th Bomb Wing retained its B-52Hs (and gained more from the disappearing units) and the 2nd Bomb Wing re-equipped with B-52Hs. Both wings established additional squadrons to handle the extra bombers. As a result of the reorganisation, the 94-aircraft B-52H fleet was consolidated at two bases, Barksdale AFB in Louisiana and Minot AFB in North Dakota, with a test/trials/evaluation detachment at Edwards AFB, California. The Barksdale wing includes an Air Force Reserve B-52H squadron.

Block	Serial	Quantity
B-52H-135-BW	60-0001/0013	13
B-52H-140-BW	60-0014/0021	8
B-52H-145-BW	60-0022/0033	12
B-52H-150-BW	60-0034/0045	12
B-52H-155-BW	60-0046/0057	12
B-52H-160-BW	60-0058/0062	5
B-52H-165-BW	61-0001/0013	13
B-52H-170-BW	61-0014/0026	13
B-52H-175-BW	61-0027/0040	14
	Total:	**102**

Memphis Belle IV *was actually the first B-52H built. The aircraft was decorated when assigned to Brigadier General David L. Young at Barksdale AFB. Young was a keen supporter of the Memphis Belle Memorial Association.*

Boeing B-52H Senior Bowl

Two B-52Hs were converted to carry the D-21B ramjet-powered reconnaissance drone under Operation Senior Bowl, funded by the CIA.

The D-21 drone was powered by a Marquardt RJ-43-MA-11 ramjet giving it a cruising speed of Mach 3.3 at an altitude of 90,000 ft (27432 m), and a range of about 1,250 nm (3,726 miles; 2315 km). The D-21 flew a pre-programmed flight profile, guided by an onboard inertial navigation system. Its mission profile ended at a pre-programmed recovery point, where the camera unit (palletised on a hatch) was ejected to descend by parachute. The camera unit with the exposed film was to be 'snatched' during its descent by a specially-equipped JC-130 Hercules (probably by radio command) using an explosive charge.

The D-21 had originally been designed for launch from the trisonic Lockheed A-12, but difficulties (culminating in a fatal accident) forced Lockheed to find a new carrier. The D-21 drone was modified to D-21B standards and had a 90-second burn, solid-fuel rocket booster slung below its body. This was used for accelerating the

drone to its 80,000-ft (24384-m) operating altitude and Mach 3 operating speed. The first of the two B-52H launch aircraft (B-52H-170-BW 61-0021, the second being B-52H-150-BW 60-0036) was sent to Palmdale on 12 December 1966, where it was fitted with massive inboard underwing pylons from which the D-21s could be suspended, and the gunner and EWO positions in the cockpit were replaced by new launch control officer stations. The aircraft were fitted with camera ports in the fuselage sides and the launch pylons to allow the launch to be filmed, but the normal tail armament was retained, contrary to some reports.

After training at Groom Lake with a unit known simply as A Flight, the two D-21 carriers were operated from late 1968 by the 4200th Support Squadron at Beale. They were kept on a virtual alert status, with two 'birds' on each B-52H. After a night launch from Beale to maintain security, the B-52Hs flew to Andersen AFB in Guam, Hickam AFB, Hawaii, or Kadena AFB, Okinawa, from where operational missions were mounted. Operational missions were flown on 9 November 1969,

16 December 1970, 4 March 1971, and 20 March 1971. The first and last drones were lost over enemy territory, and the second and third missions were fruitless because the vital palletised camera hatches containing the mission film were not recovered after being ejected from the drones. All the D-21Bs launched from the 4200th Support Squadron B-52Hs were dropped from the starboard pylon. The port station carried the backup D-21B, which was never used operationally.

The Senior Bowl programme was terminated on 23 July 1971 as a result of the technical and operational difficulties that had been encountered, as well as the very high cost of these fruitless missions, and political considerations, especially after the fourth drone was lost and China made an official complaint about US overflights. The two B-52s remain operational today, and both served as attrition reserve aircraft with the 5th BW's 23rd BS at Minot AFB during the late 1990s.

Boeing B-52H Rapid Eight

The retirement of the B-52G left the USAF's B-52 fleet without certain conventional capabilities. B-52Hs could carry some weapons without modification, but could not use either

the AGM-84 Harpoon anti-ship missile or the AGM-142 stand-off attack missile. The conversion of B-52Hs under the Conventional Enhancement Modification (CEM) programme would clearly restore

these capabilities, but CEM B-52Hs would not be available until mid-1995. To 'plug the gap', eight aircraft were converted under the Rapid Eight programme: four gained AGM-142 capability as Rapid Have Nap B-52Hs (60-0014, 60-0025, 60-0062 and

61-0004), and four gained AGM-84 capability as Rapid Harpoon aircraft (60-0013, 61-0013, 61-0019, and 61-0024). The programme was completed during 1994, and the aircraft were demodified before going through the full CEM modification.

Boeing B-52H CEM

At one time it had been intended that the B-52G would serve in the conventional role, while the B-52H remained committed to the stand-off nuclear role using cruise missiles. When inventory reductions led to the retirement of the B-52G, the B-52H force quickly had to adopt the conventional role alongside its nuclear responsibilities.

Like all previous B-52 Stratofortress variants, the B-52H could carry a limited range of conventional ordnance in its capacious bomb bay, and on the old underwing AGM-128 Hound Dog missile pylons fitted with an I-beam, rack adaptor and two multiple ejector racks. The aircraft lacked the maritime capability and specialised weapons capabilities of some modified and upgraded B-52Gs, however.

As they began to adopt conventional taskings, the B-52Hs were fitted with the AS-3858/AAR-85T(T) miniature receive terminal (MRT) – a five-channel LF/VLF receiver and an AN/ARC-171(V) USF/AFSATCOM radio – allowing the

B-52 to access the US military's satellite communications system.

The Conventional Enhancement Modification (CEM) programme for the B-52H began in 1994, adding a new HSAB underneath each wing, as used by conventional B-52Gs. This allowed the modified B-52Hs to carry a slightly smaller load of Mk 82, M117, or CBU-52/58/71/89 bombs (18 instead of 24) than the standard H with converted Hound Dog pylons, or a larger number of CBU-87s (22 instead of 18). It also allowed the aircraft to carry GBU-15 glide bombs, and weapons that were too long or too heavy to be accommodated on the original I-beam rack adaptor such as the AGM-142A Have Nap, AGM-84 Harpoon, AGM-84E SLAM, Joint Direct Attack Munition (JDAM) and Joint Stand-Off Weapon (JSOW), as well as Mk 60 CapTor mines, Mk 55/56 mines, and Mk 40 DST mines. HSAB-equipped aircraft can also carry the GBU-10 and GBU-12 Paveway II LGBs, and a range of

larger free-fall dumb bombs, including the 2,000-lb (907-kg) Mk 84 and British 1,000-lb (454-kg) bombs. With the HSAB fitted, the B-52H can carry up to nine large weapons on each pylon, depending on which weapon type is being carried.

The CEM programme also covered the provision of a GPS navigation receiver, a modern AN/ARC-210 VHF/ UHF radio with SATCOM (satellite communications) capabilities as well as secure voice encryption, Have Quick II UHF and SINCGARS VHF anti-jam/secure capability. A MIL STD 1760 databus was introduced for support of the new generation of smart weapons typified by JDAM and JSOW.

In addition to the HSAB, B-52Hs originally needed special HACLCS equipment in order to use Harpoon. From 1997 this was replaced by a Harpoon stores management overlay (SMO) system, software that could be added to any CEM- and OAS-modified B-52H.

Maritime-roled, Harpoon-armed B-52Hs were originally used only by the 2nd Bomb Wing's 96th BS. These carry

up to 12 examples of the AGM-84D Block 1C Harpoon underwing, although a Block 1D version with enhanced range is planned for the future. The first live Harpoon launch by a B-52H took place on 25 July 1996. The B-52H can target its own Harpoons, but usually relies on a third-party targeting platform in the shape of a US Navy S-3 Viking or P-3 Orion.

Budgetary cuts in 1996 led to force reductions that were once expected to reduce the B-52H fleet from 94 to 66 or 76 flying examples. Instead of being retired to the AMARC 'Boneyard' (from where few large aircraft have ever been returned to service), the 28 surplus B-52Hs were held in attrition reserve at Minot AFB.

The B-52H made its combat debut on 3 September 1996, when a pair of B-52Hs from the USAF's 2nd BW launched 13 conventionally-armed AGM-86C cruise missiles against Iraqi targets. Since then the aircraft has been heavily used in the conventional bombing role during Operation Enduring Freedom, using a variety of weapons, and in Operation Iraqi Freedom.

The 10 or so B-52Hs deployed to Diego Garcia for Operation Enduring Freedom had all been modified under the $US108 million Avionics Midlife Improvement (AMI) programme launched in January 2000 to modernise the B-52H offensive avionics system to accept special SMO software packages. They had also had their AN/ARC-210 radios modified to DAMA standards.

Further modifications in the pipeline include provision of a SATCOM data terminal, which could give an operator real-time weather information and

During Operation Allied Force CEM B-52s launched AGM-142 Have Nap EO-guided missiles. Three were carried on the wing HSABs, with a datalink pod on the rear of the starboard pylon.

Right: The autonomous delivery of laser-guided weaponry became possible thanks to the adoption of the Litening designation pod, carried on a pylon between the engine nacelles on the starboard wing. After a test in the US (illustrated), two GBU-12 Paveway II bombs were successfully guided during Operation Iraqi Freedom by a crew operating with the 457th Aerospace Expeditionary Group from RAF Fairford in England.

targeting data plus automatic updates of weapon systems in flight, and the $US48-million B-52 situational awareness defensive improvement (SADI) programme, being flight tested during 2003. This will see the replacement of the ancient AN/ALR-20A ECM receiver system and the AN/ALR-46 radar warning receiver. Air Force Reserve Command's 93rd Bomb Squadron has completed the conversion of one aircraft with NVG covert external lighting, a modification that may then be extended fleet-wide if it proves worthwhile.

Right: New weapons which have become available to the B-52H include Boeing's JDAM, as seen during an Iraqi Freedom mission. The JDAM uses GPS guidance, receiving accurate GPS positional updates from the aircraft prior to release. At present the B-52 is cleared to carry the GBU-31, the 2,000-lb (907-kg) version of the weapon. Another weapon introduced during the campaign was the CBU-105 Wind-Corrected Munition Dispenser, a cluster bomb with semi-precision capability.

Boeing 'EB-52H'

Early this century, as the B-52H's future was being examined, there were proposals to use surplus Hs as 'stand-off jammers' to compensate for the ongoing shortfall in ECM capabilities. These 'EB-52Hs' would carry very high-power countermeasures gear, allowing them to stand off while blinding enemy radar.

Boeing JB-52H

About five B-52Hs used in test roles were temporarily known as JB-52Hs. These aircraft included 60-0002 which tested the A/A42G-11 automatic flight control system and AN/ASG-21 fire control system, and 60-0003 which undertook Cat II performance tests and tested AN/ALE-25 chaff rockets.

As a JB-52H, 60-0004 carried out testing of the AN/ASQ-38(V) bombing and navigation systems, as well as ECM and low-level testing, while 60-0005 tested the cabin air conditioning and cooling systems, and the AN/APS-105.

60-0006 may not have been designated as a JB-52H, but it was used for the type's 1,000-hour

inspection and evaluation, and for GAM-87 testing at Eglin. Other B-52Hs used in the test role included 60-0023 (which conducted structural and aerodynamic testing in the maximum asymmetric GAM-87 configuration) and 61-0023 (which conducted vibration and flutter tests, and a dynamic load survey). Neither was redesignated as JB-52Hs.

Boeing NB-52H

The eventual replacement for NASA's ageing NB-52B will be a B-52H on loan from the USAF. The aircraft, 61-0025 from the 23rd BS at Minot, will eventually be modified to the same standards as the older aircraft, with a new underwing launch/drop pylon and extensive camera and observation ports. The new pylon on the NB-52H will be set farther forward than that on the NB-52B, which will allow the NB-52H to retain its flaps, thereby reducing landing speed – unlike the NB-52B, which had to have its inboard flaps removed to accommodate drop payloads.

Funds for the redesigned, relocated and improved pylon for the NB-52H are not yet available, which limits the aircraft's usefulness to NASA. It arrived at Edwards AFB on 30 July 2001 and was ceremonially handed over at NASA's Dryden Flight Research Center on 1 August, still wearing its USAF Gunship Gray colour scheme, albeit with its squadron and wing badges and 'MT'

tailcodes hastily blocked out.

The 'new' NB-52H was quickly repainted in the modern NASA gloss white-and-blue paint scheme, and was flying in this guise by April 2002. Hopes that it would have been fully

modified for its first NASA drop mission as early as December 2002 proved over-optimistic.

The USAF has retained ownership of the aircraft, as it did with the NB-52B, but it is unlikely to be returned.

Following its repaint in NASA house colours, the NB-52H wheels over its new home at the Dryden Flight Research Facility, part of the Edwards complex. The aircraft remains on USAF charge.

Boeing B-52J 'Super Stratofortress'

Although much more modern, more reliable and more efficient than the J57 engines used by the B-52G and previous variants, the B-52H's TF33 turbofan is not comparable to the modern high-bypass-ratio turbofans used on today's jet airliners.

There have been serious proposals to re-engine the B-52H fleet since the mid-1970s, and even before. The aircraft's TF33 turbofans were acknowledged as being old-fashioned and increasingly difficult and costly to support. The two JB-52E engine testbeds demonstrated that the aircraft could easily be powered by four modern high-bypass turbofans, and that such a re-engining with more fuel-efficient and reliable powerplants would dramatically reduce the maintenance burden and fuel costs,

while simultaneously extending the aircraft's range.

In 1996, a Boeing-led team (including Rolls-Royce, Allison and American Airlines) issued an unsolicited proposal to the USAF to re-engine the B-52H fleet with four 43,100-lb st (191.68-kN) Rolls-Royce RB.211-535E4-B turbofans; one RB.211 would be mounted on each underwing pylon, replacing a pair of TF33 turbofans, and there would be cockpit engine control displays as developed for the Boeing 757. The RB.211 turbofan is widely used by a number of jetliners, including versions of the Boeing 747, 757 and 767.

The analysis presented by the Boeing-led consortium showed that replacing the existing TF33 engines with RB.211s would save billions of dollars over the expected remaining life of the

B-52H fleet, quickly paying for itself while also providing greater power and increased range. The re-engining was even offered as a lease on a 'power-by-the-hour' basis, but the proposal was not pursued.

This Boeing illustration shows how the RB.211-engined B-52 might have looked. At the time that the proposal was rejected, the projected career of the B-52 was shorter than is the case today.

Thunder over Italy

The AMI in the Golden Age of the Jet

In the aftermath of World War II the initial re-equipment of the Aeronautica Militare Italiana relied on wartime types – Spitfire, P-38, P-47 and P-51 – with jet equipment arriving in the form of the Vampire in 1950. Far more important, and allowing a massive exapnsion of the force, was the acquisition through the Mutual Defense Assistance Program (MDAP) of hundreds of F-84s and F-86s from 1952 onwards. These types formed the backbone of the AMI for the remainder of the 1950s and into the 1960s, until the F-104 Starfighter entered service.

F-84G Thunderjet

The AMI received a total of 254 F-84Gs, the first of which arrived by sea at Brindisi on 19 May 1952. Service life was brief as they were rapidly replaced by the F-84F. 178 F-84Gs were returned to the USAF for reallocation, the majority going to Turkey. The type equipped three fighter-bomber wings and a reconnaissance wing.

3° Stormo
18° Gruppo, 132° Gruppo
Having previously operated the F-51 Mustang and P-38 Lightning, two Gruppi of the 3° Stormo at Verona-Villafranca received camera-equipped F-84Gs for the reconnaissance role from 1954. They served briefly until replaced by the RF-84F.

5° Stormo/5ª Aerobrigata (right)
101° Gruppo, 102° Gruppo, 103° Gruppo
The 5° Stormo Caccia Terrestre moved to Verona-Villafranca to receive its first F-84G on 1 June 1952, the Thunderjet equipping the 101° and 102° Gruppo. The newly formed 103° Gruppo joined the Stormo in July. On 1 February 1953 the unit was redesignated as the 5ª Aerobrigata Caccia Bombardieri. The wing parented the 'Getti Tonanti' aerobatic team, and its aircraft were marked with the 'Diana Cacciatrice' badge. Conversion to the F-84F began in June 1956.

6ª Aerobrigata (left)
154° Gruppo, 155° Gruppo, 156° Gruppo
The 'Diavoli Rossi' (red devils) of the 6ª Aerobrigata Caccia Bombardieri at Ghedi began giving up their Vampires for F-84Gs on 10 February 1953, and by November all three Gruppi had converted to the new type. F-84Fs began arriving in January 1956 and the wing had completed its transition by the end of the year.

51° Stormo/51ª Aerobrigata (above and left)
20° Gruppo (below left), **21° Gruppo** (below), **22° Gruppo**
Having operated F-47D Thunderbolts, the 20° and 21° Gruppo moved to Aviano in August 1952 to begin conversion to the F-84G. On 1 February 1953 the 51° Stormo, with its famous badge depicting a black cat chasing three green mice, became the 51ª Aerobrigata, and on the same day the 22° Gruppo joined the wing. This initially had F-47Ds, but by November all three Gruppi had converted to the Thunderjet, shortly before moving to Treviso-Istrana. The F-84G was operated into 1957, when the F-84F was taken on charge.

F-86E(M) Sabre

Italy's 179 Sabres were ex-RAF Canadair CL-13 Sabre Mk 4s – also known as the F-86E(M). After refurbishment in the UK, the aircraft were returned to US ownership (having been supplied under MDAP) and reallocated RCAF serials before shipment to Naples, from where they were distributed to the AMI and the Yugoslav air force. In Italian service the Sabre equipped two day-fighter wings.

313° Gruppo Autonomo (left)
Better known as the 'Frecce Tricolori' (three-coloured arrows), the 313° Gruppo Autonomo Addestramento Acrobatico was formed at Rivolto on 1 March 1961 as the Pattuglia Acrobatica Nazionale (national aerobatic team). It was initially assigned 10 F-86E(M)s donated by the 4° Stormo. These served for only two seasons, before being replaced by the long-serving Fiat G91PAN, the first of which was delivered to the team on 28 December 1963.

2ª Aerobrigata (above)
8° Gruppo, 13° Gruppo, 14° Gruppo
In October 1956 the 8° Gruppo was detached to Brescia-Montichiari to begin conversion to the F-86E(M). On 1 March 1957 the remainder of 2ª Aerobrigata Intercettori Diurni followed. In July the wing moved to Cameri-Novara, although the 8° Gruppo stayed behind at Montichiari. In January 1959 the 14° Gruppo moved on to Rimini-Miramare, where it later converted to Fiat G91s. Meanwhile, the 8° Gruppo was disbanded on 1 September 1962, and 2ª Aerobrigata followed suit a month later. This left 13° Gruppo to operate as a semi-autonomous unit for a couple of years under the Cameri base command.

4ª Aerobrigata (above and left)
9° Gruppo, 10° Gruppo, 12° Gruppo
While still a Vampire operator at Napoli-Capodichino, the 4° Stormo was raised to the status of 4ª Aerobrigata Intercettori Diurni on 1 February 1954. In March 1956 it moved to Pratica di Mare where it converted to the F-86E(M). In March 1959 the 12° Gruppo moved to Grosseto, and was followed by the 10° Gruppo in March 1961, and the 9° Gruppo in June 1961. Despite all three squadrons being at Grosseto, the wing headquarters remained at Pratica di Mare. Later in the year, in November, the 10° Gruppo moved on to Grazzanise. In early 1963 the wing was gearing up for the introduction of the F-104G Starfighter, which equipped the 9° Gruppo at Grosseto from 13 June, and the 10° Gruppo at Grazzanise from early 1964. The third squadron, the 12° Gruppo, converted to the F-86K in September 1963. Sabres initially wore the original RAF-style camouflage (left), but in later years were returned to natural metal (above). The 'Cavallino Rampante' (rearing horse) badge was worn in black on a white background.

F-86K Sabre

Fiat built 221 F-86Ks for NATO air arms, of which 63 were for the AMI. The 'Kappone' equipped one night-fighter wing, which was later renumbered and then dispersed among several Stormi.

1° Stormo (right)

6° Gruppo, 17° Gruppo, 23° Gruppo

With a white archer as its badge, the 1° Stormo Caccia Ogni Tempo was formed at Istrana on 1 May 1956. The first unit was the 6° Gruppo, which had formed on the F-86K under the 51ª Aerobrigata in November 1955. It was joined by the 17° Gruppo in August 1956, while the 23° Gruppo stood up at Pisa on the F-86K in March 1957. On 1 May 1959 the 1ª Aerobrigata was formed, and all 1° Stormo aircraft were reassigned to the 51ª Aerobrigata.

4° Stormo

12° Gruppo

The break-up of the 51ª Aerobrigata's F-86K fleet began in September 1963 with the transfer of the 21° Gruppo's aircraft to the 12° Gruppo at Grosseto. The squadron moved to Gioia del Colle on 19 September 1963.

5° Stormo

23° Gruppo

The 23° Stormo at Rimini-Miramare was partially converted to the F-104S when it was transferred to the 5° Stormo. The last F-86K flight was on 27 July 1973.

36° Stormo (left)

12° Gruppo

Having moved to Gioia del Colle in 1963, the 12° Gruppo was reassigned to the 36° Stormo CI/CB on 1 August 1966. In 1969 it began to pass its aircraft to the 23° Gruppo, a process completed in 1971 by which time the F-104S was being delivered.

51ª Aerobrigata

21° Gruppo (above), 22° Gruppo (below), 23° Gruppo (right)

The 51ª Aerobrigata's first association with the 'Kappone' was in November 1955, when the 6° Gruppo was formed at Istrana, although it was soon placed under 1° Stormo control. Later, on 1 May 1959, the 51ª Aerobrigata was re-roled as a night-fighter unit, acquiring all the aircraft of the 1° Stormo. The 21° Gruppo received ex-17° Gruppo aircraft and the 22° Gruppo received F-86Ks from the 6° Gruppo. The 23° Gruppo at Pisa also joined the wing. The F-86K wing was slowly broken up from September 1963, when the 21° Gruppo began conversion to the F-104G, followed by the 22° Gruppo in 1969. The 23° Gruppo moved to Rimini in July 1964, and was transferred to the 5° Stormo in 1973.

F-84F Thunderstreak

Republic's pugnacious Thunderstreak was delivered from 1956 to become the AMI's standard fighter-bomber. A total of 150 was delivered, and they served with six fighter-bomber wings. Unlike its predecessor, the F-84G, the 'Streak' was to have a long life in the AMI, and many were retained after their flying careers had ended to serve as decoys, ground instructional airframes and gate-guards. The type also proved popular in the 1950s as an aerobatic mount, and at least two wings had display teams using the aircraft.

5ª Aerobrigata

101° Gruppo (above)**, 102° Gruppo** (left)**, 103° Gruppo**
With the 'Diana Cacciatrice' (Diana the Hunter) badge proudly displayed, the 5ª Aerobrigata Caccia Bombardieri acquired F-84Fs to replace F-84Gs from 21 June 1956. The first squadron to get the type was the 102° Gruppo. The following month, this unit and the parent wing moved to Rimini-Miramare. The other two squadrons followed, and all had converted to the F-84F by 1957. The 103° Gruppo only operated the F-84F for a short while, beginning its transition to the G91 in June 1958. The 101° Gruppo converted to the RF-104G in early 1964, and the 102° Gruppo converted to the F-104G in May. In June 1966 the 101° Gruppo reverted to the F-84F, being transferred to the 8° Stormo a year later.

6ª Aerobrigata

154° Gruppo (left)**, 155° Gruppo, 156° Gruppo**
On 3 January 1956 the 154° Gruppo received its first F-84F to begin the 6ª Aerobrigata's transition from the F-84G, and the other two squadrons had converted by the end of the year. The F-84F wing was broken up from June 1963, when the 154° Gruppo began to convert to the F-104G. In the same year the 155° Gruppo was sent to Piacenza, and in 1966 the 156° Gruppo was dispatched to Gioia del Colle. Both operated under base control until assignment to the 50° and 36° Stormi, respectively.

8° Stormo

101° Gruppo (below)
When the 5ª Aerobrigata was broken up in 1967, the 101° Gruppo was reassigned to the 8° Stormo at Cervia (having already moved there on 28 June). The unit operated the Thunderstreak until 10 January 1970, by which time the last F-84F had been passed to the 155° Gruppo. The 'new' 101° Gruppo then converted to the Aeritalia G91Y.

36° Stormo

156° Gruppo (left)
The 156° Gruppo first acquired the F-84F in 1956 when part of the 6ª Aerobrigata at Ghedi. On 15 June 1966 it moved to Gioia del Colle, and briefly operated as a semi-autonomous unit under the Comando di Base Aerea de Gioia del Colle. On 1 August 1970 it was placed under the control of the 36° Stormo. In April 1970 the Gruppo began transition to the F-104S, but the last F-84F was not passed on to the 155° Gruppo at Piacenza until September 1971.

50° Stormo

155° Gruppo (above)
On 1 October 1963 the 155° Gruppo moved to Piacenza-San Damiano. Although still technically under 6ª Aerobrigata control it was effectively parented by the Comando di Base Aerea di Piacenza. On 1 April 1967 the squadron was officially placed under the command of the 50° Stormo. In the spring of 1972 the unit began conversion to the F-104S (unlike most other fighter-bomber units which converted to the F-104G), and the F-84F was phased out in 1973. In September that year the 50° Stormo was disbanded and its one Gruppo moved to Istrana, where it was reallocated to the 51° Stormo.

51ª Aerobrigata

21° Gruppo (above)
22° Gruppo (right)
In 1957 the 51ª Aerobrigata at Treviso-Istrana began to receive F-84Fs to replace the F-84G. Only the 21° and 22° Gruppi converted, as the 20° Gruppo had been disbanded in October 1955. Thunderstreaks were flown until 1 May 1959, when the 51ª Aerobrigata was re-roled as a night-fighter unit, and received F-86Ks. The F-84Fs were distributed among the 5ª and 6ª Aerobrigati.

RF-84F Thunderflash

The AMI took delivery of 78 RF-84Fs to equip a single tactical reconnaissance wing at Verona-Villafranca. The first arrived in September 1956, and the type served until 1974, by which time it had been fully replaced by the RF-104G.

3ª Aerobrigata

18º Gruppo, 28º Gruppo, 132º Gruppo
Known initially as the 3ª Aerobrigata Ricognitori Tattici, and from 1 July 1967 as the 3ª Aerobrigata Caccia Ricognitori, the AMI's tactical reconnaissance wing at Villafranca transitioned from the F-84G (with cameras in the port wingtip) to the RF-84F in 1956. The first RF-104G was taken on charge in in February 1970.

The 18º Gruppo (above, left and right) had been newly formed in the 3ª Aerobrigata to operate the F-84G, transitioning to the RF-84F in 1956. Its aircraft were coded between 3-02 and 3-19. The squadron was the last AMI operator of the Thunderflash, not converting to the RF-104G until the first half of 1974, by which time just six RF-84Fs remained in serviceable condition.

Soon after the 3º Stormo was raised to Aerobrigata status in 1956, the 132º Gruppo (left) began to receive RF-84Fs to replace F-84Gs. The Thunderflash was operated until 1972, when the 132º Gruppo converted to the F-104G (initially operated in fighter and visual reconnaissance roles until Orpheus camera pods were delivered). The 3ª Aerobrigata's other squadron, the 28º Gruppo (below), also received the RF-84F in 1956, and became the first squadron to transition to the Starfighter. It received its first RF-104Gs in February 1970, becoming the 28º Gruppo Aerofotografico in the process.

Aerobatic teams

During the 1930s the Regia Aeronautica had developed an excellent reputation for the quality of its formation display teams. This was interrupted by World War II, and it was not until 1950 that the first post-war team was formed – the 'Cavallino Rampante' flying Vampires. The F-84/F-86 era was to become the heyday of the Aerobrigata display teams, which were provided by the wings on a rotational basis. The success of these teams led to the creation of a permanent national team – the 'Frecce Tricolori' – in 1961.

'Getti Tonanti' (Thunder Jets)
Provided by 5ª Aerobrigata, this team followed on from the 'Cavallino Rampante' Vampires and flew four F-84Gs between 1953 and 1955.

'Tigri Bianche' (White Tigers) – above right
For the 1955 and 1956 seasons 51ª Aerobrigata flew a team of four F-84Gs.

'Cavallino Rampante' (Rearing Horses) – below
The second incarnation of the 4ª Aerobrigata team flew F-86E(M)s in 1956/57. The aircraft were painted yellow, with red nose trim. Wings and tail were in blue with white stars.

'Diavoli Rossi' (Red Devils) – above and below
First of the Thunderstreak teams, the 'Diavoli Rossi' came from the 6ª Aerobrigata at Ghedi. It flew six F-84Fs between 1957 and 1959, and displayed in the US. Trim was red with tricolours on the wings and tailplanes.

'Lanceri Neri' (Black Lancers) – above
The 2ª Aerobrigata team flew six Sabres between 1957 and 1959. It undertook several foreign appearances, including one in Persia (Iran).

'Frecce Tricolori' (Tricoloured Arrows) – right
The national aerobatic team flew nine Sabres between 1961 and 1963. This was the second scheme to be worn.

'Getti Tonanti' (Thunder Jets) – below
5ª Aerobrigata's second team flew five F-84Fs between 1959 and 1960. Rome hosted the Olympic games in 1960, hence the markings.

Tupolev Tu-16 'Badger'
Maid of all work

Like the Boeing B-47 Stratojet – its contemporary and Cold War rival – Tupolev's Aircraft '88' carved a unique niche for itself as its country's first successful swept-wing bomber and, like the B-47, was built in huge numbers to swell the ranks of the medium-range strategic bomber units. It also proved very adaptable to the needs of the reconnaissance and electronic warfare communities. More importantly, it adopted the role of missile-carrier, becoming for many years the backbone of the Soviet Union's answer to the 'blue-water' threat posed by the US Navy. Amazingly, 50 years after its first flight, the 'Badger' is still in widespread use in China, and still posing a significant threat to navies operating in the region.

Main picture: In its Tu-16K-26P form the 'Badger' maintained a place in the front line into the 1990s, and in 1991 several were brought out of mothballs and put back into service to cover a shortfall in capability caused by the grounding of the Tu-22M 'Backfire'. Representing the last of these missile-carriers is this pair of aircraft with Rogovitsa and Ritsa antennas, and each carrying two KSR-5s.

Left: Electronic reconnaissance was a major role for the Tu-16, and was responsible for a large percentage of the nearly 100 variants and sub-variants that have been identified. This is a Tu-16RM-1, at rest on an AVMF airfield.

Unmistakeably a Tupolev design, the Tu-16 was an outstanding aircraft which far outlived its initially intended career span. As well as the basic soundness of the design, the key to this longevity was the airframe's tractability, allowing it to be tailored to a wide variety of roles. The wing pylons and Ritsa antennas above the flat-pane bombardier's window identify this as a Tu-16K-26P. The small antenna protruding from the lower port side of the nose was one of two which served the Rogovitsa formation-keeping system.

In 1954 and 1955, together with my crew, I ferried a number of Ilyushin Il-28 aircraft from Moscow, Voronezh and Irkutsk to the Pacific and Baltic Fleets, and to China. When we flew over or landed at the airfield of aircraft plant no. 22 at Kazan, we looked enviously at the big silver-coloured aircraft with an unusual swept-back wing. This was the Tu-16 – known in the West as 'Badger' – and at the time it was pouring from the production line at a prodigious rate.

In the mid-1940s the design bureau of general designer Tupolev started a careful study of 'foreign' (captured German) material, and undertook its own research to explore the characteristics of the swept-back wing. The aim was to evaluate the expediency and possibility of using it on large aircraft. There were numerous problems to solve. Paramount was the need to provide good aerodynamics at high subsonic speeds to meet flight range requirements and also satisfactory take-off and landing characteristics.

Location of the engines was another important consideration, and there were many more.

The concept of the swept-back wing was tested first on the experimental aircraft '82' powered by two VK-1 engines. Early work was begun on the initiative of the design bureau, but from June 1948 the activity came under government decree. In March 1949, '82' made its first flight, piloted by A. Perelyot. During the tests at the plant the aircraft reached a speed of 913 km/h (567 mph). The aircraft was not tested further and a proposed combat version was not built. However, the swept-back wing concept was proven.

During that period the Il-28 went into serial production. It perfectly met the requirements of both the Air Force and the Navy. Nevertheless, the need for a long-range aircraft existed, and a decree, dated 10 June 1950, assigned the Ministry of Aircraft Industry the task of creating a long-range bomber. The challenge was issued to design bureaux OKB-240 (Ilyushin) and OKB-156 (Tupolev).

An order issued by the ministry on 14 June 1950 required the Tupolev design bureau to develop an aircraft fitted with the domestic AL-5 turbojet engines, designed by A. Lyulka with a thrust of 53.96 kN (12,125 lb), and to meet the following characteristics: maximum flight speed – 900-1000 km/h (560-620 mph); range – 7000 km (4,350 miles) with a combat load of 2000 kg (4,410 lb); service ceiling – 11000-12000 m (36,090-39,370 ft). Defensive weaponry was to comprise seven guns. An option was agreed for installing the promising AM-03 engines – designed by A.A. Mikulin – which offered a thrust of 78.48 kN (17,636 lb).

Taking into consideration the time limit for building the aircraft, the relatively modest requirements and the lack of flight test data of swept-back wing aircraft, the OKB-240 design bureau decided to split the work into two stages. During the first stage it was to build the aircraft with a straight wing, like that of the Il-28, and subsequently replace it with a swept-back wing. Tests of the experimental Il-46 began on 3 March 1952 and were completed on 31 July of the same year. The tests proved the characteristics anticipated by OKB-240, and development of the swept-back wing variant accelerated.

Tupolev's approach

Tupolev's OKB-156 chose another way. After considering a number of intermediate concepts, in July 1951 a mock-up and a draft design of aircraft '88' were approved. Four months later, in compliance with the decree, the aircraft was fitted with two AM-3 engines, by then producing 85.84 kN (19,290 lb) thrust each. The prototype, with engines, was completed the following March.

Accommodating the AM-3 engines entailed a wide scope of alterations, due to its larger size as compared to the AL-5. Front-end diameter was 1.47 m (4 ft 10 in), length was 5.38 m (17 ft 8 in) and weight was 3700 kg (8,157 lb). Because of its size, the engine had to be located closer to the fuselage and attached to frames, partly sunk in the wing. Putting the engine in the root of the wing required placing the air intake in front of the wing near the side of the fuselage and then separate the flow into two channels,

Above: A unique feature of the Tu-16 was its elaborate and cumbersome wingtip-to-wingtip refuelling method. Here a Tu-16R reconnaissance aircraft prepares to refuel from a Tu-16Z.

Right: For much of the aircraft's career the bulk of the 'Badger' fleet was assigned to missile-carrying tasks. Large numbers of Tu-16K-10s were produced with the huge YeN ('Puff Ball') nose radar.

Perhaps more than any other Soviet aircraft, the Tu-16 confused Western analysts, resulting in some anomalies in the way reporting names were allocated by the Air Standards Co-ordination Committee. This was only to be expected, given that the assignation of such codenames was primarily based on visual and photographic intelligence reports from interceptor pilots and other human resources. Because they looked the same externally, several distinct variants with widely differing roles were grouped under the 'Badger-A' codename, while a string of letter suffixes were applied to the reconnaissance aircraft, despite the fact that they represented an ongoing modification process to the same aircraft. Continual equipment upgrades and additions throughout the Tu-16's long career, especially concerning electronics, further clouded the issue, as did the fact that AVMF and VVS Tu-16s often differed in their equipment fit.

'Badger-A': 'basic' Tu-16, covering Tu-16 and Tu-16A bombers, Tu-16N and Tu-16Z tankers, Tu-16PL ASW platform and Tu-16T torpedo-bomber

'Badger-B': Tu-16KS – first missile-carrier version armed with AS-1 'Kennel' missile

'Badger-C': Tu-16K-10 – missile-carrier with centreline AS-2 'Kipper' and large nose radome for 'Puff Ball' radar

'Badger-C Mod': Tu-16K-10-26 – 'Badger-C' modified to carry AS-6 'Kingfish' missiles under wing pylons

'Badger-D': Tu-16RM-1 – 'Badger-C' converted for reconnaissance duties with three additional radomes under belly

'Badger-E': Tu-16R – reconnaissance version with two radomes under belly

'Badger-F': Tu-16R – similar to 'Badger-E' but with Elint sensors in underwing pods

'Badger-G': Tu-16KSR/Tu-16K-11-16 – several similar versions converted from earlier aircraft (including Tu-16KS) to carry AS-5 'Kelt' under each wing. Undernose radar enlarged ('Short Horn')

'Badger-G Mod': Tu-16K-26/Tu-16KSR-2-5-11 – 'Badger-Gs' converted to carry AS-6 'Kingfish' missile under wing pylons. Tu-16K-26 retained undernose radome and is sometimes referred to as 'Badger-G'. Some aircraft had large belly radome

'Badger-H': Tu-16 Yolka – chaff-laying platform with chaff dispensers in belly

'Badger-J': Tu-16P – electronic warfare version with ventral canoe fairing covering jammers. Earlier Tu-16SPS jammer does not appear to have been given separate codename

'Badger-K': Tu-16R – reconnaissance update with two sizeable teardrop radomes under belly

'Badger-L': Tu-16R – reconnaissance version with two underbelly radomes and underwing sensor pods. Sometimes fitted with enlarged tailcone housing jammers and nose 'pimple' antenna

one being in the structural box of the wing and the other one below it. The AM-3-powered aircraft was expected to demonstrate the following performance: range – 6000 km (3,728 miles), speed – 1000 km/h (621 mph) and service ceiling – 14000 m (45,930 ft).

After completion, the experimental aircraft '88' was redesignated as the Tu-16 and passed over to the test team. On 27 April 1952 test pilot N. S. Rybko made the first flight that lasted just 12 minutes. In ensuing flights the aircraft reached a speed of 1020 km/h (634 mph) and demon-strated a range of 6050 km (3,760 miles). The performance advantage of the '88' over the Il-46 was readily apparent.

On 13 November 1952 the prototype was submitted for state testing which continued through late March 1953. Despite the aircraft initially failing to pass these state tests, as early as July 1952 Tupolev managed to obtain an approval for launching mass production in December 1952. In July 1953 plant 22 was to make the first series aircraft and work on the rival Il-46 was stopped.

The second prototype of project '88' was 3900 kg (8,598 lb) lighter and required 2000 kg (4,409 lb) less fuel to achieve its range requirement, reducing the maximum take-off weight from 77430 kg (170,701 lb) to 71560 kg (157,760 lb). For future use the outer wing panels were

fitted with additional fuel tanks, which increased capacity of the fuel system from 38200 to 43900 litres (8,403 to 9,657 Imp gal). Other modifications included extending the nose by 20 cm (8 in), reinforcing the spar caps, widening the engine nacelle, and installing defensive weapons with sighting stations and radar sights. The crew headed by N. S. Rybko flew the second prototype on 6 April 1953. This aircraft passed the state tests that concluded in April 1954 and was recommended for fielding operationally, endorsed by a decree dated 28 May 1954.

During the state tests the following characteristics were obtained: maximum speed at 6250 m (20,505 ft) – 992 km/h (616 mph); flight range with a 3-ton payload – 5670 km (3,523 miles); service ceiling – 12800 m (41,995 ft); maximum take-off weight – 72000 kg (158,730 lb) with a take-off run of 2000 m (6,562 ft).

Full-scale production

The first serial Tu-16 aircraft No. 3200101 was completed on 29 October 1953 at Kazan-based aircraft plant 22. In 1954 the output counted 70 aircraft. The aircraft number allocated by the plant indicated the following: the first digit – the year of manufacturing; the second one – the plant; the third – any digit; the fourth and the fifth – serial number, and the last two – the number in the series.

Electronic warfare variants of the 'Badger' were produced in sizeable quantities, and also exhibited many different configurations as equipment was upgraded. The main jamming variant was the Tu-16P Buket, with jammers housed in its canoe fairings, surrounded by heat exchanger inlets and exhausts to cool the equipment in the stores bay. Jamming platforms were procured either new from the factory or by conversion of other variants. This aircraft was probably a tanker previously.

Below: Without its deadly missile cargo, the Tu-16K-10 was often intercepted during navigation training sorties, or on missile launch simulation exercises.

As with most other Soviet types of its era, the Tu-16 was exported readily to client states: China, Egypt, Indonesia and Iraq all received Soviet-built 'Badgers'. Clouding the issue in the United Arab Republic (Egypt) was the use of AVMF aircraft operating from Egyptian bases in spurious UARAF markings. This, however, is one of Egypt's Tu-16KSR-2-11s, purchased when the AVMF's 89 OMRAE departed the country in 1972.

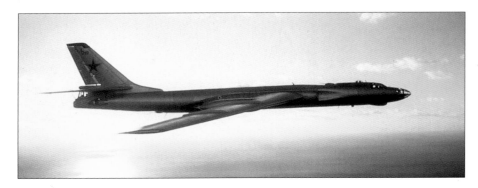

'Badger-A' bombers were delivered in roughly equal numbers to both Long-Range and Naval Aviation. The AVMF also received a fleet of Tu-16T torpedo-bombers, although they did not differ externally. The Tu-16T had a very brief career in the late 1950s, as this mode of warfare was proved to be obsolescent. The Tu-16T fleet became available for conversion, and a handful were fitted with sonobuoy equipment to form an interim anti-submarine warfare capability (as the Tu-16PL) pending the arrival in service of the dedicated Il-38 maritime patrol aircraft. Another naval speciality was the Tu-16S rescue aircraft, which was designed to drop a boat to survivors.

Sometimes a short form of numeration was used. The first manufactured aircraft were widely used for test and research purposes. In this way, after undergoing check tests that increased its gross weight to 76000 kg (167,548 lb), production aircraft No. 4201002 was taken as a standard.

During the period 1953 through 1959 plant no. 22 at Kazan produced the Tu-16 and the Tu-16A bomber, the Tu-16KS missile-carrier and the Tu-16E jammer. The number of aircraft per series varied: up to series 11 – five aircraft per series; up to series 20 – 10 aircraft per series; from series 21 – 20 per series; and from 31 to 41 – 30 aircraft per series. In the period of 1958 to 1960 there was no production of the Tu-16, and in the period of 1961 through 1963 30 series of five aircraft each were built, comprising 150 Tu-16Ks for naval aviation. In total, plant no. 22 manufactured 800 Tu-16 aircraft of different versions.

On 19 September 1953 a decree was issued followed by the order of the Ministry of Aircraft Industry to launch production of the Tu-16 at plant no. 1 at Kuibyshev (now Samara). The first aircraft was assembled from components delivered from the Kazan factory. The following nine aircraft also used components received from plant 22. The numbers assigned to aircraft produced by the Kuibyshev plant indicate the following: the first digit – plant number; the second and the third ones – type of the article ('88'); the fourth and the fifth – series number; and the two last digits – number of the aircraft in the series.

Kuibyshev produced the Tu-16K, Tu-16SPS, Tu-16E, Tu-16 Yolka and the Tu-16R. Altogether, 40 series were made. The first series (up to the 11th) consisted of five aircraft each, series 11 – 20 ten aircraft each and from series 21 onwards 20 aircraft each. In total, plant no. 1 built 543 Tu-16s.

In July 1954 a decree was issued to start production of the Tu-16T for naval aviation at plant no. 64 at Voronezh. Tu-16s and Tu-16 Yolkas were also built there. The first aircraft, Tu-16T No. 5400001, was completed in May 1955. The first digit in its number indicated the year of manufacturing; the second one – plant number; the third – any digit; the fourth and the fifth – series number; the sixth and the seventh – number of the aircraft in the series. In 1957, production at Voronezh was stopped after 166 aircraft had been built,.

Combined, the three plants built 1,509 Tu-16, comprising 747 Tu-16 and Tu-16A, 76 Tu-16T, 107 Tu-16KS, 216 Tu-16K, 233 Tu-16 Yolka and Tu-16SPS, 75 Tu-16R and 52 Tu-16E (ASO-16). However, these far from represent the number of versions of the Tu-16, since the aircraft were modernised and modified by the VVS and AVMF aircraft overhaul plants.

According to their assigned tasks, the Tu-16s were divided into combat and special-purpose aircraft. The combat aircraft included bombers, missile-carriers, torpedo bombers, ASW aircraft and reconnaissance platforms. ECM jammers, tankers, rescue aircraft and some other variants are referred to as special-purpose aircraft.

'Badger-A' bombers

The basic variant of the Tu-16 bomber was under continuous upgrade, improving its airborne equipment and self-protection gear. The OPB-11 optical vector/synchronous bomb sight was subsequently replaced by the better OPB-15. Some aircraft were equipped with the A-326 Rogovitsa radar system, which provided station-keeping information for formation flying at a range of up to 20 km (12.4 miles), and gave warnings of potential collisions.

Prototypes for the Tu-16A nuclear bomb carrier were created through modifying the second and the third series Tu-16 aircraft nos 3200102 and 4200103. By 1958 453 similar aircraft had been built, 59 of which were equipped with an inflight refuelling system, becoming the Tu-16ZA. An almost equal number of aircraft was divided between long-range and naval aviation units. Unlike the standard bomber, the stores bay of the nuclear carrier contained an electric heating system and featured an improved air-tightness system. The crew cockpit was equipped with blinds. The bottom part of the fuselage had a special thermal protection skin.

Production of Tu-16 bombers stopped in 1958, but modifications continued and many were subsequently converted for other tasks. In the late 1960s the modification of aircraft no. 7203829 increased its capability to carry FAB-250 bombs from 16 to 24, and FAB-500s from 12 to 18. Some aircraft were equipped with SPS-100 jammers, while some individual aircraft carried Rubin radars required for the OPB-112 optic sights. In the 1960s, with the accent moving rapidly away from free-fall bombers, 156 Tu-16As were retrofitted to carry KSR-2 and KSR-11 missiles, becoming Tu-16KSR-2s or Tu-16K-11-16s in the process.

In compliance with a decree dated 28 March 1956, the

Few Western air arms saw more of the 'Badger' than the Svenska Flygvapnet (Swedish Air Force), for whom intercepts of 'Badgers' over the Baltic were an almost daily occurrence. Above is a pair of Tu-16K-10 'Badger-Cs', while at right is a Tu-16P Buket equipped with Geran nose jamming antenna. A rarely-seen variation on the Buket was the Fikus aircraft, which had five small radomes under the stores bay.

Tu-16B aircraft was developed. It was an attempt to improve the performance of the aircraft through the adoption of the 107.91-kN (24,250-lb) RD 16-15 engine that had been developed by OKB-16. Two aircraft were modified and were tested until 1961. The results were satisfactory but serial production was never launched.

As well as normal free-fall weapons, Tu-16s carried guided bombs. After a careful study of the characteristics of German guided bombs, from the early 1950s the UB-2F (Tchaika) and UB-5 (Kondor) bombs were developed. They were of two different diameters and weighed 2240 kg (4,938 lb) and 5100 kg (11,243 lb), respectively. The first one was intended for the Il-28 and the Tu-16, the second one only for the Tu-16. They utilised a three-point method of guidance – the navigator used the OPB-2UP optic sight to adjust the trajectory, watching the target and the tracer.

A thermal imaging guidance system was also tested. As a result, the fielding of Tchaika-2 with a thermal homing head was endorsed. In 1956 a UBV-5 5150-kg (11,354-lb) bomb with both high explosive and armour-piercing warheads was developed to fit the Tu-16. Gradually, a disadvantage of the guided bomb became obvious – it was necessary to get close to the target. The idea to equip the bomb with a missile engine for longer range complicated its design and made it almost as expensive as a normal missile.

Tu-16 bombers in service

Tu-16s served in many variants with dozens of aviation regiments of both long-range (DA – Dalnyaya Aviatsiya) and naval aviation (AVMF – Aviatsiya Voenno Morskogo Flota). Regiments could either be an independent part of heavy bomber divisions (TBAD – Tyazhelaya Bombardirovochnaya Aviatsionnaya Diviziya) or mine-torpedo aviation divisions (MTAD – Minno-Torpednaya Aviatsionnaya Diviziya). In 1961 the latter were renamed naval missile carrier divisions (MRAD – Morskaya

Above: This 'Badger-A' rests at Zhukovskiy. The undernose radar has been removed.

Right: A 'Badger' refuels at Kamennyi Ruchey air base in the Soviet Far East during the Cold War. This was a major base for patrol and attack missions over the Pacific.

Left: A Tu-16 being shadowed by a US Navy F-4 Phantom highlights the undercarriage nacelles which were introduced by the 'Badger' and which became a Tupolev hallmark. The reasons for the nacelles were simple enough: to cater for the high weight of the aircraft a bogie undercarriage was required, and the high-speed wing was too thin to accommodate such a structure. To minimise the cross-sectional area of the nacelles, the bogies were designed to somersault so that they lay flat in the nacelle when retracted.

Raketonosnaya Aviatsionnaya Diviziya). The AVMF also had independent reconnaissance air regiments (ODRAP – Otdelnyi Dalniy Razvedyatelnyi Aviatsionnyi Polk) and squadrons (ORAE – Otdelnyi Razvedyatelnyi Aviatsionnyi

Below: This regular Tu-16 bomber is on display at the VVS museum at Monino, complete with an array of bombs. The Tu-16 could carry a 9000-kg weapon.

Eskadrilya). A TBAD or MRAD consisted of two to three heavy bomber regiments (TBAP – Tyazhelyi Bombardirovochnyi Aviatsionnayi Polk) or naval missile carrier regiments (MRAP – Morskoy Raketonosnyi Aviatsionnyi Polk). Both AVMF and DA aviation regiments consisted of three squadrons. Two squadrons operated combat aircraft while the third flew tankers and electronic countermeasures aircraft.

The first Tu-16s were received in February-March 1954 by 402 TBAP at Balbasovo and 203 TBAP at Baranovichi. Later they were fielded at Engels, while several Tu-16As deployed to Bagerovo.

Re-equipping the AVMF with Tu-16s was preceded by numerous appeals by the navy command for a long-range aircraft. On 24 June 1955 Minister of Defence of the USSR Marshal Zhukov, who did not favour the Navy much, forwarded a letter to the Presidium of the CPSU Central Committee reporting a decision to start re-equipment in the following year. It was planned to send the first 85 aircraft to the Northern Fleet, 170 aircraft of the 1957 delivery plan to the Black Sea and Pacific Fleets and only in 1958 provide 170 aircraft for the Baltic Fleet. In this way the AVMF was to receive 425 Tu-16s within three years.

The first four AVMF Tu-16s were delivered on 1 June 1955, and on 25 June the aircrew of 170 MTAR started flying at Bykhov airfield. In April 1956 Tu-16s were deployed with 5 MTAR of the Black Sea Fleet, and with the Northern Fleet. The Pacific Fleet received them only in 1957. In the initial service period the AVMF mastered bombing and, after receiving the Tu-16T, torpedo dropping, whose technique did not differ from bombing. In the middle of 1957, Tu-16s were delivered to the 4th naval aviation flight training centre, subordinate to the naval mine-torpedo air college at Nikolaev, where a squadron was organised. There were plenty of volunteers to fly the Tu-16 and the selection of candidates focused on age, education, health, level of training, morale and feedback from commanding officers. Pilots with experience of the Tu-4, Tu-14 and Il-28 at the level of class one with a total of not less than 600-700 flying hours were chosen as aircraft commanders, while pilots trained at the level of class two with a total of not less than 200 flying hours became assistant aircraft commanders.

For the Soviet Union the Tu-16 opened an era of heavy jets. It had a lot of peculiarities and a heavy emphasis was placed on the quality of flight training. As initial flying skills were acquired, the mastering of combat employment

followed. With the exception of bombing, the focuses of the DA and AVMF combat training were quite different.

Crews of nuclear weapon-carrying aircraft were selected in a special way. Besides being examined by different commissions, they made a signed statement of non-disclosure of top-secret information. Having been granted admission, the crews studied the physical properties of nuclear weapons and crew manuals, especially those covering emergency procedures. Then, endless training and flights with dummy bombs followed. The crews were trained to precisely reach the target at an appointed time and carry out a single bombing. Crews from the DA's 420 TBAP were the first to receive nuclear weapons for the Tu-16A. They were on combat duty at different airfields, including the naval base at Vesyolyi. With the coming of the Tu-16A, the Spetsnaz squadrons of Il-28s were disbanded.

Nuclear tests

DA Tu-16As were involved in nuclear tests on the ranges near Semipalatinsk and on Novaya Zemlya. In the period from 1957 to 1962, before the treaty banning nuclear tests came into force, 37 nuclear bombs had been dropped at Semipalatinsk. On 22 November 1955 the crew of a Tu-16A dropped the first RDS-37d nuclear bomb with half-filler (1.6 MT yield) over Semipalatinsk. Later, the Tu-16A participated in refining the bombs at Novaya Zemlya. Tu-16A crews not only dropped bombs, but in the interests of short-sighted scientists and dim-witted state-mongers passed through radioactive clouds to collect samples. On 30 November 1961, when the Tu-95V dropped a 'super-bomb' with a TNT equivalent of 57 MT, insulating sleeves of electric wire bundles of the escorting Tu-16 started smoking and the aft gunner received burns, while the aircraft was 55 km (34 miles) away.

In the early 1960s, with ICBMs not yet in service, both DA and AVMF units were ordered to be ready for nuclear retaliation, with targets designated and distributed. However, the limited capabilities of the Tu-16 did not let it reach the USA even from airfields in Amderm, Severomorsk and on Wrangel Island. By the early 1960s no fewer than 16 airfields had been built within the Arctic circle, including ones at Vorkuta and Tiksi. They were regularly used for inter-theatre manoeuvres, as well by the Tu-16s of naval aviation. The earlier experience of using ice airfields by

Il-28s was deemed to be suitable for the Tu-16. However, the first attempt of landing a Tu-16 on an ice airfield in May 1958 resulted in the loss of the aircraft.

After ICBMs took over combat duty, the significance of the 'pure' bomber evaporated and they began to be turned into missile-carriers and other variants. A few bombers were left in the long-range aviation combat training centre at Ryazan. However, major difficulties were to plague the rearming of the 'Badger' fleet missile systems as almost all of them, with the exception of the KSR-5, required considerable improvement and refinement.

As a postscript, during the war in Afghanistan the Tu-16A and Tu-16KSR-2-5 (a missile-carrier with bombing capability) once again became pure bombers to deliver attacks against congregations of Mujahideen, dropping bombs of 250 to 9000 kg (551-19,841 lb) in weight. Bombing was carried out in daylight with optic sights and support was provided by Tu-16Rs and Tu-16Ps for reconnaissance and jamming of Pakistani radars. The aircraft flew from Khanabad, Mary and Karshi airfields in either groups of three or four aircraft or in squadrons of eight to 10 aircraft. On 28 April 1984 a strike was delivered by a group of 24 aircraft, each carrying 24 to 40 FAB-250 bombs.

Tu-16T torpedo bomber

While the first Tu-16s were laid down as standard bombers, a decree issued on 12 July 1954 required that part of the production be converted into torpedo bombers. Plant no. 64 at Voronezh was charged with this task. The first Tu-16T, no. 5400001, left the factory in May 1955. In total, 76 Tu-16Ts were built.

Tu-16Ts did not look any different from their bomber counterparts. As well as bombs, provisions were made for carrying six 45-54 VT and four RAT-52 torpedoes, and six to 12 mines of different types. The entire load was accommodated in the stores bay. For suspending and releasing torpedoes a pneumatic torpedo control system was installed, which fed air to the torpedoes to start their gyros 0.6 seconds prior to release. There were circuits for charging capacitors and warming the weapons. To provide a clean release with minimum yaw, the weapons compartment of the aircraft was fitted with a central guiding girder and four rocking yokes that guided torpedoes from the moment of their release from the locks to the moment of exiting the bomb compartment.

A special circuit to ensure the torpedo's travel depth was added, drawing on the experience of the Pacific Fleet in operating the Il-28 aircraft with RAT-52s. In the incident in question, the torpedo exited the water, climbed 500 m (1,640 ft) and flew for 3.5 km (2.2 miles).

Deploying the Tu-16T added problems, while bringing little real advance to the capabilities of naval aviation other than torpedo bombing from altitude. Good results had been demonstrated during tests with the Tu-2 when torpedoes were released with a divergence angle ('fan' release). Calculations indicated that, if the RAT-52 torpedoes were released with a divergence angle, hit probability could increase by 40-50 percent. However, the optimists were outnumbered by the doubters, and attention shifted its focus to missiles. Torpedoes were not developed any further.

With the Tu-16KS entering naval service, Tu-16T training continued for some time, and trials exercises were conducted. One of them took place in June 1959 with the Black Sea Fleet's 943 mine torpedo air regiment. Three aircraft participated in the exercise, each aircraft packed with six 45-54 VT torpedoes. They undertook three group sorties but the exercise did not carry any practical value and the findings were only of theoretical use. Thus, the Tu-16T turned out to be the last torpedo bomber in the inventory.

Tu-16PL ASW platform

While anti-submarine systems were being developed and hopes for them were high, the situation in the fleets

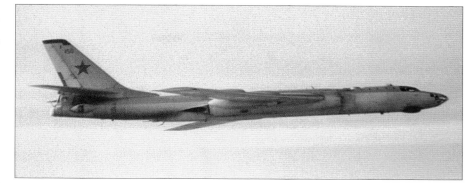

Above: 'Badger-As' continued in service as trainers and for general duties long after the front-line nuclear mission had passed to the missile-carriers. Many served at Ryazan to provide conversion training before crews moved to their operational version.

Left: Tu-16Ts were armed with up to 10 torpedoes, but they could also carry mines, like these AMD-2s.

was quite difficult. There were no assets capable of searching for submarines. Naval aviation possessed a considerable number of relatively new (by the date of manufacturing) and redundant Tu-16Ts, and it was decided to fit them with ASW equipment. The Northern Fleet was the priority, where a public design bureau was established. In 1962 a number of proposals were put forward as to how to equip the Tu-16T for the ASW role. The upgrade was to be co-ordinated with the Tupolev OKB.

In Soviet service the Tu-16 saw its moment of action during the occupation of Afghanistan in the 1980s. 'Badger-A' trainers – like this aircraft – were sporadically involved, as were some missile-carriers. The Tu-16KSR family carried its missiles on wing pylons, and consequently retained a bomb bay.

The black and white attachment projecting from the starboard wingtip of this 'Badger' identifies it as a Tu-16Z tanker, intercepted over the Baltic by a Swedish Saab Viggen in the 1980s. The aircraft is fitted with the SPS-5 Fasol ('kidney bean') jamming equipment, denoted by the small 'pimple' radome under the belly – a common sight on later tanker and Tu-16KSR aircraft. The 'shadow' is a JA 37 from 2./F13, the second squadron of the Brávalla Flygflottilj at Norrköping.

Here a Tu-16R 'Badger-F' refuels from a Tu-16Z. Once the connection had been established between tanker and receiver the process asked only good formation-keeping of the receiver pilot, but getting hooked up in the first place involved a very demanding manoeuvre undertaken in close quarters to the receiver. By necessity the wingtips had to overlap for the hook-up to be successful.

Dwarfing even the sizeable 'Badger' is the immense bulk of a Typhoon SSBN. Tankers like this Tu-16Z usually served in mixed squadrons with EW aircraft, one squadron supporting two of missile-carriers in a regiment.

In 1962 Tu-16 aircraft from 9 MRAP (Naval Missile Air Regiment) were fitted with the Baku radio sonar system while the stores bay was configured to carry 40 sonobuoys and small bombs in cassettes, becoming Tu-16PLs in the process. In April 1966 the equipment of the Tu-16PL was refined to allow the employment of the AT-1 anti-submarine torpedo. Converting the Tu-16 into an anti-submarine platform was a forced measure. The Tu-16PL squadrons existed over five years before they were disbanded in 1967 in the Northern Fleet, and the following year in the Pacific Fleet.

Tu-16S: search and rescue

A decree of the Government dated 25 December 1955 committed the Ministry of Defence to create a search and rescue service in the Armed Forces, and to form search and rescue squadrons in all naval theatres by 1958. In particular, Minsudprom (the Ministry of ship building industry) was to build in the fourth quarter of 1956 two prototypes of self-propelled boats to equip the Tu-16. The Ministry of aviation industry was to equip the Tu-16 with a boat and present it for testing in the first quarter of 1957.

On 17 June 1958 a prototype evaluation commission examined a model of the project 647 rescue boat (code of the subject was 'Arkhangelsk'), featuring remote radio control. The model received an approval. However, a boat of different design entered service.

The Tu-16S was built on the basis of the Tu-16T, and was named 'Fregat naval aviation rescue system'. The system also included a transportable Fregat air-droppable boat which could be controlled manually or remotely from the aircraft. The aircraft was additionally equipped with boat attachment units, boat guidance and control equipment. The boat was made from duralumin and was of open design with folding tents and a watertight platform which ran along the entire length. The boat was fitted with a 25-kW (33-hp) petrol engine. The endurance of the boat (measured by food and water supply) was up to three days, and it could travel 1480 km (920 miles) in still water with a speed of 5 kt (9.2 km/h) and navigate sea states up to 5. The boat carried airbeds, dry clothes, canned water, fishing tackle, medicine, signalling equipment and a food supply.

Some 14 Tu-16S Fregats were produced by conversion, but they were scrapped in the early 1980s. They had never been used for their intended purpose.

Tankers

Fielding the Tu-16 brought with it the problem of how to increase its range: the only option was air refuelling. As well as range increase, it offered the advantage of an aircraft being able to make a lightweight take-off from a limited-size or muddy airfield, followed by a top-up refuelling. On 17 September 1953 the Ministry of Aviation Industry issued an order prescribing the Tupolev OKB-156 and other aircraft builders to design all bombers with air refuelling systems. At the same time, OKB-918 was ordered to continue working on the tanker for the Tu-16, and submit it for testing in the third quarter of 1954. The first Tu-16Z with the wing refuelling system arrived for testing in 1955 and was tested and refined for more than 18 months. From 1957 the refuelling system was installed on all Tu-16s. In total, 571 Tu-16s had the receiver kit installed, and 114 were converted into Tu-16Z tankers.

Any Tu-16 with a feed tank (10500-10800 litres/ 2,310-2,378 Imp gal) and additional equipment could be used as a tanker. From the starboard wing was extended a 37-m (121-ft 5-in) hose with an inner diameter of 76 mm (3 in) which was attached to a cable. For night refuelling, the right landing gear fairing had searchlights, three lamps and a manual searchlight operated by the gun operator. The left wing panel of the receiver had a contact unit with automatic control and monitoring devices. The unit gripped the feed hose and connected it to the fuel line.

Tanking was possible at an altitude of up to 11000 m (36,090 ft) and a speed of 480-500 km/h (298-311 mph). The fuel feed ratio was 2000 litres (440 Imp gal) per minute. Up to the automatic shutting of the fuel cocks, the refuelled aircraft could receive 17500-19500 litres (3,850-4,289 Imp gal), which increased the flight range by 1500-2000 km (932-1,243 miles).

It must have been hard to invent something as imperfect as the wing refuelling system, which ultimately killed dozens. The tanking process was complicated by the fact that the commander of the aircraft in his left-hand seat could not see the wing of his own aircraft and had to be guided by the marks on the hose and by commands from the gunner.

Both long-range aviation and naval aviation began mastering the art of tanking in 1958, and in 1961 the AVMF began night refuelling, although the procedure was identical. After it had established visual contact with the tanker, the receiver assumed an initial position 2-3 m (6.5-10 ft) above the hose and with 4-6 m (13-20 ft) between the wing and the hose. Relative speeds were then equalised. The pilot then stabilised at a distance some 45 m (148 ft) aft of the tanker, using the hose marks as a guide. Maintaining elevation and position aft, the receiver aircraft was then edged to the left so that the line of its port wingtip over-lapped that of the tanker by about 3-5 m (10-16 ft). Next, the receiver's port wingtip was lowered slightly over the hose to snag it, and then the receiver moved out to about 6-8 m (20-26 ft) separation to grip it.

Above: Over 100 tankers were converted to Tu-16Z or ZA standard (the latter converted from Tu-16A nuclear bombers). There were also a few Tu-16N conversions, which had hose-drogue refuelling systems.

Left: This view shows the limited separation between the two aircraft. Night refuelling greatly increased the risk of collision.

Below left: Laying the wing on top of the trailed hose was the first stage of refuelling, as demonstrated by a Tu-16K-10. The pilot then moved the aircraft out to snag the hose under the receiver's port wingtip.

Below: Having snagged the cable, the receiver had to close the distance between it and the tanker for fuel to flow. Here the receiver is a Tu-16R 'Badger-L'.

Above: A KS is seen at the moment of launch from a Tu-16KS, the Kobalt-1M guidance radar being clearly visible in its retractable 'dustbin'.

Above right: Tu-16KS aircraft were built as such from new, and featured strong pylons for carriage of the 2735-kg (6,030-lb) KS missile. They were subsequently modified to carry KSR-2s and, later, KSR-5s.

During the winching of the hose from the tanker until contact was established with the nipple, the gunner kept a running report, chiefly concerning the remainder of the rear end of the hose. Upon establishment of contact the refuelling system was activated, and the distance between the two aircraft was reduced by 10 m (33 ft). After that fuel feed started. Altitudes over 8000 m (26,250 ft) made refuelling complicated because of the aircraft's lack of manoeuvrability. Consequently, an altitude of 6000-7000 m (19,685-22,965 ft) was considered ideal.

There were other tanker versions: the Tu-16N was designed for refuelling the Tu-22 'Blinder' using a standard hose-drogue unit and probe receiver system. This was tested on Tu-16N no. 1882401 and several aircraft were retrofitted in 1963. The Tu-16NN was based on the Tu-16Z but had a fuselage HDU and the wing refuelling system removed. It differed from the Tu-16N in having aerodynamic plates on the wingtips. Long-range aviation retrofitted a total of 20 aircraft. The Tu-16D was a Tu-16 retrofitted for receiving from a drogue-equipped tanker system. Tests were successful.

Tu-16K – missile-carriers

Of all the many tasks assigned to the Tu-16 throughout its long career, the most important was that of missile-carrier, leading to a wealth of variants. Work to create the Kometa aircraft missile system started in 1947. The cruise missile was developed by the A.I. Mikoyan OKB-155 headed by M.I. Gurevich, the carrier aircraft was the responsibility of the OKB headed by A.N. Tupolev, and SKB-1 headed by S.L. Beriya was charged with the development of the control and guidance system. The missile system was intended to attack large surface ships at a range of 2000 km (1,243 miles) from the base airfield with the carrier aircraft beyond the reach of hostile air defences.

Mikoyan's KS missile (AS-1 'Kennel') was a smaller replica of the MiG-15 aircraft, with a maximum speed of 1100-1200 km/h (684-746 mph), range of 70-90 km (43-56 miles), length of 8.29 m (27 ft 2 in), launch weight of 2735 kg (6,030 lb) and a warhead of 1015 kg (2,238 lb). It was fitted with a single-mode (short-duration) RD-500K engine.

The control system consisted of equipment installed both on the carrier aircraft and the missile. Aircraft-mounted equipment included the Kobalt target detection, lock-on and auto-tracking radar, and the K-3 missile radio control equipment. The missile carried the K-2 radio control equipment, which was matched with the APK-5V autopilot. The K-2 receiving antenna was fin-mounted, and the K-1 homing equipment with its antenna was located in the forward part of the missile. Kometa was a radio command system, with semi-active homing in the terminal phase. When the target was detected and the carrier aircraft radar changed over to lock-on and tracking mode, the missile was launched and the operator guided its flight. As the signal reflected from the target was received by the aircraft's radar, the missile switched to homing mode using the K-1 system.

KS missile

Mikoyan's pug-nosed KS-1 missile was clearly based on the MiG-15 fighter. Guidance equipment consisted of the K-1 homing equipment, which was housed in the nose radome, and the K-2 radio command guidance equipment, the antenna for which was located in a bullet fairing at the tip of the fin (right). The Kometa system was originally tested on the Tu-4KS, but the operational missiles were fielded by the Tu-16KS aircraft of 124 TBAP in 1957.

Both Egypt and Indonesia received Tu-16KS aircraft and their associated KS missiles. These two are Indonesian machines, the aircraft above showing the Kobalt-1M radar extended on the ground for maintenance access.

A directive of the Main Staff of the Navy, dated 30 August 1955, was the starting point for the organisation of the 124 heavy bomber air regiment (TBAP) as a unit of the Black Sea Navy aviation. It was equipped with 12 Tu-4KS and two MiG-15SDK, plus sundry MiG-15UTIs, Li-2s and Po-2s. It was the first missile-carrier aircraft regiment in the air force. The MiG-15SDK (and subsequently the MiG-17SDK) was fitted with the equipment of the KS missile system, an autopilot and recording equipment.

For training, one or two MiG-15SDKs were suspended from the Tu-4KS and the latter took off and headed for the test range. Flying along a combat course, the crew of the Tu-4KS operated in the sequence identical to the missile launch procedure, although the engine of the converted fighter was started by the pilot. After separation from the Tu-4KS the receiving unit of the MiG-15SDK received guidance signals that were translated into control commands fed to the autopilot. The pilot did not interfere with the control of the aircraft until it dived and approached the target to a distance of 500-600 m (1,640-1,970 ft). Then he switched off the autopilot, took over control of the aircraft and flew to the landing airfield. The performance of the guidance system was analysed on the basis of data registered by the recorders in both the Tu-4KS and MiG-15SDK.

Tu-16KS 'Badger-B'

Taking into consideration the disadvantages of the Tu-4, it was decided that the missile system needed another carrier. The Tu-16 was the only alternative and, in 1954, along with some upgrade work, aircraft no. 4200305 was fitted with part of the Tu-4KS equipment. An improved Kobalt-1M radar was fitted and two external wing racks with locks and fairings were installed. Until the moment of their release the missiles were supplied with fuel from a 2300-litre (506-Imp gal) No. 1 tank, which was isolated from the aircraft's own fuel system.

At the same time, the crew gained one more member – a navigator-operator who was responsible for all operations related to the employment of the missiles. A suspended cabin was provided which was fitted to the lower attachment points of the bomb racks in the stores bay. The cabin had everything required for life support. Operating inside the cabin was quite difficult, especially in the summer. Besides, the operator in his box could not see anything and felt separated from the crew.

First deliveries of Tu-16KS aircraft with the Kometa missile system started in 1953, to naval aviation. In 1954, at the Black Sea Fleet aviation airfield at Gvardeiskoe in the Crimea, the 27th Separate Aviation Unit was formed. Live missile test launches were carried out at Bagerovo, the base of the Air Force R&D Institute. They ended in 1955 with the adoption of the missile. The tests indicated that the maximum flight speed of the Tu-16KS carrying two missiles dropped by 100 km/h (62 mph) and with one missile by 30 km/h (18.6 mph), while the take-off run became longer. The range of the Tu-16KS carrying two missiles and returning empty was 3250 km (2,020 miles) and 3560 km (2,212 miles) with one missile. The take-off weight of the aircraft carrying two missiles increased to 76000 kg (167,548 lb).

'Badger-Gs' carrying the KSR-2 missile were converted from either Tu-16KSs (Tu-16KSR-2) or from redundant free-fall bombers (Tu-16KSR-2A). The upgraded aircraft and its new missile entered service in 1963, but it was soon supplanted by the much better KSR-5 system.

This Tu-16KSR-2-11 has the SPS-5 Fasol EW system mounted in its belly. When upgraded to carry the KSR-11 anti-radar missile, the 'Badger-G' was known as either the Tu-16KSR-2-11 or the Tu-16K-11-16, depending on its former configuration. Aircraft with KSR-11 capability were fitted with the Ritsa radar-targeting system with characteristic nose antenna array.

Left: This Tu-16KSR-2 carries two missiles. From the calibration marks on the forward fuselage it can be assumed to be a test aircraft. KSR-2 tests were undertaken in 1960/61, although not without problems. KSR-11 tests were accomplished at much the same time. In 1968 the system was 'tweaked' and recertificated as the KSR-2M to allow launches from low altitude.

Series production of the KS missile with folding wings and increased fuel capacity was launched in 1957. With allowance made for the capabilities of the carrier aircraft's equipment, the missile upgrade provided a launch range of up to 150 km (93 miles).

Tu-16KSR

The KSR programme was created in accordance with an order of the Ministry of Aviation Industry dated 29 April 1957. It charged OKB-273 with the task of improving the Kometa system using new Rubikon airborne equipment, which combined the functions of the K-2M and the then up-to-date Rubin-1 radar (NATO: 'Short Horn'). In compliance with a decree issued 2 April 1958, a design bureau headed by A.Ya. Bereznyak was preparing the KSR version of the KS missile, equipped with a two-chamber liquid propellant engine designed by A.M. Isaev. The aim was to increase the altitude and launch speed range. The voracious engine required more fuel and the capacity of the tank for TG-02 fuel reached 666 litres (146.5 Imp gal). For the AK-20F oxidiser there was a 1032-litre (227-Imp gal) tank made of stainless steel. The fuel ignited when the two components reacted with each other. The warhead was taken from the K-10S missile developed by OKB-155. Since there was no air intake, the diameter of the KSR missile was reduced to 1 m (39.4 in) and the wings were made to fold. The missile radio control equipment included a K-1MR station and an APK-5D autopilot.

Between 1 July and 15 November 1959, 22 test sorties were undertaken, and KSR missiles were launched during nine of them. Out of six launches at a range of 90-96 km

KSR-2/11 missile

Although its roots lay in the KS, the KSR-2 missile was much cleaner aerodynamically thanks to the use of a liquid-propellant rocket motor. The motor had two chambers: one for boost and one for cruise. This example is seen on display under the wing of a Tu-16R reconnaissance aircraft.

(56-60 miles), four resulted in direct hits. The tests near Feodosia revealed low reliability of the guidance system and the missile was not endorsed for fielding. Nevertheless, the KSR missile entered service with the 5th Mine-torpedo Air Regiment of the Black Sea Fleet aviation and crews carried out launches.

Tu16K-16/Tu-16KSR-2

As a further development of the Tu-16KSR system, the Tu-16K-16 was developed in accordance with a decree dated 22 August 1959. OKB-283 updated the homing head of the missile by installing a larger antenna, covered by a radome which was almost the diameter of the missile, while the stabiliser was moved from the fin to the fuselage. The missile suite was designated K-16, while the system itself acquired the designation of Tu-16K-16. It included the Tu-16KSR-2 aircraft, the KSR-2 missile (AS-5 'Kelt'), Rubikon control equipment fitted to the aircraft and the missile, testing and measuring equipment, and servicing assets. The system was designed to attack targets such as large ships (with a displacement over 10,000 tons), railway bridges and dams. The guidance system used the missile's active radar from launch to strike, a principle later to be called 'fire-and-forget'.

Tu-16KSR-2 aircraft were fitted with the following equipment: upgraded radar designated Rubin-1k (Rn-1k), external store racks, and an AP-6E autopilot instead of the AP-5-2M. Changes were made in the fuel system. Missile guidance equipment consisted of the Tu-16's Rn-1k radar, the KS-2M homing radar of the cruise missile and its AP-72-4 autopilot. The Rn-1k searched for the target and commanded the KS-2M radar and the AP-72-4 autopilot before the release of the missile. During ground preparation of the Rn-1k radars it was possible to change the frequency range of their transmitters by up to 3 percent, which prevented interference during multiple launches. The navigator operated the Rn-1k radar in all its modes.

The KSR-2 missile was a mid-wing monoplane with a length of 8.65 m (28 ft 5 in), wingspan of 4.5 m (14 ft 9 in) and a sweep angle of the wing and the tail unit of 55°. The weight of the missile was 4000 kg (8,818 lb). In its forward part the missile accommodated the KS-2M radar and the warhead compartment, which could be fitted with either a conventional or nuclear charge. The 850-kg (1,874-lb) FK-2 high-explosive hollow-charge warhead was capable of penetrating steel armour 30 cm (12 in) thick. For attacking ground targets there was an FK-2N high-explosive warhead.

In the detachable tail section was the autopilot and a C5.6 liquid propellant engine which weighed 48.5 kg (107 lb). The engine developed 11.9 kN (2,674 lb) thrust operating in the boost mode and 6.93 kN (1,559 lb) in the cruise mode with the fuel consumption 5.4 and 2.3 kg (11.9 and 5.1 lb) per second, respectively. It started automatically 7 seconds after release. Power was supplied by a self-activating storage battery.

In June-July 1960 production Tu-16KS no. 720368 was fitted with additional equipment and acquired the designation of Tu-16KSR-2. Tests carried out from 20 October 1960 to 30 March 1961 revealed the need for further refinement and testing. A decree issued on 30 December 1961 endorsed the system for fielding by both naval and long-range aviation. With two missiles at a take-off weight of 76000 kg (167,548 lb), the Tu-16KSR-2 had a range of up to 1850 km (1,150 miles). The range of the missile's radar in detecting seaborne and ground targets was 220-290 km (137-180 miles), and launch range was 160-170 km (99 to 106 miles).

Beginning in 1962 some 205 aircraft were converted to 'Badger-G' configuration. These comprised 50 Tu-16KS, which became Tu-16KSR-2s, and 155 redundant Tu-16A bombers, which became Tu-16KSR-2As. Their wings were reinforced to accommodate external racks, the maximum flap setting was reduced by 5° and cutouts were made in the flaps for the missile fin. As well as the ability to carry

the KSR-2 missile, the 'Badger-G' retained its free-fall bombing capability. In the early 1970s some of the Tu-16KSR-2As were equipped with SPS-5 Fasol and SPS-100 Rezeda jammers. The latter was located in a special compartment in place of the tail gun.

Tu-16K-11 anti-radar system

Developed in accordance with a decree issued on 20 July 1957, the Tu-16K-11 system was intended as a development of either the KS or KSR missiles for the anti-radar role. Since the KSR provided for the installation of a larger diameter antenna it was selected as the preferable platform.

By the end of 1959 development of the Ritsa reconnaissance and target designation radar was completed. Components were housed in the nose landing gear bay, while a T-shaped direction-finding antenna was located on the upper framing of the navigator's canopy. To maintain the aircraft's centre of gravity the AM-23 nose gun was removed.

The KSR-11 missile was based on the KSR-2 and had much in common with it. The nose section of the missile was occupied by the 2PRG-10 (passive radio target seeker) with a gyrostabiliser and elements of the radar. Behind them there was a warhead, integral tanks with fuel and oxidiser, while the tail section housed the AP-72-11 autopilot, storage batteries and other equipment. A liquid propellant engine was installed under a detachable tail fairing. The weight of the armed missile was 3995 kg

(8,807 lb) with 1575 kg (3,472 lb) of oxidiser and 530 kg (1,168 lb) of fuel. The 840-kg (1,852-lb) warhead was of two types: FA-11 high-explosive fragmentation or FK-2 high-explosive hollow charge.

After the 2PRG-10 had completed tuning to the frequency of the target radar and had locked on to it, the missile was launched. The aircraft released the missile at a distance of 160-170 km (99 to 106 miles) upon receiving its readiness signal. Depending on the launch altitude, the missile dropped by 400 to 1200 m (1,312 to 3,937 ft) within the first 5 seconds. After 7 seconds the liquid propellant engine started operating and reached afterburning mode.

Forty seconds later an electronic time relay connected the output of the 2PRG-10 radar course channel to the autopilot, and from that moment its signals exercised directional control. Another 20 seconds later the engine changed

Although the full warload comprised two missiles, 'Badger-Gs' were mostly intercepted carrying one on the port wing. Above is an aircraft showing evidence of having been a Tu-16KS in a former incarnation. It is most likely a later Tu-16K-26 or Tu-16KSR-2-5, which retained the ability to launch the older KSR-2 missiles despite being equipped for the much better KSR-5.

Above left: This 'Badger-G' lacks the nose-mounted Ritsa antenna, and is probably a Tu-16KSR-2 or KSR-2-5. As well as anti-ship attacks, the KSR family of weapons could also be used against high-value and land targets with large radar returns such as bridges, dams or large buildings.

Tu-16KSR-2-5-11/Rn-1M

When the KSR-5 missile (AS-6 'Kingfish') was added to the Tu-16's arsenal, its increased range caused some operational problems. Chiefly, the range of the missile's active seeker was greater than that of the standard Rubin-1k radar fitted to Tu-16KSR aircraft. An initial attempt to overcome the shortfall involved fitment of Berkut radar (from the Il-38 'May') to the 14 aircraft of a single Baltic Fleet squadron. A more widespread upgrade, however, was the installation of a Rubin-1m radar in a large belly radome, while the undernose Rn-1k and nose cannon were removed. Rn-1m radar was retrofitted to both Tu-16KSR-2-5 and Tu-16KSR-2-5-11 aircraft, the numerical suffixes denoting which missiles the aircraft could carry. On at least one occasion a Tu-16KSR-2-5-11 with Rn-1m radar was spotted while carrying a KSR-5 and KSR-11 together.

'Badger' details

Wings and tail

The rigid two-spar wing had a torsion box structure, with high aspect ratio, and consisted of a wing centre section and wing panels. Sweep angle was 35° which increased critical Mach number and reduced shock stall. Since the aerofoil section of a swept wing is at an angle to the line of flight, local speeds along it are determined by the component of the airflow rather than the full velocity vector, as is the case with a straight wing.

Set at an anhedral angle of 3°, the wing had an area of 164.65 m² (1,772 sq ft). For greater angles of attack, the wing tips (from rib seven) had profiles with stall-resistant characteristics. The trailing edge of the wing was equipped with single-slot Fowler extension flaps and sealed-type mass-balanced ailerons. The tail unit was designed with a higher critical Mach number than the wing. This was achieved through low aspect ratio and a higher sweep angle of 42°.

Fuselage and emergency egress

The fuselage of the aircraft was an all-metal semi-monocoque. The nose and aft sections housed pressurised cabins for the crew. The entrance to the front cabin was located under the seat of the navigator-operator, the entrance to the rear compartment was under the seat of the tail gunner. Jettisonable doors above and below allowed emergency escape. The pilots were ejected upwards after their seats reached the rearmost position, while other crew members escaped downwards. The pilots' control columns were pushed by the pneumatic system to the foremost position. Maximum g-force developed by the pilot seat ejection system was 18, with an initial seat velocity of 22 m (72 ft) per second. G-force of ejecting downwards reaches -3 to -5.

Powerplant and fuel

Two Mikulin AM-3 turbojets, with axial-flow compressors and two-stage turbines, formed the initial powerplant, each producing a maximum thrust of 85.84 kN (19,290 lb). From 1958 the more powerful RD-3M (93.79 kN/21,076 lb thrust) engines were fitted, and from 1961 the upgraded RD-3M-500 was installed, offering the same thrust but with a maximum take-off rpm of 4,700.

The engines were started by an engine-mounted starter using B-70 gasoline. The engines burned T-1, TS-1 or RT kerosene held in 27 flexible fuselage and wing fuel tanks separated into 10 groups (five per engine) with a total capacity of 43800 litres (9,635 Imp gal). With a normal take-off weight of 72000 kg (158,730 lb) the fuel load was 34360 kg (75,750 lb). Fuel consumption was controlled automatically and did not require attention from the crew apart from in an emergency. Fuel could be dumped from the wing tanks and nos 1, 2 and 5 fuselage tanks.

Control system

The aircraft featured a dual control system with mechanical push-pull control linkages without hydraulic actuators. The system included AP-5-2M autopilot servo units (subsequently replaced by AP-6E and AP-6B). The flap and trimmer drives were electromechanical with an alternate mechanical elevator control.

Undercarriage

The tricycle landing gear had a nose gear with twin wheels able to turn through 40° on either side. The nose wheels were controlled by the steering control wheel in the pilots' cockpit. The main landing gear, with four-wheel bogies, retracted back into the undercarriage nacelles. The tailskid was extended and retracted by an electric actuator. A brake chute was provided, roughly doubling the drag of the aircraft on landing.

Aircraft systems

The hydraulic system extended and retracted the landing gear, opened and closed the doors, and controlled the brakes. The hydraulic brake control system served as an alternate system for the primary when the stores bay doors were closed. The crew life support system included a liquid oxygen unit and individual oxygen masks. Wing anti-icing system was by hot air, bled from the engine compressor. Air was ducted through the wing and exited through gills at the wing tips. The leading edges of the tail unit were equipped with electric thermal heaters.

Above: The navigator could perform visual bomb drops from the glazed nose panel. This Tu-16KSR-2-11 has the Ritsa antenna installed.

Right: Access to the front crew compartment was made through this hatch (with drop-down ladder) between the nosewheel and the Rubin-1k radar.

Below: Detail of the two tail positions. Above the turret was the PRS-1 Argon gun-laying radar.

Aircraft converted or built to carry missiles on the wings had a cut-out in the flaps to accommodate the missile's fin, as well as two upper surface fences.

Many 'Badger-Gs' already had an anti-radar capability thanks to the KSR-11 missile, but improvements in this field were introduced by the KSR-5P version of the AS-6 'Kingfish'. When modified to carry this weapon the aircraft were designated Tu-16K-26P, which became the final major variant of this missile-carrying family, and the last 'Badger' version to see widespread Soviet front-line service. All K-26Ps had the Ritsa direction-finding antennas on the nose.

power for the deployment of the lower fin and other operations.

Missile launch altitude varied between 500 and 11000 m (1,640 and 36,090 ft) and aircraft speed during launch ranged between 400 and 850 km/h (248 and 528 mph). Launched from 500 m the missile range was 40 km (25 miles), while from 11000 m it could reach 280 km (174 miles). The missile was 10.6 m (34 ft 9 in) long with a wingspan of 2.6 m (8 ft 6 in) and a body diameter of 0.92 m (3 ft). The fully tanked missile weighed 3952 kg (8,712 lb). It was equipped with either a 9A52 high-explosive hollow charge or a TK38 nuclear warhead.

The missile guidance system included VS-KN radio guidance equipment, autopilot, Mach switch and altitude switch. The VS-KN was a long-range active homing seeker which locked on to and automatically tracked the target while under the wing of the aircraft, following identification of the range and direction to the target. The missile was released at a rated distance. Two seconds later the engine ignited and the missile accelerated. After 15 seconds it began to climb and, upon reaching Mach 3, the booster chamber of the engine was cut off. At an altitude of 18000 m (59,055 ft) the missile began to level off and the engine switched to sustainer mode. After that the missile remained in a stabilised flight at an altitude of 22500 m (73,820 ft). As the slant range to the target reached 60 km (37 miles), a dive command was issued. At a distance of 400-500 m (1,312-1,640 ft) to the target, direction and pitch radio guidance was disabled. In case of jamming, the missile magnetron frequency was retuned and operation of the logic circuitry switched to the memory mode.

As time went by, a number of modifications were made to the basic KSR-5, such as the low-altitude KSR-5N (the K-26N system with a low-altitude radar) and the KSR-5M (the K-26M system for countering complex small-size targets).

Flight tests of the K-26 system started in October 1964 with two aircraft, a converted Tu-16K-11-16KS (no. 8204022) and a Tu-16KSR-2A no. 5202010, which was redesignated Tu-16KSR-2-5. Some flights of the programme were made by Tu-16K-26 no. 420073 and Tu-16K-10-26 no. 1793014. Tests were suspended several times, and it was not until 12 November 1969 that a decree approved the K-26 system for fielding.

From 1969 the overhaul plants of the Air Force and the Navy modified 15 Tu-16K-11-16KS into the Tu-16K-26, 125 Tu-16KSR-2A into the Tu-16KSR-2-5-11 and 110 Tu-16KSR-2A into the Tu-16KSR-2-5. By comparison with the Tu-16KSR-2-5-11, the latter lacked the nose-mounted antenna of the Ritsa system and could not use the KSR-11 anti-radiation missile.

KSR-5 missile

The KSR-5 missile appeared in the 1970s, and was far more capable of penetrating heavy defences than the KSR-2/11. After launch it accelerated and climbed to altitude, before diving at around Mach 3 on to its target. For the anti-ship mission a large conventional warhead was usually employed, while for land targets a nuclear TK38 was the warhead of choice. The KSR-5M was a low-altitude launch version.

Below: A KSR-5 is displayed on a Tu-16. The ventral fin would normally be folded until launch, but is seen here in flight position.

Above: Like its forebear, the KSR-5 employed a two-chamber motor. The large chamber accelerated and climbed the weapon to Mach 3 at around 18000 m (59,055 ft), while the smaller chamber sustained it in level flight.

Tu-16K-26 s/n 1883704
1st Squadron, 99th Guards Independent Reconnaissance Regiment (GvORAP)
Priozyorsk, Sary-Shagan air base, Kazakhstan, late 1980s

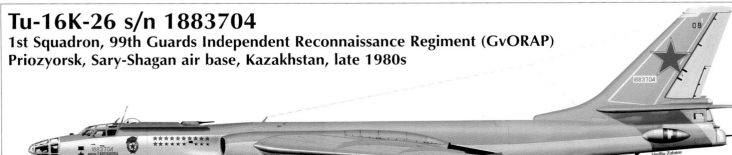

The Tu-16K-10 represented a major advance in Soviet missile warfare, for it introduced the first supersonic weapon in the form of the K-10S. The missile drew on elements of the MiG-19, including a version of its engine, and its guidance system was tested using MiG-19SMK surrogates. Installing the massive YeN radar in the Tu-16 necessitated moving the navigator from his nose compartment.

Above: This dramatic photograph captures the moment of launch of a K-10S. The missile was introduced to service in 1960, and was chiefly viewed as an anti-ship weapon.

Right: Crew inspect an armed Tu-16K-10. The missile was carried on a pylon which raised the weapon into the belly recess to provide ground clearance for take-off and landing, and to reduce drag in the cruise. For firing the pylon was lowered so that the missile could fly free from the aircraft.

Problems emerged with retrofitting the Tu-16KSR-2 aircraft. It was discovered that the KSR-5 missile's radar surpassed the aircraft's Rn-1k in detection range. As an alternative, 14 aircraft of the naval aviation were equipped with the Berkut radar of the Il-38 aircraft, with a slight

Right: A Tu-16K-10 makes a pass with its K-10S missile in the lowered launch position.

Below: This view of a Tu-16K-10 at rest shows the missile in its raised position. The radome of the YeN radar actually improved the aerodynamics of the 'Badger'.

change in the shape of the radome. In 1973 some of the Tu-16KSR-2-5-11s were fitted with a Rubin-1m radar optic system with a longer detection range, increasing the launch range of the KSR-5 missile to 450 km (280 miles). The radome of the Rubin-1m radar was located under the forward part of the stores bay, and required the removal of the nose gun. Tu-16KSR-2-5s were also upgraded with Rubin-1m radar.

Tu-16K-26P

This system was developed in accordance with a decree issued on 7 February 1964. It was a modification of the K-26 system and was intended to destroy radar contrast targets with either KSR-2 and KSR-5 missiles, and radars operating in pulse mode with either KSR-11 and KSR-5P missiles. The KSR-5P missile radio guidance system consisted of the VSP-K equipment and an autopilot mounted in the missile, and elements of the aircraft-carried ANP-K radar and Ritsa direction-finding system. Two KSR-5Ps could be launched against one or two targets, provided they were within a 7.5° sector from the longitudinal axis of the aircraft. After the release of the missiles the aircraft made a turn.

State tests of the system began in April 1972. A decree issued 4 September 1974 approved it for fielding to the Navy. In the following year the naval overhaul plants started converting the Tu-16KSR-2-5-11 into the Tu-16K-26P. The aircraft retained its ability to launch the older missiles, and to deliver free-fall weapons.

Tu-16K-10 'Badger-C'

Development of missile-carrying 'Badgers' proceeded along two parallel lines. While the Tu-16KS/KSR family was developed, the Tu-16K emerged as a separate track with a very different mission fit. In the fullness of time, the two tracks were to move closer together with the common adoption of the KSR-5.

Initially intended for the '105' aircraft (Tu-22 'Blinder'), the K-10 system was intended to attack surface vessels with a displacement over 10,000 tons. Construction of the '105' aircraft was delayed, so it was decided to test the K-10 system on the Tu-16 instead. A decree issued on 3 February 1955 initiated development of the Tu-16K-10, with completion of tests scheduled for the third quarter of 1958. On 10 December 1957 plant no. 22 at Kazan completed the first Tu-16K (no. 7203805) and the second (no. 7203806) the following year.

Meanwhile, the first K-10 missile was sent to the test range at Vladimirovka in October 1957. From January 1957 two Tu-16Ks, two MiG-19SMKs and 36 missiles (34 with telemetry equipment and two live) were employed in tests. Initially, the test series encountered numerous unforeseen difficulties, and in the middle of 1958 a dilemma emerged as to whether it was expedient to proceed with the K-10 or not. In the event, a compromise was reached in which a new missile was to be developed for the Tu-22 (Kh-22) while the K-10 continued its development as a weapon for the Tu-16.

Following the guidelines of this decision, taken by the government commission for the military industry on 28 July

1958, another 10 aircraft equipped with telemetry equipment and two production Tu-16Ks were allocated for tests. In the course of testing 44 launches of missiles were made, including four launches against ground targets with corner reflectors and 34 against a target vessel on the air force's range no. 77 in the Caspian Sea. The ship was the half-submerged 9,100-ton tanker *Chkalov*, also equipped with corner reflectors.

From 5 September 1959 to 5 November 1960 the guidance system of the K-10 was tested. For that purpose ships of the Black Sea Fleet were employed as targets and the MiG-19SMK aircraft were used as surrogate missiles, having been equipped with the K-10's guidance system. In 1959, before tests were complete, large-scale production of the Tu-16K-10 was launched.

During state tests 20 K-10S missiles with production equipment were launched, of which eight missiles failed to reach their targets because of malfunctions and defects in design and manufacturing, and two missed their targets because of mistakes by the crews. Despite objections by the manufacturer, the results of the tests stated in the Act indicated that the technical serviceability of the system was rated as only 0.55. Later, on the basis of results of live launches carried out by the naval aviation, the hit probability in the Act of state tests was increased to 0.8. A decree of the Government of the USSR issued on 12 August 1961 endorsed the fielding of the K-10 system to naval aviation.

Production threatened

Overall, the Tu-16K-10 complex comprised the Tu-16K carrier aircraft, the K-10S air-to-ship missile, control system, ground preparation and servicing equipment. The Tu-16K differed from the Tu-16KS to such an extent that retrofitting the latter did not seem expedient. The first serial Tu-16K, no. 8204010, was built at Kazan in April 1958, but by the end of 1959 only seven Tu-16Ks equipped with refuelling systems had been built, and production was stopped. The destiny of the entire system was under threat as production of the Tu-16 was closed. Plant no. 64 at Voronezh was making the An-10, Kazan was preparing for production of the Tu-22, while plant no. 1 at Kuibyshev was transferring to the production of missiles.

On 6 June 1958, to save the situation, the Air Force commander-in-chief, deputy chairman of the Council of

Ministers D.F. Ustinov and the chairman of the State Committee for aviation, turned to the Central Committee of the CPSU with a proposal to continue building the Tu-16 at the Kuibyshev plant and make 173 aircraft (13 Tu-16s and 40 Tu-16Ks in 1958 and 60 Tu-16Ks during each of the following two years). By June 1960 the Kuibyshev plant had built 59 Tu-16Ks. In June next year Kazan began the manufacture of 150 Tu-16Ks, to add to seven produced earlier. A total of 216 Tu-16Ks was built.

To employ the K-10S missiles the Tu-16 was considerably upgraded: the nose section was redesigned with a large radome housing the antenna system for a target search and tracking channel, and the YeN radar (NATO: 'Puff Ball') transmitter. The radome slightly improved the aerodynamic characteristics of the aircraft and was necessarily large to accommodate the YeN's large antenna, which also traversed mechanically through 120° azimuth to continue guiding the missile even after the aircraft turned away. The missile guidance channel antenna was located in the fairing under the forward fuselage in the place of the standard Tu-16's radar.

The Tu-16K-10 was mainly operated by the AVMF, and it was commonly intercepted by Western air arms over international waters. Despite unreliability, the K-10 system offered significant advantages over the KS – notably the ability to launch up to 18 missiles simultaneously and the ability of the YeN radar to support the missiles at an angle up to 60° from the direction of flight.

K-10 missile

This K-10S is seen on its loading trolley, with wings folded. The RD-9FK turbojet engine had afterburning to propel the missile to supersonic speeds. The fin folded for carriage. Variants of the basic missile were the nuclear K-10SB, K-10SN low-level launch weapon, K-10SD with increased range, K-10SND low-level, increased-range weapon, and K-10SP jammer missile.

Production of Tu-16K-10s reached 216, making it the most numerous of the missile-carrying versions. Construction was handled at both Kuibyshev and Kazan to satisfy the demands for the type. However, according to plans the aircraft should not have been built at all: the K-10 system was destined to be fitted to the Tu-22 'Blinder', although delays with that programme meant that it was applied to the Tu-16 instead. The 'Blinder' eventually matured with the K-22 system with the Kh-22 (AS-4 'Kitchen') missile.

The capabilities of the Tu-16K-10 expanded over the years, mostly in the areas of increased missile range (K-10SD) and reduced launch altitude (K-10SN). The latter was of particular importance as all forms of aerial warfare moved to low-level to defeat ever more sophisticated defences. A novel idea was the use of the K-10SP missile, which had jammers instead of a warhead. It blinded hostile defences to allow armed missiles to penetrate.

The crew of the Tu-16K consisted of six men. The navigator was moved to the navigator-operator station, and the stores bay was made longer. Tank no. 3 was removed, and the aircraft was fitted with a pressurised bay cabin for the navigator-operator of the YeN radar and missile guidance equipment. An additional 500-kg (1,102-lb) fuel tank provided a supply for the missile's engine while it was on the rack. Bombing capability was removed.

K-10S missile carriage was provided on an external rack under the fuselage. For take-off and cruise the rack was raised so that the upper half of the missile was inside the stores bay, but before launch the rack and missile were lowered by 55 cm (22 in) into the firing position. Despite somewhat better streamline characteristics than other variants, the cruising speed reduced to 780-820 km/h (485-510 mph) and combat radius to 1900 km (1,180 miles).

Known to NATO as the AS-2 'Kipper', the K-10S had a 7-m² (75-sq ft) wing with 55° sweep, and a folding tail unit. Power came from a short-endurance RD-9FK turbojet engine with an afterburner chamber, a version of the RD-9B engine used in the MiG-19 fighter. When not in use the engine was covered by a drop-down fairing. The engine provided a speed of 2030 km/h (1,261 mph), making the K-10S the USSR's first supersonic missile. Dimensions were 9.75 m (32 ft) length, 4.18 m (13 ft 8 in) wing span, 0.92 m (3 ft) body diameter and weight of 4500 kg (9,921 lb) with 780 kg (1,720 lb) of fuel. The missile could carry either a conventional or a nuclear warhead.

Missile guidance equipment was located in the extremities of the fuselage: the forward part housed the ES-2-1 antenna and elements of the ES-2 homing head, the tail part housed the ES-1 radio guidance equipment. The main tank was made from steel and was located between frames 15 and 19. The total capacity tank was 1573 litres (346 Imp gal). Behind the main tank there were elements of the ES-3 autopilot, hydraulic accumulator and hydraulic pump.

The high-energy YeN radar was used for searching for and tracking the target while the missile was on its trajectory. Frequency spacing of the radar allowed the simultaneous launch of 18 missiles. The K-10S missile was launched at a distance of 180-200 km (112-124 miles) from the target. Before launch, the navigator-operator locked on the target for auto-tracking, guidance equipment ran through self-test procedures, and the missile engine was run up to full afterburner.

Missile trajectory was divided into three phases: self-contained flight, stabilised altitude flight, and terminal dive. Depending on the launch altitude, 40 seconds after release the missile could have dropped by 600-1500 m (1,969-4,920 ft). It then climbed and flew horizontally at a stabilised cruise altitude. At a distance of 100-110 km (62-68 miles) to the target the carrier aircraft commanded the missile to dive at an angle of 12-17° and the missile attained its maximum speed. At an altitude of 2000 m (6,562 ft) the dive angle changed to 5-6° and the missile flew at its second stabilised altitude of 1200 m (3,937 ft). Some 15-20 km (9.3-12.5 miles) out from the target the missile homed in direction and altitude. 6.5 km (4 miles) short of the target the missile dived at an angle of 15-18°. The warhead exploded on impact with the target.

Meanwhile, 100 seconds after missile launch the Tu-16K's crew could turn 60° either side with a roll of 9-12°, the bank angle being limited by the stabilisation of the YeN antenna, which continued to guide the missile towards the target and monitored its flight by the signals from its transponder.

In 1963 the Tu-16K-10N with the low-altitude K-10SN missile was developed. A number of engineering solutions helped reduce launch altitude from 5000 to 1500 m (16,404 to 4,921 ft) and the second stabilised altitude from 1200 to 600 m (3,937 to 1,968 ft). Subsequently, the altitude of the last phase was successfully brought down to 90-150 m (295-492 ft).

Right: This cockpit view was taken aboard a Tu-16K-10 during a mission in 1961. In the centre of the dashboard is the missile control panel.

Below: Landing the Tu-16 with the missile retracted was no problem, as can be judged from this view. However, if the retraction mechanism failed there was very little ground clearance to land back safely with the weapon.

Tu-16K-10SD

This upgrade employed an increased-range missile with an upgraded engine fuel system, which reduced fuel consumption per kilometre and added 65 km (40 miles) in range. The TK-34 warhead was replaced by a smaller TK-50 and it provided the possibility to install an additional 200-litre (44-Imp gal) tank. This gave a 40-km (25-mile) increment in range. Therefore, the maximum missile launch distance grew by 105 km (65 miles) to 325 km (202 miles). The launch altitude range expanded from 5000-10000 m (16,404-32,808 ft) to 1500-11000 m (4,921-36,090 ft). The retrofitted missile was designated K-10SD. The low-altitude missile, when retrofitted for a wider range of employment, was designated as the K-10SND.

Growing missile range in turn required increased capabilities of the carrier aircraft's YeN radar, a task which presented a challenge. Basic research was conducted at the

33rd Naval Aviation Centre. Detection range was increased through changing frequency and the length of outgoing pulses. The range of detecting typical sea-borne targets grew from 320 to 450 km (199 to 280 miles). These new advances meant that, after launching the missile from maximum distance, by the moment of impact the aircraft was 265 km (165 miles) away from the target, rather than 140 km (87 miles) as it had been before.

Tu-16K-10SP

This system was developed on the initiative of naval aviation, and featured a K-10SP jammer drone that protected armed missiles while they approached their targets. Navy overhaul plant no. 20 (in Leningrad) was charged with the job of retrofitting missiles, while tests were conducted at Centre no. 33. In fact, the K-10SP missile opened a completely new direction – ECM drones.

K-10SPs were fitted with one of the Azaliya jammers (SPS-61R or SPS-63R operating on a centimetric wavelength). Jamming characteristics could be changed before the flight, depending on the tactical situation. The jammer was first used during the Kvant electronic warfare exercises in the Pacific in 1976 and received positive responses. In accordance with the order of the Minister of Defense of the USSR dated 11 April 1979, the K-10SP jammer missile was endorsed for fielding for service.

K-10SPs were used again during the Ekran electronic warfare exercises by the Northern Fleet in 1981. That time six K-10SPs were covering an K-10SN missile, which the ship-borne Volna and Shtorm air defence systems were tasked to destroy. To avoid possible trouble, and taking into account the low effectiveness of the two systems, the K-10SN missile descended only to the second stabilised altitude of 1200 m.

Target designation for naval fire was provided by the MR-310A centimetric wave radar. It was discovered that, from a distance of 130-140 km (81-87 miles), the jammers produced a solid flash on the screens of the ship's radars and target designation did not seem possible. Only the MR-600 radar, operating in a different waveband, provided target designation and the K-10SN was destroyed 15 km (9.3 miles) short of the ships. However, the crews were in a high state of turmoil as they had not been informed that the K-10SN would descend.

Tu-16K-10-26 'Badger-C Mod'

This system was developed in compliance with a decree issued on 23 June 1964 and essentially added the KSR-5 missile to the K-10 system. It was intended for destroying ground and sea-borne targets with the KSR-5, KSR-2 and K-10SND missiles. The aircraft could carry two KSR-5Ns

The Tu-16K-10-26 introduced the KSR-5 missile, greatly expanding its capabilities. The ability to carry the K-10 was retained, and on rare occasions all three missiles were carried, albeit for short-range missions.

The Tu-16 was essentially a pack hunter. Firing one or two missiles each, a group of aircraft launching from different directions hoped to saturate the defences of a large ship and its air defence support vessels.

Tu-104Sh trainer for the Tu-16K-10

Two ex-Aeroflot Tu-104 airliners were converted with YeN radar and its large radome to serve as trainers for Tu-16K-10 'Badger-C' navigators and weapons officers, with operating stations in the cabin. They were designated Tu-104Sh and were delivered to the AVMF in 1964. In 1969 both aircraft were upgraded with the Ritsa radar direction-finding system, and had an antenna mounted on top of the nose forward of the windscreen. Subsequently, both Tu-104s were given Tu-22M 'Backfire' radar and had wing pylons added, becoming Tu-104Sh-2s in the process. They continued as 'Backfire' trainers until replaced by the Tu-134UBK. This aircraft – SSSR 42342 – was photographed at Pushkin in the early 1980s after it had been upgraded to Tu-104Sh-2 standard.

Above: The wing pylons were the principal external clue to the identity of the Tu-16K-10-26. The YeN radar had sufficient range to match that of the missile's radar, alleviating the problems encountered by 'Badger-Gs' when they were upgraded to the new weapon.

Right: This group of stored Tu-16K-10-26s awaits the scrapman's torch. Visible are the flap cut-outs and wing reinforcements required to accommodate the KSR-5 missile. Also on display is the protruding rod on top of the nacelle which provided a visual indication of the 'down and locked' position of the main landing gear.

and one K-10SND. Additional missile carriage required reinforcing the wing, installing external store racks and limiting maximum flap setting to 25°. Addition of the KSR-5 greatly increased striking capabilities, although combat radius was cut to a third.

Factory tests of the system were carried out from November 1966 to March 1967 with two Tu-16K-10-26s

As well as active radar-guided KSR-5s, the 'Badger-C Mod' was modified to use the radar-homing KSR-5P, allied to elements of the Ritsa system. With this equipment installed, the designation became Tu-16K-10-26P. The Taifun upgrade expanded anti-radar capability and allowed the aircraft to identify hostile radars at long range. This system was applied to the Tu-16K-10-26P in the 1980s, but was of limited value.

converted from Tu-16K nos 1793014 and 2743054. State tests began in late 1968 and ran through to April 1969. A decree dated 12 November 1969 approved the fielding of the Tu-16K-10-26. Retrofitting was conducted at Navy overhaul plants from 1970. A total of 85 aircraft was retrofitted, essentially identical to the Tu-16K-10 with the exception of the wing racks for the KSR-2 and KSR-5 missiles.

As with the Tu-16KSR series, an anti-radiation capability was added through the use of KSR-11 and KSR-5P missiles. The resulting Tu-16K-10-26P was developed in accordance with a decision of the military industrial complex issued on 21 January 1976. The KSR-5P radio guidance equipment included the VSP-K equipment installed on the missile, ANP-K equipment on the aircraft, and monitors for the I-41 and I-42 units of the Ritsa system.

Taifun upgrade

In 1980 the system was upgraded with the Taifun equipment, which had been installed in the production Tu-16K-10-26P no. 3642035 of naval aviation. The aircraft was fitted with LO-67 Taifun reconnaissance and target designation equipment, which was designed to operate together with the direction-finding element of the KSR-5P's homing head and identify characteristics of target radars. A naval overhaul plant was responsible for retrofitting. Check tests took place at Akhtubinsk and Nikolaev from 29 December 1979 to 18 June 1980. They indicated radar signals could be detected at a distance of 380-400 km (236-248 miles) with the aircraft flying at 9000-11000 m (28,528-36,090 ft), with lock-on range some 20-30 km (12.5-18.5 miles) less. The Taifun VSP-K system operated in two modes: reconnaissance and target designation.

Despite indifferent results, work on expanding the combat performance of the Tu-16K-10-26P with the Taifun equipment continued at Akhtubinsk and Nikolaev from 23 September 1984 to 14 January 1985. The final conclusion

stated that launch of the KSR-5P was possible in limited situations with the YeN-D radar monitoring distance to the target.

Tu-16K-10-26B

This system was prepared by naval aviation and developed at the naval overhaul plants during the period of 1974-1976 and gave the missile-carriers a bombing capability. Tu-16K-10-26 aircraft were fitted with 12 fuselage and wing external store racks, allowing the carriage of bombs from 100-500 kg (220-1,102 lb) in weight, to a total of 9000 kg (19,841 lb). A modernised OPB-1RU sight was fitted. An aircraft armed to the teeth with bombs looked quite impressive but, with hindsight, the Tu-16K-10-26B could hardly be seen as a great achievement, since bombs could only be delivered if the target was visible. Furthermore, bombing accuracy was low since the sight had its roots in the 1930s.

Missile-carriers in service

The Tu-16KS was first delivered to 124 TBAP of the Black Sea Fleet in June 1957 (from 3 October redesignated 124 MTAP). Its inventory included 16 Tu-16KS, six Tu-16SPSs, six Tu-16Zs and an An-2. The first missile was

launched by the Tu-16KS in December 1957 against a target in the Caspian Sea. After 124 MTAR, the Tu-16KS rearmed 5 MTAP, also part of the Black Sea Fleet, and in 1958 joined the Northern and Pacific Fleets. The rate of assimilation of the new missile system can be demonstrated by 124 live launches in 1958, and 77 sorties by the Northern Fleet using the MiG-17SDK surrogate.

KS missiles had only one radio guidance frequency and, until 1959, launches had been practised from one, two or three directions to eliminate mutual interference. Under such conditions a combat formation of 12 Tu-16KSs was

The Tu-16K-10-26 'Badger-C' was built as a pure missile-carrier, and had no bomb bay. Despite the fact that free-fall bombing had little place in the Cold War maritime scenarios in which the 'Badger-C' was designed to operate, the AVMF desired a conventional bombing capability for its aircraft. Fuselage BD4-16-52 bomb racks could be added (above), while further racks could be mounted on the KSR-5 pylons, as seen on the aircraft at left (with bombs) and empty on the aircraft below. These could accommodate up to 12 from a range of small/medium FAB general-purpose bombs. With this upgrade the aircraft became known as the Tu-16K-10-26B.

*Above: The 'Badger-F'
codename was applied to
the Tu-16R equipped with
SRS-3 Elint pods under the
wings. The underfuselage
radomes were retained.*

*Right: Photographed by
another Tu-16R, this DA
aircraft is fitted with a
Geran nose jammer and
Rezeda tail jammer.*

*Above: On 25 May 1968 this
Tu-16R, captained by Major
Pliev, crashed into the
Norwegian Sea while
snooping round a US Navy
carrier.*

*Below: The large teardrop
radomes under this Tu-16R
signify fitment of the SRS-4
Elint suite.*

900-1200 km (560-745 miles) deep and the strike dragged on for 1.5 hours. The study of the missile system discovered a possibility to increase the number of guidance channel frequencies to six and launch missiles from directions differing by 45-60°.

In 1958 launches and guidance of two missiles from one carrier aircraft were practised. In 1960 the option of launching missiles from four to six directions was tested. Four aircraft launched eight missiles and seven of them hit the target. An attack from six directions also turned out to be a success, and the strike lasted for just 40 seconds.

Improved systems

The Kometa system was slowly refined over the next few years. The improvement succeeded in reducing launch altitude from 4000 to 2000 m (13,123 to 6,562 ft), but despite all the tinkering the most reliable guidance of missiles was provided at a speed of no more 420 km/h (260 mph). The extending radome of the K-2M radar antenna was also a speed-limiting factor. In 1961 the Kometa system armed five air regiments of the AVMF: 9 MTAP of the Northern Fleet, 5 MTAP and 124 MTAP of the Black Sea Fleet, and 49 and 568 MRAP of the Pacific Fleet.

In December 1959 the flying and technical personnel of 924 and 987 MTAP of the Northern Fleet began mastering the Tu-16K-10. In May 1961 574 MRAP of the Northern Fleet and 170 MRAP of the Baltic Fleet's 57 MRAD began their retraining.

In 1959 the Tu-16K was delivered to the Northern Fleet and preparation for launches got under way. In July 1960 the crews of Colonels Myznikov and Kovalev were appointed to carry out the first AVMF launches of the K-10S. Myznikov's crew made the first launch against a target 165 km (102 miles) away, but the missile undershot.

Then the second crew followed with a direct hit. Within two weeks a total of five launches had been made: one of them was considered to be abnormal (crew mistake), and one missile hit a wave crest 200 m (656 ft) short of the target.

In 1960-1962 the Tu-16K-10 was fielded to seven air regiments and the number of launches grew from 79 in 1960 and 126 in the next year, to a record figure of 147 in 1962. This year was marked by another event. On 22 August, in the course of the Shkval exercise involving the Air Force, Navy and Strategic Missile Forces, the crew of Lieutenant Colonel V.P. Krupyakov, commander of 924 MRAD, made the first launch of the K-10SB missile with a 6-kT nuclear warhead against a target (a barge with corner reflectors) from a distance of 250 km (155 miles) on a firing range near Basmachny bay (Novaya Zemlya).

In the process of mastering the Tu-16K-10, crews were often challenged with unusual situations. For example, in 1961 when the crew of captain G.A. Zimin flew a Tu-16K-10 to carry out simulated launches, the external rack did not retract the missile and left it in the lowered, firing position. The crew manual did not provide any recommendations as to what to do in such a situation. There was real concern that during landing the missile would brush against the runway, as the normal landing angle of attack was 8°. Thankfully, all ended well. After several research flights with extended missile landings (five of them were made by the crew of Zimin), the crew manual was amended with a relevant instruction.

Experience gained during the first years of operating the Kometa missile system and the Tu-16K-10 indicated that the latter was highly unreliable, though it had been considerably refined by 1961. Almost 50 percent of launches carried out in 1961 failed, 32 percent of which were through technical or manufacturing faults. Against that background, the design of the Kometa system looked more successful.

Sometimes, missiles chose targets on their own. In 1964 a crew from 169 MRAP of the Pacific Fleet launched a live K-10SND missile over a firing range at Cape Tyk. The missile homed in on a Japanese timber ship, the *Sine-Maru*, that had sailed from Nikolaevsk-na-Amure and had entered a prohibited area. The timber ship was saved due to the fact that the missile had been set for self-destruction 400 m (1,312 ft) short of the target. Only some fragments reached the ship. One crewmember was wounded and the timber ship called at the port of Holmsk for medical aid. The accident was followed by an investigation.

Naval aviation began training with the Tu-16K-16 (KSR-2) missile system in February 1963, the first units being 540 squadron the 33rd Naval Aviation Centre and 12 MRAP of the Baltic Fleet. In 1964 one squadron of 568 MRAP was retrained and in 1967 all of the Pacific

Fleet's 49 MRAD was converted. The first live launches were performed by the crews of 12 MRAD from 25 October to 23 November 1963, at a firing range in the Caspian Sea. The target was acquired for steady tracking at a distance of 200-210 km (124-130 miles), launch following two or three minutes later.

With the fielding of the new system, trials were undertaken by Centre no. 33 to decrease launch altitude to 2000 m (6,562 ft). An order of the Minister of Defence of the USSR dated 22 April 1968 approved an upgrade for the missile allowing its employment in the 500-10000 m (1,640-32,808 ft) altitude range. They were designated KSR-2M. Research flights indicated that the minimum launch range at an altitude of 500 m was 70-80 km (43-50 miles), limited by the radar visibility of the target and pre-launch preparation time. The Tu-16K-16 had only a short career in the AVMF, being overtaken by the Tu-16K-26 system and its KSR-5 missile.

Fielding the Tu-16 aircraft offered the possibility of large-scale inter-theatre manoeuvres. In 1966 seven Tu-16Ks of the Northern Fleet deployed to the Pacific Fleet along the Northern route, with two intermediate landings. The trip took 26 hours. The return journey required only one landing and lasted for 11 hours 30 minutes.

Mastering group air refuelling dramatically reduced flight time. The tankers were flown to airfields along the route beforehand. This allowed Tu-16K missile-carriers from the Northern Fleet to simulate short-notice strikes against appointed targets in the Sea of Japan, including live launches at the firing range. In the beginning of 1970 nine Tu-16K-10-26s from 143 MRAD simulated a nocturnal tactical strike against a group of ships in the Pacific Ocean, with air refuelling from five Tu-16Zs. Missile-carrier units occasionally carried out tactical air exercises two to three days long, using US and NATO ship formations for simulated tactical launches.

From 1966 theatre exercises also involved non-nuclear operations, and directives were issued requiring missile-carrier crews to rediscover bombing skills. In August 1967 943 MRAP of the Black Sea Fleet, along with forces from Bulgaria and Romania, took part in a theatre exercise involving the 'organisation and execution of landing operations aimed at destroying the enemy in a coastal area'. In one of the scenarios the Tu-16s delivered a strike dropping 72 FAB-250 bombs. Centre no. 33 also obtained positive results from using the Tu-16 to bomb ships.

In April 1970 the large-scale Okean manoeuvres were undertaken. 13 MRAP undertook six successive launches of missiles on the firing ranges. On 20 April 10 Tu-16Ks of the Pacific Fleet's 169 MRAP deployed to the Northern Fleet, making an intermediate landing at Olenya and launching K-10S missiles on the Kolsky peninsula firing range.

'Badgers' abroad

Following the earlier deployment of Tu-16Rs, at the end of 1970 a decision was taken to send an AVMF missile-carrier unit to the United Arab Republic. The squadron was named 89 Otdelnaya Morskaya Razvedovatelnaya Aviatsionnaya Eskadrila (OMRAE – independent naval

missile air squadron). It consisted of 10 Tu-16KSR-2-11s from 9 MRAP. The aircraft were painted in Egyptian camouflage with UAR national insignia. On 4 November the first aircraft landed at Asuan airfield. The other aircraft arrived in accordance with the schedule, and an An-12 brought in technical personnel. Four Tu-16SPS from the Baltic Fleet came to provide ECM support for the squadron.

The squadron was charged with the task of retraining Egyptian crews (who had been flying the Tu-16KS) for the Tu-16KSR-2 aircraft. The Egyptians' previous experience and high level of general education made the task easy. Combat training flights were carried out by mixed crews: aircraft commander, navigator and the 'aft' were Arabic, while the pilot on the right seat and the second navigator were AVMF naval aviation instructors.

Initially, training required interpreters, but subsequently communication turned into a mixture of Russian, English and Arabic, aided by hand signals. By June 1972 all 10 Arab

Above: A US Navy Tomcat keeps close watch on a pair of Tu-16R 'Badger-Fs' as they prowl around a US Navy carrier battle group. The initial Tu-16R variant, with SRS-1 Elint equipment, gave way to a new configuration which carried SRS-3 Elint antennas in wing-mounted pods. Similar pods were used later to house other intelligence-gathering equipment, including air samplers.

Above left: As part of the ongoing improvement of reconnaissance 'Badgers', some aircraft had the gun removed, SRS-4 Elint antennas fitted in two equally-sized radomes, and Rubin-1k radar installed. These were designated Tu-16RM-2, and were known to NATO as 'Badger-Ks'.

Above: A close-up of a Tu-16RM shows the port-side window which may have covered an oblique camera. As the Tu-16R family developed, optical sensors decreased in importance.

Left: In original guise (called 'Badger-E' by NATO) the Tu-16R retained the bomber's original RBP-4 radar. The main features were the two widely-spaced radomes for the SRS-1 Elint system.

Right: Operated by the AVMF's Northern and Pacific Fleets, 24 Tu-16RM-1s were produced by conversion of the Tu-16K-10. Here one overflies HMS Ark Royal at typically low level. Reconnaissance 'Badgers' aimed to approach ships under the radar to get the best intelligence.

Below: Primary mission equipment of the Tu-16RM-1 consisted of the two unequal-size SRS-1 Elint antennas found on early Tu-16Rs and a single SRS-4 in between them. The radar was 'tweaked' to give better performance in the reconnaissance role, and was known as the YeN-R.

Below and below right: Intercepted by the Swedish air force over the Baltic, this 'Badger' is almost certainly a Tu-16R of some sub-variant, but could also be an electronic warfare aircraft. While EW-dedicated aircraft carried Elint sensors like the Tu-16Rs, so Tu-16Rs carried ECM equipment and chaff dispensers, clouding the division between the two. The distinctive fairing above the flight deck is part of the A-326 Rogovitsa system, which was used for station-keeping in formation flying and to alert of impending collisions with other similarly-equipped aircraft. This equipment was installed primarily for missile-carriers flying carefully structured mass attacks, but was also fitted to reconnaissance and EW machines.

crews had been trained for live missile launches. The following month, orders for the withdrawal of the Soviet staff arrived. The Egyptians purchased the Tu-16KSR-2-11s, along with ammunition, technical material and equipment.

The last overseas 'tour' for the Tu-16 took place in 1980 when an independent mixed air regiment (169 OSAP) was formed at Cam Ranh airfield in Vietnam. The regiment included 16 Tu-16Ks and Tu-16Rs, which periodically flew along prescribed routes, accumulating valuable data about operating the aircraft in conditions of high humidity. Further Tu-16s deployed to the base in 1982 and 1988, arriving direct (using inflight refuelling) or via Sunan in North Korea.

In 1989 the conclusion that basing the Tu-16 in Vietnam had little practical value was drawn. On 28 August 1989 the Minister of Defence of the USSR announced that 169 OSAP would disband in 1990, leaving an independent mixed squadron (OSAE). The Tu-16s left in two groups: one group landed at Pyongyang airfield for refuelling and the other group proceeded to the airfields of the Pacific Fleet.

Reconnaissance aircraft

Long range and a tractable airframe made the 'Badger' an ideal platform for reconnaissance, and a dedicated variant was planned from the outset. The Tu-16R was built in accordance with a decree of 3 July 1953. Several design

bureaux were charged with developing equipment for it at the same time. Tu-16 no. 1880302 was additionally fitted with nine cameras for strip survey and mapping, and the SPS-1 jammer. State tests resulted in a decision to launch series production of the Tu-16R ('Badger-E'). In 1957 plant no. 1 at Kuibyshev manufactured 44 Tu-16Rs and 26 more with different sets of equipment the following year. In particular, the weapons bay accommodated a three-container automatic chaff-dispenser for ASO-16 and ASO-2b type chaff.

During production the equipment suite of the Tu-16R changed more than once. In the beginning, the aircraft was fitted with the SRS-1 and SRS-3 electronic intelligence stations, along with RBP-4 (RBP-6) radars. The SRS-1 was capable of detecting and identifying basic parameters of radars operating in the range of 10 to 500 cm (4 to 197 in). All operations were manual and a seventh crew member was required. He was called a special operator and occupied a suspended pressurised cabin, fitted with an ejection seat and the main units of the Elint suite, in the stores bay.

The SRS-3 automatic radio registered operations of 3- to 30-cm (1- to 12-in) band radars on a film that was decoded after flight. Cigar-shaped containers for the SRS-3 were attached to the pylons under the wings, leading to the NATO codename 'Badger-F'. Subsequently, the SRS-1 was replaced by the SRS-4 with teardrop-shaped radomes bigger than those of the SRS-1. The Tu-16R had several cameras with lenses of 20 to 100 cm (7.9 to 38 in) focal length. For night time photography the NAFA-MK-75 camera was carried.

In the late 1970s some Tu-16Rs were upgraded to Tu-16RM configuration. Besides higher resolution cameras, they were fitted with the SRS-4 Elint suite and the Rubin-1k (Rn-1k) radar. The SRS-1 and SRS-3 were removed. Flight radius of the Tu-16R with a take-off weight of 75800 kg (167,108 lb) reached 2500 km (1,553 miles). Inflight refuelling increased the figure by 1300-1400 km (808-870 miles).

Tu-16RM-2

In 1962 the Tu-16K-16 (KSR-2) missile system entered service with naval aviation and the Tu-16RM-2 was developed to support it in 1965, on the initiative of the AVMF. Using the Tu-16R as a basis, the nose gun, bombing equipment and suspended pressurised cabin were removed, and

a container with a 7000-litre (1,540-Imp gal) flexible tank was installed in the cargo compartment, thus increasing capacity of the fuel system to 51000 litres (11,219 Imp gal). The RBP-4 (RBP-6) radar was replaced by the Rn-1k, capable of detecting large ships at a distance of 200-240 km (124-149 miles).

A platform with the receiving antennas of the SRS-4 system was installed under the entrance door of the pressurised cabin, while omnidirectional sector antennas of the SRS-4 were attached to the doors of the cargo compartment. Navigation equipment was improved with the addition of the DISS-1 ground speed and angle of drift Doppler computer, while the RSB-70 transmitter was replaced by the R-836 Neon. Only two cameras remained. The navigator-operator was responsible for the SRS-4 in flight.

Combat radius of the Tu-16RM-2 grew by 700 km (435 miles) and reached 3200 km (1,988 miles) at a 79000-kg (174,162-lb) take-off weight. One inflight refuelling raised it to 4200 km (2,610 miles). Naval aviation plant retrofitted a total of 12 aircraft.

Tu-16RM-1 'Badger-D'

In such a way, the Tu-16RM-2 aircraft was capable of reconnoitring, providing guidance and target designation for the Tu-16KS, Tu-16K-16/KSR-2 and Tu-16K-11-16. But the number of those aircraft remaining in service with naval aviation continued to shrink, and most of the units were being retrained for the Tu-16K-10 with a missile launch range of 220 km (137 miles), subsequently increased to 325 km (202 miles). The energy of the Tu-16K's YeN radar also grew, providing detection of large ships from as far as 450 km (280 miles).

Consequently, once again a disproportion arose between the capabilities of the Tu-16K and Tu-16RM-2 radars to detect seaborne targets, and the ability to provide target designation for missile systems became apparent. Thus, there was a need for an aircraft able to detect surface targets and offer target designation for employment of the Tu-16K-10SD and Tu-16K-26 missile systems. These reasons brought about a decision to develop a reconnaissance aircraft on the basis of the missile carrier.

The resultant Tu-16RM-1 was converted from the Tu-16K at a naval overhaul plant. All missile-related equipment was removed and an additional container with a flexible tank, like the one in the Tu-16RM-2, was installed in the cargo compartment. The capacity of the fuel system increased to 48000 litres (10,559 Imp gal). Two SRS-1M radomes were installed in the area of the stores bay, with one SRS-4

radome situated between them. The navigator-operator, located in the suspended pressurised cabin, controlled their operation.

Elements of the YeN-R search radar were fitted in the forward compartment at the navigator's station. The radar, with a pulse power of about 180 kW, was capable of detecting large surface ships at a distance of up to 480 km (298 miles). It was equipped to determine the largest ship for targeting prioritisation. The unrefuelled flight radius reached 3200 km (1,988 miles) and flight duration increased to 7 hours 45 minutes. The crew consisted of six, including the navigator who also operated the YeN-R. In 1966-1967 the naval overhaul plants converted 24 Tu-16K-10 aircraft into Tu-16RM-1s. They entered service with the Northern and Pacific Fleets.

A total of 75 Tu-16Rs was produced, shared by the AVMF and DA. Ongoing modifications and new systems dramatically altered the look of the aircraft throughout their long careers – none more so than the adoption of SPS-100 Rezeda jammers which replaced the tail gun turret of DA Tu-16Rs in the latter part of their service lives.

Above: From 1968 six AVMF Tu-16Rs operated from Cairo-West wearing spurious UARAF markings. The aircraft and crews were drawn from the Northern Fleet's 967 ODRAP but were detached to the specially formed 90 ODRAE for the deployment, which was later bolstered by Tu-16SPS jammers. Here an Egyptian-marked Tu-16R is seen in Russia.

Above left: The Egyptian detachment was a constant thorn in the side of the US Navy's Sixth Fleet in the Mediterranean. This Tu-16R with SRS-1 ('Badger-E') passes an 'Essex'-class boat. This may not be an AVMF aircraft, for in October 1967 Egypt had acquired two Tu-16Rs (with SRS-1) of its own.

Left: Intercepted by a US Navy fighter over the eastern Mediterranean, this 90 ODRAE Tu-16R carries SRS-3 Elint pods under the wings. The blade antenna on the aircraft's spine just behind the wing trailing edge was one of the antennas for the SRS-2 system. A similar antenna was carried under the fuselage in a position slightly aft of the dorsal aerial.

Radiological reconnaissance

Under a decree issued on 22 November 1967, the Tu-16RR was developed, designed to collect airborne samples of radioactivity. In 1969 Tu-16R no. 1883305 was fitted with two filter pods on the wing pods instead of the SRS-3 Elint pods, and dosimeters. Early in 1970 eight Tu-16Rs were converted into Tu-16RRs.

Another variant was the Tu-16RC, which was fitted with the MRSC-1 Uspekh airborne equipment suite to provide target designation for the missile systems of ships and submarines. Three aircraft were used to refine individual elements of the Uspekh system.

Reconnaissance 'Badgers' in service

Deployment of the Tu-16 boosted the capabilities of the DA and allowed the AVMF to reach wide expanses of ocean. The Tu-16 was to be arguably most important as a missile-carrier, but its successful employment in this role greatly depended on the reconnaissance aircraft, whose crews were characterised by high professionalism.

Tu-16Rs entered service with independent long-range reconnaissance air regiments (ODRAP). In the DA every corps had an ODRAP attached. In the AVMF the first Tu-16Rs were delivered to 967 ODRAP of the Northern Fleet aviation at the end of 1957, and to 50 ODRAP in the Pacific Fleet aviation in October 1958. In 1960 one squadron of 30 ODRAP of the Black Sea Fleet was retrained for the Tu-16R.

Tu-16Rs were delivered to the regiments that formerly operated the Il-28R, which made mastering equipment easier. In fact, AVMF Tu-16Rs were widely used in the surveillance of the fleets of neighboring countries almost from the very beginning of conversion, flying in pairs. The range/duration increase made possible through inflight refuelling became very important.

In the summer of 1961 the crew of the Tu-16R commanded by Major A.V. Uzlov, commander of 967 ODRAP, flew to the North Pole with two refuellings. Flight duration was 11 hours 48 minutes. With air refuelling, the crews of the Tu-16R from 317 independent mixed air regiment (OSAP – Otdelnyi Smeshannyi Aviatsionnyi Polk) flew from Yelizovo airfield (Kamchatka) and operated in the Pacific beyond the islands of Japan. In 1962 the crew of Colonel I.S. Pirozhenko, commander of 317 OSAP, flew a Tu-16R to Midway Island in the Pacific and back with two refuellings. Only those who understand that the crew had no alternate airfield can estimate the impact of this flight.

Crews were quite creative in developing techniques for the reconnaissance of carrier battle groups to discover their composition. The first contacts with the CVBG usually came with the interception of the Tu-16 by shipborne fighters, long before they managed to reach visual contact with the ships. To counter this, the Tu-16R crews descended to very low level after detecting the ship's radar, approaching under the radar lobe and often directly overflying the battle group.

Quite often AVMF and DA units worked together on reconnaissance missions. The following is an example, in which Northern Fleet aviation took initial information and, using its own assets, carried out additional reconnaissance, provided target designation and simulated a strike delivered by a regiment of missile-carrying aircraft against NATO ships participating in the Folex-64 manoeuvres.

The events unfolded in the following sequence: before entering the Norwegian Sea the group of NATO ships was under the surveillance of a Tu-95 'Bear'. Then a

Tu-16RM-1 from 967 ODRAP took over. The crew detected the ship's radars using Elint sensors and relayed positional information to a Tu-16R. At a distance of 450-460 km (280-286 miles) from the ships the crew of the Tu-16RM-1 established contact with the aircraft-carrier group using the YeN-R radar. Having ascertained the initial data, this was passed to the leaders of tactical (additional reconnaissance and strike) groups to set up the specified directions and appointed time of an attack. The crew of the Tu-16RM-1 simultaneously guided five Tu-16Rs and two groups of Tu-16Ks to the ships. The missile-carrying aircraft simulated tactical launches.

Tu-16Rs abroad

Following the mauling received by the Arab side in the 1967 Six-Day War, interest in Soviet aid increased. At the time the Soviet navy's 5th squadron of ships was in the Mediterranean. Its headquarters was housed in mother ships absolutely inadequate for command purposes, or on the command and control cruiser *Zhdanov*, which swarmed with rats and cockroaches. The squadron HQ encountered great difficulties in obtaining intelligence about the maritime situation.

In March 1968 the governments of the USSR and United Arabic Republic (Egypt) signed an agreement which would allow the deployment of an air group of six AVMF Tu-16Rs to conduct reconnaissance in the interests of both countries. The strength of the unit, named the 90th special-purpose separate long-range reconnaissance squadron (90 ODRAE) was determined as 170 men. The Tu-16Rs were drawn from the 967 ODRAP and were given UAR national insignia, the location being referred to as 'Object 015'. The 'Badgers' flew to 'Object 015' via Hungary and Yugoslavia, observing all international rules.

Cairo-West airfield was to become the base for the squadron. Flying and technical officers were accommodated in well-furnished cottages with conveniences that could only be dreamt of back in the Motherland. A systematic series of reconnaissance flights obtained a wealth of information about the activities of ships in the Mediterranean. Later, a detachment of Tu-16SPS ECM aircraft from the Baltic Fleet beefed up the group.

Electronic warfare aircraft

Electronic countermeasures provides both active and passive interference affecting enemy radioelectronic assets. A considerable number of Tu-16s were produced to perform this role, in several variants.

Some Tu-16P Buket aircraft featured extended wingtips. These may have been additional antennas, but more likely indicate that the aircraft were converted from tankers, which have the extended tips added as part of the tanker conversion process.

Known to NATO as the 'Badger-H', the Tu-16 Yolka was primarily a chaff-dispensing platform, equipped with a bellyful of chaff cutters and dispensing chutes. As with the reconnaissance aircraft, the EW family was reconfigured and upgraded several times, and Yolka aircraft were fitted with active jamming equipment to complement the passive chaff. The aircraft right has a large stick antenna aft of the stores bay, while the aircraft above has a small radome.

Officially designated Tu-16Gs (Grazhdanskiy – civil), but publicly known as Tu-104Gs, Aeroflot had a small number of disarmed 'Badgers' for crew training pending the arrival of the Tu-104 airliner. Nicknamed 'Krasnaya Shapochka' ('Little Red Riding Hood'), the aircraft also flew newspaper matrices (and other high-priority items) around the country until airliners assumed the role.

Below: Among the many Tu-16LL test aircraft which have served with the LII at Zhukovskiy, this former Tu-16K-10 was one of the most unusual, fitted with the nose section from the Myasishchev M-17 'Mystic'.

The Tu-16SPS was designed to create active noise interference of particular frequencies for jamming hostile radars. In 1955-1957 Kuibyshev built 42 Tu-16s equipped with SPS-1 jammers, and 102 Tu-16s with SPS-2 jammers. The first equipment created a 60-120 W interference in the 30-200 cm (12-79 in) band, while the SPS-2 produced 250-300 W interference in the 9.5-12.5 cm (3.7-5 in) band. The operator tuned to the frequency of the target radar using data provided by the SRS-1BV and SRS-1D high-precision radio reconnaissance systems. The SPS-1 and SPS-2 could jam only the radars selected by the operator. Tuning took 2-3 minutes.

In the Tu-16SPS the operator sat in a removable pressurised cabin in the rear part of the cargo compartment. Two SPS-2 jammer antennas under teardrop-shaped radomes were installed in the bottom part of the fuselage ahead and aft of the stores bay. Whip antennas of the SPS-1 jammer were mounted on top of the fuselage behind the operator's blister or in the bottom part of the fuselage forward of the cargo compartment. Subsequently, the Tu-16SPS was fitted with three ASO-16 chaff-laying systems, whose presence was indicated by dispensers in the bay doors. From 1962 the SPS-1 and SPS-2 jammers began to be replaced by the Buket active jamming system.

Tu-16P Buket

This EW variant served the same purpose as the Tu-16SPS, but was equipped with the Buket system developed in the second half of the 1950s. This included SPS-22, SPS-33, SPS-44 and SPS-55 jammers, whose transmitters covered a wide frequency range, operated in automatic mode and were able to simultaneously jam several radars. Operation of the Buket equipment was powered by four additional PO-6000 AC converters and one PT-6000. The system was installed in the weapons bay, whose doors were replaced by a platform with units equipped with pressurisation and cooling systems. In its bottom part along the axis of the aircraft there was a long canoe radome covering the antennas.

Beginning in 1962, over 90 Tu-16s were equipped with the Buket suite. Ten years later, when it became clear that the Buket interfered with the aircraft's own operations, the radiation sector was narrowed.

Several Tu-16Ps were equipped with the Fikus system with five rotating directional antennas. They were installed under the fuselage and were covered by a radome. The equipment was tested on Tu-16P nos 1882409 and 1883117.

Under a decree dated 21 July 1959 the Tu-16P was developed with the RPZ-59 system. This used a K-5 air-to-air missile body to form the basis of a passive electronic countermeasures system. Six launchers in the cargo compartment fired missiles to create a cloud of chaff in front of the aircraft. Tests ran till 1964, and for safety reasons 12 launchers were placed under the wing. This led to the Pilon system which comprised 12 RPZ-59 missiles under the wing of a Tu-16P Buket. The system was not developed any further.

Yolka chaff-layers

Developed at the same time as the Tu-16SPS, the Tu-16 Yolka (fir tree) aircraft were fitted with an active electronic

M-16 Mishen drone

As 'Badger' airframes reached the end of their fatigue lives, many were converted to serve as full-scale targets for missile tests and practice launches. The Kazan factory performed the first conversions, which were known as the M-16 Mishen (target), or Tu-16M. All EW and navigation equipment was removed, and radio guidance equipment was installed. The first became operational in April 1965. From the early 1980s, as Tu-16s were retired in large numbers, conversions were performed by the VVS's 12 ARZ at Khabarovsk.

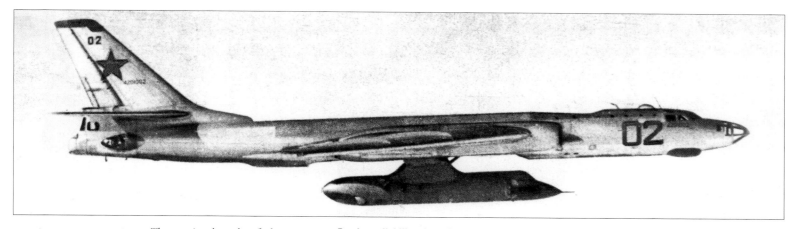

countermeasures system. The entire length of the stores bay housed seven ASO-16 chaff dispensers, with ejectors in the bay doors. Also, the aircraft were fitted with the SPS-4 Modulyatsiya jammer. The chaff dispenser used different types of chaff, covering most bands apart from 75-cm waveband radars. With a one-second dispensing interval, one load of ASO-16 was sufficient for 15-20 minutes operation (depending on the type of chaff). As well as new-build aircraft, some Tu-16Ts were modified to Yolka standard. In 1957-1958 naval aviation received 71 Yolka aircraft. They were modernised more than once, and carried both passive and active electronic countermeasures systems.

Some Tu-16 Yolkas were fitted with the SPS-61, 62, 63, 64, 65 and 66 jamming sets, which shared the common name of Azaliya. Aircraft carrying SPS-61, 62 and 63 were additionally equipped with SPS-6 Los, while those with SPS-64, 65 and 66 had SPS-5 Fasol. The antenna for Azaliya was located in the forward section of the stores bay.

Other equipment was also fitted to the jamming aircraft. At the end of the 1970s the Tu-16Ps were fitted with SPS-151, 152 and 153 jammers from the Siren series. With the fielding of heat-seeking missiles, the Tu-16P started to carry ASO-2 and ASO-7ER thermal noise equipment (IR flares), with dispensers located in the landing gear nacelles and rear fuselage fairings. There were two Tu-16Es designed for photographic, electronic, radiological and chemical reconnaissance, as well as ECM duties. They were equipped with the SPS-5 and SPS-151 jammers, and carried up to four containers of ASO-16.

Non-military Tu-16s

Tu-16s were widely used for purposes other than their originally intended ones. Many were converted for trials duties. Following are descriptions of the more important programmes.
■ In the 1950s an aircraft designated as the Tu-16KP was fitted with equipment for monitoring and adjusting the flight trajectory of missiles designed by S. Lavochkin. Equipment was housed in a suspended pressurised cabin as on the Tu-16R. In 1960 all work was stopped.

Ready availability of surplus aircraft and a capacious internal bay made the Tu-16 a sought-after trials platform, especially for powerplant trials. At least nine Tu-16LLs have been used for this task, carrying the test engine on a pylon which can be extended in mid-air (below). The most unusual of these platforms was the aircraft used to carry a complete Yak-38 fuselage section (above) to test the complicated propulsion system of the V/STOL fighter.

Above: H-6s take shape in the Xian factory, which has produced at least 150 'Badger' copies. The first wholly-Chinese aircraft was first flown by Xu Wenhong on 24 December 1968. The aircraft in the foreground is an H-6D missile-carrier.

Left: As well as the baseline H-6 and H-6A nuclear bomber, a HZ-6 reconnaissance version was developed. This has wing and fuselage fairings similar to those fitted to Tu-16Rs.

During the course of its production, a number of improvements have been made to the basic H-6A, including the fitment of extended, rounded wingtips, as displayed by this aircraft. In PLAAF service the type equips several bomber regiments, which form part of bomber divisions that also operate the Harbin H-5 (Il-28 'Beagle'). Regiments are believed to be based at Shijiazhuang in the Hebei Military District, Datong (Shanxi MD), Xian and Wugong (Shaanxi MD), Nanchang (Jiangxi MD), Harbin (Heilongjiang MD) and one base in Guangdong MD. As well as free-fall bombers, the PLAAF is believed to have missile-carriers in service, possibly armed with the ramjet-powered YJ-16 missile. H-6s also perform reconnaissance and EW tasks.

Above: H-6s conduct free-fall bombing. An unusual role for the type is breaking up the ice dams which form annually on the Yellow River.

Below: At least one H-6 serves as an engine testbed, using the same pylon arrangement as the Tu-16LL.

■ In the beginning of 1960 two Tu-16Ks were retrofitted for filming, and for a long time were used for monitoring the course of flight tests.

■ The Tu-16G and Tu-104G were variants of the Tu-16 outfitted for fast delivery of mail and matrices of central newspapers. With the arrival of airliners the aircraft were returned to the military.

■ The Tu-16 Tsyklon-N was used for weather modification and research, including the study of thermodynamic properties of the atmosphere. Naval aviation allocated two Tu-16KSR-2-5 aircraft, nos 6203203 and 6203208, which were modified by overhaul plant no. 20 at Pushkin in 1977. The aircraft were fitted with cassette holders for firing silver iodide, or containers with cement, for cloud-seeding. The external racks were also expected to carry containers with reagents. The crew included two meteorologists. The tests ran as long as 1980 and the aircraft, wearing Aeroflot colours, were handed over to the Air Force at Chkalovskaya airfield. They were used for rain-making over Moscow during the Olympic games of 1980. In 1986 the aircraft were used for dissipation of radioactive clouds in the wake of the Chernobyl disaster.

In November of the same year it was decided to use the aircraft in international programmes and to turn them into meteorological laboratories. In early 1990, during the course of modification at Pushkin, no. 6203203 was fitted with equipment for flying international airways. Tests were not completed, and the aircraft were later phased out as their service lives expired.

■ In 1970 two Tu-16Ns were modified for spraying carbonic acid. They flew from Chkalovskaya airfield.

■ The Tu-16AFS was retrofitted from a Tu-16 in 1970 for aerial photography of the Baykalo-Amurskaya trunk railway. Several photo cameras were located in the cargo compartment.

■ The Tu-16LLs were flying laboratories. They were converted in 1954 and were used for testing jet engines in an underslung pod. From 1957 for two years Tu-16 no. 1881808 was tested with engines featuring thrust reversal, which shortened the landing run by 30-35 percent. One of the aircraft was used for adjustment of the RD36-51V high-altitude engine. More recently, a Tu-16LL has flight-tested the GTRE GTX-35VS Kaveri engine developed for India's LCA fighter.

Tu-16s for export

People's Republic of China

Four years after the DA and AVMF received the Tu-16, it was decided to export the type. An agreement with the People's Republic of China for the manufacturing of the Tu-16 was signed in early 1956, with production entrusted to Harbin. The first aircraft was assembled from Soviet-made components and first flew on 27 September 1959. Designated H-6, it was handed over to the People's Liberation Army Air Force in December. At the same time a second aircraft was being assembled.

In 1958 production of the H-6 was assigned to a new aircraft plant at Xian. Aircraft assembled at that plant from Soviet components were built as H-6A nuclear bombers. On 14 May 1966 an H-6A dropped the third Chinese nuclear bomb. On 24 December 1968 the first H-6A made entirely from Chinese-built parts took to the air. Further versions were developed, including an inflight-refuellable reconnaissance platform and a tanker. The H-6I was a one-off trials aircraft powered by four Rolls-Royce Speys – two in the standard engine trunks and two in pods slung under the wings. In 1975 development of the H-6D anti-ship aircraft was launched. It was armed with two YJ-6 radar-guided missiles, based on the ship-borne P-15. The first flight was made on 29 August 1981. The H-6D entered service in 1985, and was later exported to Egypt and Iraq.

Egypt

In 1965-1966 Oktyabrskoe airfield of the Black Sea Fleet provided training for a group of flying personnel from the UAR, followed by deliveries of different versions of the Tu-16 to Egypt. For a year from 5 July 1966, Egypt hosted a group of 19 Soviet pilots, engineers and technicians, whose task was to train the flight crew and technical personnel of the UAR Air Force and pass on practical experience of servicing and operating the Tu-16KS. Six Tu-16KS crews were to be formed. By June 1966 four pilots, three navigators, as many navigators-operators and two gunner/radio operators had completed training in the USSR.

Preliminary training took five months and only by 12 December 1966 did training and test flights begin. The Soviet group performed its mission: by March 1967 three crews had been trained for live missile launches in both day and night, and the other three by April. During the Arab-Israeli war of 1967 most of the Tu-16s were lost on the ground.

A decree of the Government issued 24 January 1966 ordered the reallocation of Tu-16Ts from the Black Sea fleet to Egypt. On 26 September 1967 six Tu-16Ts were flown by the Black Sea Fleet crews to 'object 015' (as Egypt was codenamed) via Hungary and Yugoslavia. By 6 October instructor pilots had performed several training and test flights from Beni-Sueif, and subsequently returned to the Soviet Union. Further deliveries comprised two Tu-16Rs, which were flown to the UAR in early October 1967. They were equipped with SRS-1 Elint, RBP-4, ASO-26 and Sirena-2. The cost of each was estimated at 802,508 rubles. As recounted earlier, the Egyptian Air Force subsequently received ex-AVMF Tu-16KSR-2-11 aircraft.

During the October 1973 war the Tu-16s were based beyond the reach of Israeli fighter-bombers. Reports suggest that 25 missiles were launched with five of them allegedly reaching their targets, hitting two radars and a supply unit. After military co-operation with the Soviet Union ended, Egypt turned to China for spares, and additional H-6 aircraft. In July 1977, during a four-day conflict

Above and left: The most important version to be developed by the Chinese is the H-6D missile-carrier. It has maritime search/attack radar in an enlarged chin radome, and carries two YJ-6 missiles (export designation C-601, reporting name CAS-1 'Kraken').

with Libya, the Egyptian Air Force delivered strikes against a number of targets, including Tobruk, El Adem and Al-Kufr. Allegedly, two radars were destroyed. By early 1990 the Egyptian Air Force had 16 Tu-16s remaining: by 2000 they had been retired.

Indonesia

In 1961 crews of the Black Sea Fleet ferried 25 Tu-16KS 'Badger-Bs' to Indonesia and trained flight crews. Two squadrons were formed at an airfield near Jakarta. The AVMF group was headed by Lieutenant Colonel Dervoed, and members of the group proudly wore badges to show that they had crossed the Equator. After relations with Indonesia were broken off in 1965, the aircraft languished in a semi-serviceable state through a lack of spares, and were soon scrapped.

Iraq

By the beginning of the 1967 Six-Day War, Iraq had received eight Tu-16s, based at Habbaniyah, although they did not participate in active hostilities. Six Tu-16KSR-2-11s

Known as the HY-6, Xian has developed a tanker version with wing-mounted hose-drogue units (here refuelling a pair of J-8IIDs). Based on the standard bomber, the HY-6 is intended purely for the tanker role, and lacks the glazed bombardier's nose and armament.

The most successful of the Tu-16s operated by Egypt were the Tu-16KSR-2-11 'Badger-Gs' which, armed with KSR-2 missiles, were used in anger against Israel during the 1973 war, and in a brief border conflict with Libya. The aircraft had Ritsa antennas above the nose to assist with launching the anti-radar KSR-11.

Above: Egypt's 'Badger' force survived until 1999/2000, easily long enough to participate in regular exercises held jointly with US forces. Here two Tu-16KSR-2-11s taxi for a mission from their Cairo-West base. The survivors of the original batch of 10 Tu-16KSR-2-11s (plus survivors from the other variants delivered – Tu-16R, Tu-16T and Tu-16KS) were augmented by deliveries of a few H-6s from China.

Right and below: Indonesia's KS-armed 'Badgers' had the shortest career of any user, and were not used in anger, despite the confrontation with the UK in Borneo. They swiftly fell into disuse after the end of Soviet aid in 1965.

were delivered in 1970 and, again, they did not participate in the 1973 war. However, in 1974 they were used in Kurdistan. By the beginning of the Iran-Iraq war in 1980 the Iraqi AF had eight Tu-16s, which were used for delivering strikes against the airport in Tehran and other objectives. In 1987 Iraq bought four H-6Ds and C-601 missiles from China. By 1991 the Iraqi fleet was no longer operational, the majority having been destroyed on the ground during Desert Storm. At least three were destroyed on 18 January by F-117s at Al Taqaddum.

Twilight of the 'Badger'

For 37 years dozens of modifications maintained the strike potential of the Tu-16 in both DA and AVMF service. From the early 1970s the supersonic Tu-22M 'Backfire' began to replace the Tu-16, but the process was very protracted.

At the end of 1981 long-range aviation had 487 Tu-16s in service, while naval aviation had 350. By 1988 the combat component of the naval aviation included 14 missile-carrier regiments, and six of them were equipped with 212 Tu-16Ks. The grounding of the Tu-22M in 1991 because of design-manufacturing troubles saw the Tu-16 brought back to those units affected, since it was necessary to maintain operational skills. However, that year was to be the starting point of a wholesale retirement of the 'Badger'.

By 1980 106 Tu-16s, comprising 72 DA aircraft and 34 from the AVMF, had been lost for varying reasons. Most losses were attributed to human error. Naval aviation claimed an accident rate of 2.4 per 100,000 flying hours. Reliability of the aircraft was confirmed by the fact that, for a period of 15 years, only two breakdowns were caused by design-manufacturing troubles.

On the eve of the collapse of the USSR, the European sector of the country housed 173 Tu-16s, while a further 60 were deployed in the Far East. After the Belovezhsky deal, 121 aircraft were passed into Ukrainian possession, and 18 went to Belarus. Russia's share counted 34 aircraft in Europe. By 1994 the Tu-16 had been withdrawn from combat units, although the AVMF retained 62, of which 53 were Tu-16Ks held in reserve.

Lt Col Anatoliy Artemyev (Soviet/Russian Navy, Retired)

Top: Some of the 18 'Badgers' – mostly Tu-16K-26s – that were acquired by Belarus in 1991 are seen at Bobruysk air base. All were preserved in a flyable condition in the hope that they could be sold to various aviation museums. No aircraft were bought and all were scrapped.

Above: The Tu-16 had a long service life, its front-line career in the missile role being prolonged by delays encountered by the Tu-22M. This is a Tu-16K-10.

Left: This Tu-16 guards the entrance to Dyagilevo air base at Ryazan. This airfield is the home of the DA's long-range combat training centre, and had Tu-16s assigned throughout the type's service life.

Before Barbarossa

Early in 1941 Hitler's Third Reich had reached its zenith, with German forces in control of the majority of western Europe and engaged in fighting in North Africa. While having been unable to subdue the RAF over Britain, the Luftwaffe was still able to prove itself in operations over North Africa, the Balkans, Greece and the northern Mediterranean. This photo spread is a snap-shot of the Luftwaffe around the Mediterranean prior to the watershed German invasion of the Soviet Union.

Above: Gathered in a featureless corner of the western desert, a gaggle of Junkers Ju 52/3ms sits on the hard sand of a temporary airstrip in North Africa. Some display yellow markings around the cowlings and on the rudder, a marking assigned to Balkan theatre aircraft. A single Bf 110D-3 ('3U+NS' of 8. Staffel, ZG 26 'Horst Wessel') with its engines running sits in front of a fuel dump, the type being used to protect the lumbering transports from any marauding Allied fighters.

Left: The pilot of a 7. Staffel, ZG 26 Bf 110C adjusts his flying helmet, having handed his pith hat to an adjunct standing on the wing. Under the front of the cockpit the aircraft is adorned with a penguin holding an umbrella with a pair of white arrows behind it. Another Bf 110C (right), this time with the black cat with a high-arched back on a red arrow of 8./ZG.26 on the nose, forms the back drop to a group of Luftwaffe air and ground crew.

Below: Kicking up a dust storm, a Bf 110 taxis in the desert. While the Zestörer had fared badly in the fighting over Britain, it was able to prove a useful addition to the campaign over North Africa. Having two engines proved to be a great advantage in the harsh desert environment.

Above: *Bf 110D-3 '3U+NS' of 8. Staffel, ZG 26 (also featured in the opening photo) flies protectively between the camera ship and a trio of Ju 52/3ms flying low over arid land. The Deutsches Afrikakorps existed at the end of a long supply chain and the Ju 52/3ms provided a means of rapidly re-supplying the troops in the field, but were vulnerable if caught by Allied fighters. Transports flying to North Africa had to run the gauntlet of Allied fighters based on the British fortress of Malta and along Allied-held North Africa itself. Indeed, while attempting to resupply the Deutsches Afrikakorps in April 1943, no fewer than 432 transports, mostly Ju 52/3ms, were lost within three weeks.*

Above left: *Ju 52/3ms sit on an airfield on Sicily under the shadow of Mount Etna. The aircraft in the background carry single digit numbers on their rudders.*

Left: *A BMW 132 engine is manoeuvred through the freight door of a Ju 52/3m. While being the standard transport for the Luftwaffe throughout the war, the Ju 52's small freight door limited what it could carry.*

Below: *Groundcrew lug a Rb 50/30 camera to a Bf 110C-5, the camera port visible between the wings. This C-5 is unusual because it has two nose MG FF cannons holes reportedly deleted from this sub-type.*

Operation Merkur - Crete 1941

Having been forced out of Greece, the Allies congregated on Crete in the knowledge that British sea power would make any form of sea invasion costly for the Axis forces. Thus on 25 April 1941 Hitler signed Directive No.28 for Operation Merkur, an airborne invasion of Crete. Around 150 Stukas of Fliegerkorps VIII were available for softening up attacks on Crete, based at Argos, Corinth, Mulaoi, Melos and Scarpanto in Greece. The Stukas also played a vital role attacking British shipping around Crete, helping to sink four cruisers, six destroyers and many transports, as well as damaging the aircraft-carrier HMS *Formidable*. The invasion itself was undertaken by Fliegerkorps XI, the airborne, para and glider troops arm of the Luftwaffe. In fact it had been the AOC of Fliegerkorps XI, Generaloberst Kurt Student, who had gained Hilter's permission for the assault. Using 493 Junkers Ju 52/3ms as jump-ships and about 80 DFS 230 assault gliders capable of carrying eight troops, the Germans intended to land in three waves on Crete, but clouds of dust, delays and collisions disrupted the plan, spreading the German troops far and wide. This greatly increased the price they paid for Crete: more than 7,000 men were killed and 174 Ju 52/3ms lost, greatly reducing the numbers available two months later for the invasion of the Soviet Union.

Above: Ju 87B-2s, including examples from 7. Staffel, StG 77 return to base devoid of bombs. The Stukas display the yellow nose and tail markings that were the theatre markings for the Balkans, with at least two variants of the nose markings displayed. Unusually for a Luftwaffe type, some Stukas carried an identification letter on the upper wing as displayed by 'J' (left) seen on a Greece strip in April 1941, with a 2. Staffel example in the background awaiting the loading of bombs and fuel. Operating alongside the Ju 87s in the anti-shipping role around Crete until June 1941 were KG 2's Dornier Do 17Zs (below).

Left: The roving nature of desert warfare precluded all but the most basic of equipment being taken along with the advancing (or retreating) Luftwaffe formations. Most maintenance had to be done in the open air, in temperatures that could range from the baking heat of midday to just above freezing during the night. Dust greatly reduced the reliability and life of aircraft engines, meaning that they had to be replaced in the field more often than in most other theatres of operations. These Luftwaffe maintenance personnel have built a tripod from which to hang a winch and pulley system next to the Daimler Benz DB601 powerplant of a Bf 110. Aware that an official photographer was taking their picture all three are smartly dressed in the black jackets and caps from which the nickname 'blackmen' arose. In the background are two StG 1 Ju 87s, with a cache of fuel drums near one of the aircraft. The nearest Ju 87 is a R model, with long-range fuel tanks under the wings.

Right: A Bf 110 overflies a patch of green provided by palm trees. The aircraft has the supplementary oil tank fitted to the D-3 subtype visible on the fuselage by the trailing edge of the wing, and the extended rear fuselage past the tailplane containing a life raft, but is not carrying the external fuel tanks associated with that version.

Far right: As well as the flying units the Luftwaffe's anti-aircraft organisation played a role in the desert campaign, providing short-range defence against attacking Allied aircraft. Behind this hasty AAA emplacement a Bf 110 creates its own dust storm.

Above: Posing for the camera ship, a pair of Bf 109E-4/Trops of I./JG 27 is seen somewhere over the Western Desert, wearing the camouflage scheme developed for operations in Libya. From above and at low level the scheme proved to be remarkably effective, especially nearer the coast where there were more shrubs, broken only by the Balkenkreuze on the upper wings and the white band on the rear fuselage, a theatre marking for the Mediterranean area. The Bf 109E reached North Africa in April 1941, largely proving superior to the Royal Air Force Hurricanes and Kittyhawks ranged against it.

Right: Operations in the desert were hard on men and machines and tactics had to be developed for the simplest of operations. Spacing take-off runs was necessary because of the large amount of dust kicked up by any movement on the desert surface.

Westland Wyvern

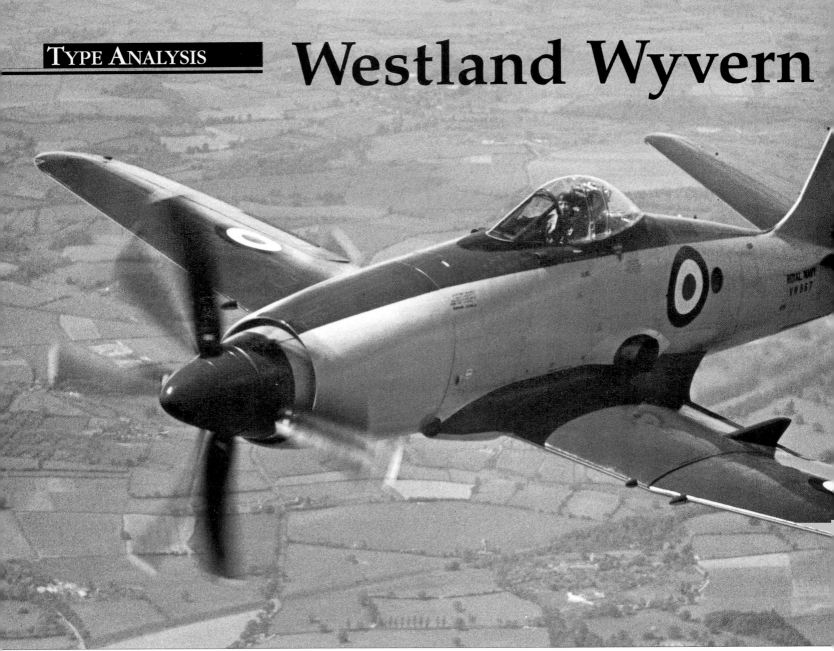

As the Fleet Air Arm's first and last turboprop strike aircraft, the Westland Wyvern saw service with the Royal Navy between 1953 and 1958. Its introduction into service provided the FAA with valuable experience, and could have represented a major step in the transition from piston-engined to jet-powered aircraft. Unfortunately, too many years were spent in development and the Wyvern entered service more than six years after the first prototype had flown. Moreover, the performance of the turboprop version was less than that of the early piston-engined Wyvern.

The Python-powered Wyvern certainly looked the part, especially in its initial TF.Mk 2 version (before the engine cowling was cut back). However, performance lagged way behind that of contemporary jet aircraft, and development of a turboprop-powered carrierborne attack aircraft proved to be a dead-end.

First thoughts on the aircraft which eventually became the Wyvern can be traced to early 1944 when W.E.W. 'Teddy' Petter, Westland's technical director, started work on the design of a single-seat long-range fighter capable of carrying a torpedo. In the initial design, the engine was behind the cockpit compartment and needed an extension shaft to drive the airscrew. Although this configuration provided exceptional forward view for the pilot, it presented several drawbacks, including weight penalty and complexity. A more conventional engine installation was finally preferred as Westland technicians were aware that their aircraft was much larger in size with greater weight than contemporary fighters.

Initially designated P.10, then W.34, the Westland project had been intended to be powered by a Rolls-Royce Eagle engine, although the Griffon, an unidentified Rolls-Royce powerplant and the Napier Sabre were also considered. Development of the Eagle, Britain's most powerful piston engine, had begun in March 1943 and it had run on the test bench in February 1944. The version selected for the W.34 was the Eagle 22 that was fitted with contra-rotating airscrews. It was a 24-cylinder flat-H sleeve valve engine intended to produce a maximum power of 3,500 hp (2611 kW) for a dry weight of 3,900 lb (1769 kg).

Interest and backing for the W.34 were expressed by Navy officials and, in the course of 1944, Specification N.11/44 was issued. As the specification was written around the Westland proposal, it was not put out to competitive tender. The general lay-out of the W.34 – completed by A. Davenport, the techni-

A salvo of eight 60-lb R/Ps is fired from a No. 830 Squadron Wyvern during firing trials off Malta. This was the standard rocket load, although it could be increased to 16 by the use of double-tier launchers.

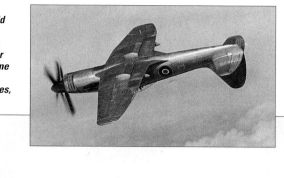

The prototype Westland W.34 Wyvern Mk I was first flown from Boscombe Down rather than the short field at Westland's Yeovil plant. The aircraft had no operational equipment or armament, and was used to assess aerodynamic qualities, during which a number of shortcomings were revealed. These resulted in some changes to subsequent aircraft. The final four prototypes were completed with full navalised features, such as folding wings and arrester hooks.

and issued Specification F.13/44 for a long-range escort fighter. When Westland received an order for six prototypes in November 1944, it took consideration of the RAF interest and included one aerodynamic test machine, three prototypes for the RN and two for the RAF. Nevertheless, the RAF order was cancelled on 7 December 1945 following the end of the war in the Pacific and the RAF's decision of introducing only jet fighters into service. On its part, the Navy was still reluctant to bring jets into service aboard its carriers. The new powerplant had yet to demonstrate its suitability for carrier operations, it had high fuel consumption, and required deck adaptation.

Eagle-powered prototype batch

Thus, the six prototypes previously ordered were all completed as naval W.34s, which were later named and designated Wyvern Mk Is. The first one, serialled TS371, was completed at the Yeovil factory in October 1946 and transferred to Boscombe Down, home of the Aeroplane and Armament Experimental Establishment (A&AEE), for its first flight to take advantage of its long runway compared to the small grass field at the Westland plant. TS371 made its maiden flight on 16 December 1946 with

Harald Penrose, Westland's chief test pilot, at the controls. The Wyvern was the first British aircraft featuring an eight-bladed contra-rotating propeller. This installation cancelled out the engine torque effect and reduced airscrew diameter. TS371 was a non-navalised version, its equipment excluded folding wings, arrester hook and armament. It was mainly used for aerodynamic and handling trials by Westland until 1947, when it crashed on 15 October following a failure of the propeller bearing. The pilot, Squadron Leader Peter J. Garner, Westland's assistant chief pilot, was killed while he attempted a forced landing in a field.

TS375, the second prototype, flew for the first time on 10 September 1947. This aircraft was similar to the first Wyvern. Nevertheless, following Sqn Ldr Garner's fatal accident with TS371, modifications were retroactively introduced on TS375, particularly an experimental ejection seat, a dihedral tailplane and upper wing surface divebrakes.

Serialled TS378, TS380, TS384 and TS387, the four subsequent prototypes were fully navalised and were fitted with four Hispano Mk V cannon. Wyvern TS378 differed from the other prototypes in having a de Havilland six-bladed contra-rotating propeller rather than the

cal director, and J.F.W. Digby, the chief designer, who succeeded Petter to be head of Westland's design team – was presented in September 1944 to meet Specification N.11/44.

This specification called for a carrierborne fighter with the ability to deliver the appropriate weapons to attack ships and land targets: an 18-in (46-cm) Mk XVII torpedo, one 2,000-lb (907-kg) or three 1,000-lb (454-kg) bombs, and eight 60-lb (27-kg) rocket projectiles. The aircraft had to be fully navalised with adequate deck landing ability, stressed undercarriage and powered wing-folding. Dimensions were limited to 18 ft (5.48 m) span folded, 40 ft (12.19 m) length and 15 ft 9 in (4.80 m) height. There was no limitation on weight except for landing, which should not exceed 17,500 lb (7938 kg). The maximum speed had to reach the figure of 500 mph (805 km/h) at 20,000 ft (6096 m) and the endurance had to be 2.5 hours, plus 15 minutes combat and 45 minutes at best speed.

Early in the design programme the RAF started to take an interest in the Westland W.34

The third prototype, TS378, was fitted with a de Havilland Propellers six-bladed contra-prop, although the standard Rotol eight-bladed unit was fitted later. This view emphasises the inverted-gull wing design, the wide and shallow radiators, and the deep trough housing the 12 exhaust stubs of the 24-cylinder engine. TS378 was used in the initial carrier trials campaign.

Resplendent in the new Royal Navy colours that were being introduced around 1950, one of the pre-production Wyvern TF.Mk 1 batch demonstrates the carriage of a torpedo. The final three aircraft from this batch were not completed.

two Rotol four-bladed airscrews, and also in receiving slightly different elevator horn balances to improve longitudinal control and stability. It was later brought to the standard of the other Wyvern Mk Is. TS378 was delivered to the A&AEE in April 1948 for airfield dummy deck landings (ADDLs). The first deck landing trials were undertaken aboard HMS *Implacable* on 9 June 1948, Lt Cdr Hickson successfully landing the Wyvern three times before all prototypes were grounded following a propeller failure on TS380.

Trials resumed on 13 July when TS378 made 15 landings and take-offs and TS380, which had joined the deck landing programme, accomplished a further seven on the next day. A second series of carrier trials was made by Wyvern TS378 aboard HMS *Illustrious* in May and June 1949. The prototypes were mainly tested by the A&AEE, although TS380 and TS387 also served with the Royal Aircraft Establishment (RAE) at Farnborough for NAD and crash barrier trials. TS384 was handed over to Rolls-Royce at Hucknall in June 1948 for engine trials but was written-off as a result of a forced landing on 14 April 1949.

Wyvern TF.Mk 1

In August 1946 Westland obtained an order for a batch of 20 pre-production Wyvern TF.Mk 1s to be built to Specification 17/46P. Nevertheless, interest in the Wyvern Mk 1 had

long since waned. Future development of the piston-engined Wyvern had been seriously considered since the end of 1944. One year later, the Ministry of Supply incited Rolls-Royce to abandon further design development of the Eagle engine and concentrate on turbojet and turboprop work. At the same time, Westland was directed to cease work on the Wyvern Mk 1 in favour of a turboprop version. In these conditions, only a small batch of about 30 Eagle engines was produced to power the prototype and pre-production Wyverns. The order for the 20 pre-production TF.Mk 1s was subsequently cut back to 10 aircraft and eventually only seven were built, receiving serials VR131 to VR137.

Although the first turboprop-powered prototypes were already flying when the pre-production TF.Mk 1s left the assembly line at Yeovil, the Mk 1s were still used for some trials as the gas turbine-powered Wyverns were plagued by engine difficulties. Besides usual handling trials at the A&AEE, more particular tests were conducted. Wyvern VR131 was handed over to de Havilland in May 1949 and to Rotol three months later for propeller trials, and the RAE also undertook airscrew development tests with VR134 in November 1950. VR133 made a series of deck landing trials aboard HMS *Eagle* in June 1952 and was sent to the RAE the next year for crash barrier trials.

The Clyde Wyvern

As interest in the piston-engined Wyvern Mk 1 faded away, Specification N.12/45 was issued in December 1945 to cover the development of gas turbine-powered Wyverns, and a

The cockpit of the Wyvern TF.Mk 1 was typical for the day, with six primary flight instruments (for air speed, vertical speed, direction, altitude, attitude and turn/slip) in the central dashboard.

contract for three prototypes was placed two months later. Two propeller turbines were considered to power the W.35 Wyvern, which later received the mark number TF.Mk 2: the Rolls-Royce Clyde and the Armstrong Siddeley Python. The first prototype TF.Mk 2 to fly was the last serial number, VP120, which was fitted with a Rolls-Royce Clyde driving a Rotol six-bladed co-axial contra-rotating propeller. The Mk 2's fuselage differed significantly from that of the Mk 1: the cockpit was raised and the fuselage depth was increased to provide space for the exhaust duct; a more curved canopy was adopted; and two windows were fitted in the aft fuselage for oblique cameras. Among other changes was the deletion of the under-wing radiators.

Harald Penrose made the initial flight of VP120 on 18 January 1949, but it lasted only three minutes as the cockpit filled with smoke just when the Wyvern left the ground. Penrose

VR133, a TF.Mk 1, was used by the RAE for ground barrier arresting trials. In this view it sports a metal fixture which mimicked the profile of the Fairey Gannet. A tricycle undercarriage was lashed up to further enhance the simulation.

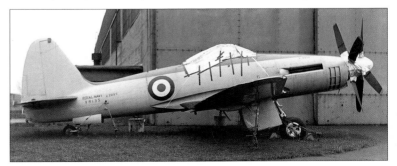

Several of the redundant TF.Mk 1s were employed as ground instructional airframes after their flight trials were completed. VR135 was initially at Yeovilton before being transferred to Bramcote, where it is seen in 1956.

immediately turned to land back at Boscombe Down where the cause was found: fuel was leaking on to the Clyde's exhaust pipe. The Clyde was still an experimental engine and the Wyvern VP120 was then the first aircraft to fly with it. After its initial test programme, VP120 was sent to Rolls-Royce in July 1949 for engine development, the Clyde becoming the first turboprop to obtain a full civil and military type certificate. It had performed a clearance test at 3,500 ehp (21611 ekW) and, by water/methanol injection, the power had been raised to 4,500 ehp (3357 ekW). Unfortunately, the Clyde suffered from various problems and, owing to pressure of work on other more promising projects, Rolls-Royce soon discontinued its development. VP120 was then handed over to Napier in April 1950 with the aim of installing a Nomad engine – a diesel piston engine coupled with an axial-flow compressor and multi-stage turbine. This project was discarded and afterwards VP120 was delivered to the RAE, where it ended its days in September 1950 during crash barrier trials.

Python power

Installation of the Armstrong Siddeley Python engine in the Wyvern airframe needed further modifications due to its greater power. The forward fuselage profile was slightly different and overall length was increased by 21 in (53 cm). The engine drove a Rotol eight-bladed co-axial contra-rotating propeller and, for a weight of 3,150 lb (1429 kg), the Python 1 engine developed 3,670 shp (2738 kW) on take-off plus 1,150 lb (5.12 kN) of jet thrust through bifurcated exhaust ducts. Compared with the Clyde-powered VP120, other external differences were the fitting of an air intake for the oil cooler in each wing root, the absence of an arrester hook, and the camera window was covered. The first Python-engined Wyvern,

These views highlight the differences between the Clyde-powered TF.Mk 2 prototype (VP120, above) and the Python-powered aircraft (VP109, below). The Python installation entailed a considerable lengthening of the forward fuselage. The Clyde-powered aircraft had a cooling grille on the lower port side of the engine cowling.

VP109, flew on 22 March 1949 with Sqn Ldr Mike Graves at the controls.

It was followed into the air on 30 August 1949 by the last Wyvern TF.Mk 2 prototype, VP113, which was also powered by a Python. VP113 featured a completely redesigned and enlarged fin and rudder to improve the aircraft's handling and it was sent twice to the RAE, in September and October 1949, for spinning tests. Tragically, it crashed on the edge of Yeovil airfield on 31 October 1949 following an engine failure at the end of an official demonstration. Sqn Ldr Graves endeavoured to make an emergency landing on the airfield but overshot and ran into a property, killing himself and three occupants of the house.

Flight handling and engine development trials continued with VP109. This Wyvern was successively modified until it crashed on 24 April 1952. In the space of three years, three different fins and rudders were installed, each time with an increase in size, and a fourth modification included the fitting of a dorsal fin and the addition of dihedral to the tailplane.

Pre-production batch

In the meantime, assembly of a batch of 20 pre-production Wyvern TF.Mk 2s had begun.

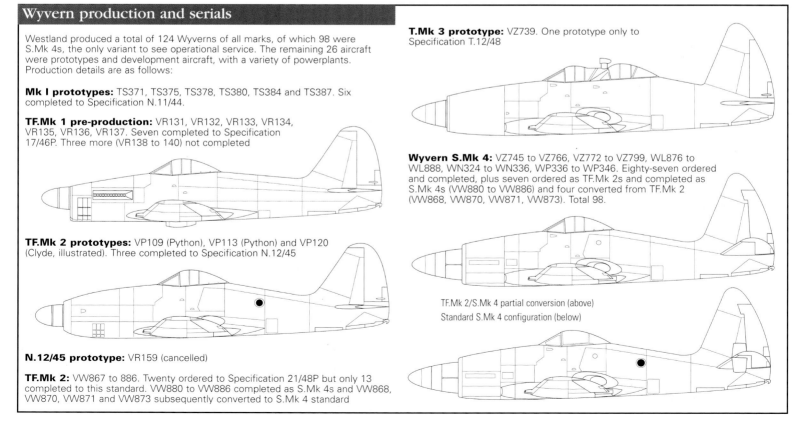

Wyvern production and serials

Westland produced a total of 124 Wyverns of all marks, of which 98 were S.Mk 4s, the only variant to see operational service. The remaining 26 aircraft were prototypes and development aircraft, with a variety of powerplants. Production details are as follows:

Mk I prototypes: TS371, TS375, TS378, TS380, TS384 and TS387. Six completed to Specification N.11/44.

TF.Mk 1 pre-production: VR131, VR132, VR133, VR134, VR135, VR136, VR137. Seven completed to Specification 17/46P. Three more (VR138 to 140) not completed

TF.Mk 2 prototypes: VP109 (Python), VP113 (Python) and VP120 (Clyde, illustrated). Three completed to Specification N.12/45

N.12/45 prototype: VR159 (cancelled)

TF.Mk 2: VW867 to 886. Twenty ordered to Specification 21/48P but only 13 completed to this standard. VW880 to VW886 completed as S.Mk 4s and VW868, VW870, VW871 and VW873 subsequently converted to S.Mk 4 standard

T.Mk 3 prototype: VZ739. One prototype only to Specification T.12/48

Wyvern S.Mk 4: VZ745 to VZ766, VZ772 to VZ799, WL876 to WL888, WN324 to WN336, WP336 to WP346. Eighty-seven ordered and completed, plus seven ordered as TF.Mk 2s and completed as S.Mk 4s (VW880 to VW886) and four converted from TF.Mk 2 (VW868, VW870, VW871, VW873). Total 98.

TF.Mk 2/S.Mk 4 partial conversion (above)
Standard S.Mk 4 configuration (below)

VP109 tested various features, including two enlarged fin shapes, of which the larger (above) was adopted for production. It also trialled a dorsal fin, and the cut-back engine cowling which appeared in the definitive S.Mk 4 (above right).

Left: VW867 was the first of the TF.Mk 2 pre-production batch. As built, the aircraft had full-length engine cowlings and straight tailplanes.

These aircraft, built to Specification 21/48P and serialled VW867 to VW886, were powered by a Python 2 that was intended to cancel the time-lag in power response to movements of the throttle. Unlike the Python 1 engine, where the power output was controlled by the throttle which operated the fuel control unit and the propeller constant speed unit (CSU), the Python 2 ran at a constant speed (7,800 rpm) and thus any throttle movement only induced a change of propeller pitch through the propeller control unit replacing the CSU. Although the first air trials proved the lag in power response had disappeared, violent engine surges occurred at speeds above 350 kt (648 km/h) and caused gear and shaft damages. Other changes from the prototypes included the fitting of small airbrakes in the upper surface of the wings, and of airflow barrier fences on the wing trailing edge to prevent aileron twitch when the airbrakes were applied. The new Martin-Baker Mk 1B also replaced the ML ejection seat.

VW867, the first pre-production Wyvern TF.Mk 2, made its initial flight on 16 February

Westland produced a single W.38 prototype of the T.Mk 3 trainer. Although it was built in answer to an official requirement, in the event the Admiralty did not proceed with a trainer version. The T.Mk 3 had a brief career as a Westland 'hack' after its initial trials, before it was damaged in a forced landing at Seaton, Devon. After three weeks spent in a salt marsh resisting all attempts to salvage it, the T.Mk 3 was written off.

1950. Following successful arresting trials at the RAE, VW867 went aboard HMS *Illustrious* on 21 June 1950. In the hands of three A&AEE pilots, 25 landings were performed and these initial deck landing trials were concluded in one day. The Wyvern was the second British turboprop aircraft to deck land, being preceded by only two days by the Fairey Gannet ASW aircraft. The 20 pre-production Wyvern TF.Mk 2s were extensively used for development trials as the handling remained difficult and the gas turbine continually required improvements to be made operationally suitable.

Two-seat Wyvern

Besides the order for 20 pre-production Wyvern TF.Mk 2s, a contract was placed to provide a two-seat dual-control version under Specification T.12/48. This aircraft was intended to ease the transition to the Wyvern turboprop for pilots having experience only in piston-engined aircraft. The prototype, known as the Westland W.38 or Wyvern T.Mk 3 and serialled VZ739, made its first flight on 11 February 1950. The airframe was similar to the Python-powered TF.Mk 2, except for a deeper rear fuselage to accommodate a second cockpit.

Both cockpits were fitted with ejection seats and their canopies could be opened independently and jettisoned. The junction between the two canopies was formed by a perspex

connecting tunnel that supported a periscopic mirror intended to give a better forward view for the instructor. As the Wyvern T.Mk 3 was built, the tailhook was removed and its small-size tailplane was replaced by a larger one, in the same way as on the prototype VP109. Due to a lack of official interest, the Wyvern T.Mk 3 was retained by Westland after the trials were completed and served as a communications aircraft. This unique aircraft was written off on 3 November 1950 following a turbine blade failure that obliged Sqn Ldr D. Colvin to make a forced landing with wheels up in a marsh.

Strike Wyvern

Although the Wyvern still needed improvements to obtain service clearance, a production order for 50 Wyvern Mk 4s was placed in 1951, the aircraft being serialled VZ745 to VZ766 and VZ772 to VZ799. Westland later received three further contracts covering two batches of 13 Wyverns, serialled WL876 to WL888 and WN324 to WN336, and a last batch of 11, serialled WP336 to WP346. Total Wyvern Mk 4 production amounted to 98 machines as the last seven pre-production TF.Mk 2s (VW880 to VW886) were completed as Mk 4s and four more TF.Mk 2s (VW868, VW870, VW871 and VW873) were converted to the Mk 4 standard. The initial role designation of torpedo-fighter was naturally changed to the strike attack role as the Wyvern had no fighter pretensions, resulting in a change of designation from TF.Mk 4 to S.Mk 4.

Production S.Mk 4s differed little from the VP109 prototype in its final configuration, almost every modification having been tested on this airframe. The main external characteristics of the Mk 4 were a reshaped engine cowling that was cut back to permit cartridge starting, modified aileron tabs, horn-balanced rudder, dihedral tailplane and the fitting of auxiliary finlets to the tailplane. Another improvement was the adoption of speared tips on the Rotol propeller blades. Through the various modifications the handling characteristics really improved and the Wyvern S.Mk 4 appeared as a pleasant aircraft to fly and an efficient weapons platform. Nevertheless, the introduction of the turboprop did nothing to improve its performance. Top speed was significantly decreased in comparison with the Eagle-powered TF.Mk 1, except at sea level where the S.Mk 4 reached 383 mph (616 km/h) and the TF.Mk 1 only 365 mph (587 km/h). The rate of climb also deteriorated, with the turboprop Wyvern limited to 2,350 ft (716 m) per minute, against 2,900 ft (884 m) for the previous version.

There were no developments of the Wyvern beyond the S.Mk 4, although plans were made for an S.Mk 5, which would have been powered by a Napier Double Eland turboprop.

Anatomy of the Wyvern

The Wyvern S.Mk 4 emerged as a single-seat, single-engined, low-wing cantilever monoplane. Fuselage and wings had an all-metal stressed skin structure. Engine and cockpit situations gave the pilot an excellent forward view

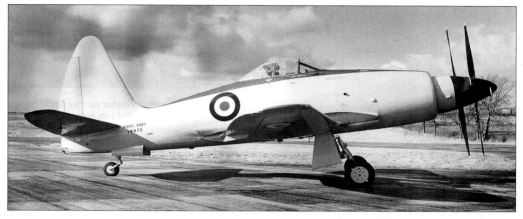

Seven TF.Mk 2s (including VW884, left) were completed to a partial S.Mk 4 standard, while four which had been completed as Mk 2s (including VW870, below) were subsequently modified as S.Mk 4s. The main external differences were the cut-back engine cowling and the dihedral tailplane. Further 'full' S.Mk 4 features, such as the horn-balanced rudder and tailplane finlets, were added later.

for an aircraft of tailwheel configuration. The cockpit was surrounded by two fuel tanks of 90 Imp gal (409 litres) and 173 Imp gal (786 litres). Total internal fuel capacity reached 511 Imp gal (2323 litres) with the two 95-Imp gal (432-litre) outer wing and two 29-Imp gal (132-litre) inner wing leading-edge tanks.

The wing centre section was built integrally with the fuselage. The Wyvern had an inverted gull-wing shape with 1° anhedral inner wings and 6° dihedral outer wings. The outer wing panels and wing tips folded upward, overall width being then reduced to 18 ft (5.49 m). The wings had hydraulically operated Youngman flaps with three positions on the inner wing, and combined dive-brakes and plain-type split flaps on the outer wing sections. The ailerons were of the sealed forward balance type with spring tabs and an electrically controlled trim tab on the port aileron. Main wheels and the tailwheel hydraulically retracted, respectively, inward into the wing centre section and forward into the fuselage, and were covered by fairings. The Dowty undercarriage legs shortened during retraction as there was not enough room for them when extended, their extended length being dictated by propeller and under-fuselage stores clearance.

Internal armament consisted of four 20-mm Hispano Mk 5 cannon with 200 rounds each. Seven attachment points allowed the carriage of a variety of external loads. The underfuselage attachment point could take a 20-in (51-cm) torpedo, a 1,000-lb (454-kg) bomb or a 150-Imp gal (682-litre) fuel tank. The two inner wing pylons could each support a 1,000-lb bomb or a 100-Imp gal (455-litre) fuel tank. Each outer wing panel had two rocket launching rails able to carry two 25-lb (11.3-kg) or 60-lb (27-kg) rocket projectiles (R/Ps). Other loads, such as 16 R/Ps or two 1,000-lb bombs, were possible with some adaptation.

Improving the Wyvern

As the first 18 production Wyvern S.Mk 4s were rolled out from Yeovil, they only had a limited clearance that prevented any carrier use. Nevertheless, these aircraft allowed the formation and work-up of the first Wyvern squadron in May 1953. During this time, Armstrong Siddeley and Rotol worked hard to find solutions to the engine control problem and to provide a fully cleared Wyvern for carrier operations. The Python 3 brought the solution. This engine was basically a Python 2 incorporating a Rotol engine control unit (ECU), which was an inertial control unit (ICU) coupled with an anticipator. Four Wyvern S.Mk 4s equipped with the ECU – VZ746, VZ750, VZ774 and VZ777 – underwent deck landing trials with the A&AEE from June to November 1953.

VZ777 was one of four production S.Mk 4s which tested the improved Python 3 engine and its Rotol engine control unit. It is seen here during carrier trials in Eagle in 1953, operating under the auspices of the A&AEE Boscombe Down.

Following the usual ADDLs, trials were successively carried out on HMS *Eagle*, HMS *Illustrious* and USS *Antietam* using both standard and mirror techniques. The trials carried out aboard the US carrier with an angled deck proved interesting. In one landing, VZ750 missed the six wires as its hook bounced. The engine response in opening up from cut power was quick and the aircraft went around again safely. The final A&AEE conclusion reported that the Wyvern S.Mk 4 was satisfactory for deck landings by using either the mirror sight or the standard approach technique. In reality, due to the high level of vertical velocity and to the nose down change of trim at the cut, and to poor elevator effectiveness after the cut, the Wyvern S.Mk 4 was considered unsuited to the standard method.

During the years following entry into service, several modifications were introduced. S.Mk 4s with Python 3s delivered from spring 1954 had new flaps with increased area and a revised wing-folding mechanism excluding the wing tips. Two years later, another series of improvements appeared. These included the adoption of a new composite canopy with a metal rear fairing in place of the fully glazed model and a flat bullet-proof windscreen, the fitting of a

Martin-Baker Mk 2B ejection seat with automatic separation, and the introduction of underwing centre-section perforated airbrakes.

Into service

Based at RNAS Ford with Blackburn Firebrand TF.Mk 5s, No. 813 Squadron was selected to become the first Wyvern unit. Early in May 1953 the pilots visited the Westland works to see the Wyvern in construction and to talk to the test pilots. On 18 May Lt Cdr C.E. Price, who had been flying the Wyvern for Westland, joined the squadron as Commanding Officer and two days later brought the first Wyvern from Stretton, where the aircraft were prepared for delivery. At the end of June 1953 all pilots were converted to the Wyvern and the full squadron complement of 12 aircraft was attained the following month.

In the meantime, No. 813 Squadron had experienced some trouble with the Python engines; bits of metal from the starter motor had been going into the engine. Although the starters were changed, the trouble happened again and low-pressure starters were sent from Merryfield for installation. The next Wyverns delivered were fitted with Armstrong Siddeley gas starters.

An early production S.Mk 4 displays a full armament load, consisting of one torpedo under the fuselage and 16 60-lb R/Ps under the wings on double-tier launchers. Above the rockets are the muzzles for the four 20-mm cannon. By the time the aircraft reached service the air-dropped torpedo had largely been consigned to history as a viable weapon, and the Wyvern would see service with bomb, rockets and guns as its principal weapons. The centreline and two inner wing hardpoints could carry 1,000-lb bombs or drop tanks.

On 11 June 1953 an incident occurred to Cd Pilot Gee when his engine stalled and the jetpipe caught fire at the end of the landing run. Although he immediately shut down the engine and operated the fire extinguisher, the Wyvern was extensively damaged and had to be sent to the base Aircraft Repair Service. The following day, Sub-Lt Cooper touched the propellers and flaps on the runway when landing, having forgotten to put the landing gear down. He managed to get the aircraft up to come round again for a successful landing.

No. 813 Squadron's first participation in an exercise occurred with Exercise Momentum from 15 to 23 August 1953, in which Fleet Air Arm aircraft attacked naval convoys defended by the RAF. The Wyverns made dummy attacks with rocket projectiles. Clearance for live R/P firing was released on 28 August, and on the same day 10 sorties were flown over Bracklesham Bay: results were varied due to troubles with the gyro sights. During Exercise Mariner Lt Farthing's aircraft crashed and sank into the sea when its engine stopped on 28 September 1953. While Lt Lawrence and Lt MacFarlane carried out a search north of Le Havre, they ran out of fuel and were diverted to Deauville in France, where they stayed until an air starter trolley arrived the next day on an RAF Valetta.

VZ746, the first S.Mk 4 that was modified and released for carrier operations, arrived on 9 October 1953 and was used during the month by the Commanding Officer, the Senior Pilot (Lt Cdr Genge), Lt MacFarlane and Cd Pilot Gee to perform Aerodrome Dummy Deck Landings. On 28 October, VZ746 was embarked aboard HMS *Eagle*. Unfortunately, the port undercarriage leg was broken off and no squadron pilot got to fly it as they returned to Ford soon after.

Six months elapsed before No. 813 Squadron received new aircraft that were not restricted to land-based operations and had the definitive ICU equipment and anticipator operative for the Python engine, the first one being delivered on 10 April 1954. With these new aircraft

Specifications – Wyvern TF.Mk 1 and S.Mk 4

	Wyvern TF.Mk 1	Wyvern S.Mk 4
Wingspan	44 ft 0 in (13.41 m) unfolded	44 ft 0 in (13.41 m) unfolded
	18 ft 0 in (5.49 m) folded	20 ft 0 in (6.10 m) folded
Wing area	355 sq ft (32.98 m²)	355 sq ft (32.98 m²)
Length	39 ft 3 in (11.96 m)	42 ft 3 in (12.88 m)
Height	15 ft 6 in (4.72 m)	15 ft 9 in (4.80 m)
Empty weight	15,443 lb (7005 kg)	15,600 lb (7076 kg)
Max. take-off weight	21,879 lb (9924 kg)	24,550 lb (11136 kg)
Powerplant	Rolls-Royce Eagle 22 24-cylinder inline engine	Armstrong Siddeley Python 3 propeller turbine
Power rating	2,690 hp (2007 kW) at sea level and 3,260 hp (2432 kW) at 18,000 ft (5486 m) at 3,500 rpm	3,670 shp (2738 kW) plus 1,150 lb (5.12 kN) thrust at 8,000 rpm at sea level
Maximum speed	456 mph (734 km/h) at 23,000 ft (7010 m)	383 mph (616 km/h) at sea level
Service ceiling	32,100 ft (9784 m)	28,000 ft (8534 m)
Maximum range	1,186 miles (1909 km)	910 miles (1464 km)
Armament (both versions, basic)	four 20-mm Hispano Mk V cannon; provision for one 20-in (51-cm) torpedo, or three 1,000-lb (454-kg) bombs, or eight or 16 60-lb (27-kg) rocket projectiles	

No. 813 Squadron carried on its training, including night flying and night ADDLs using the mirror technique from mid-June 1954. At last, on 24 September, the unit with 10 Wyverns embarked in HMS *Albion*. After all the pilots had performed five days of Deck Landing Practice they sailed for Gibraltar. The flying programme in the Mediterranean began on 12 October but on the next day a serious accident occurred. Lt MacFarlane's aircraft suffered a flame-out as a result of fuel starvation during the launch and dropped into the sea. Although the aircraft was cut in two by the ship and sank, Lt MacFarlane escaped successfully by using his ejection seat underwater, and was rescued by the waiting helicopter with minor injuries.

Enforced shore interlude

As a consequence of this accident, the unit was disembarked at Malta on 16 October 1954 and remained ashore at RNAS Hal Far for five months. While flying from the island, a fatal crash was suffered by the squadron on 20 October when Lt Banner disappeared into the sea. The accident was thought to be caused by a compressor failure as the consequence of foreign object ingestion and all Wyverns were grounded pending the dispatch of stone

guards, which were at last fitted on 7 November. A new CO, Lt Cdr R.M. Crosley, arrived at Hal Far on 20 December 1954. During its stay on Malta No. 813 Squadron performed R/P firing and bombing on the Delimara range, many photographic sorties, dummy strikes on Royal Navy ships in the area, and participated in several other exercises.

No. 813 Squadron finally embarked on HMS *Albion* on 22 March 1955 and sailed for the United Kingdom, arriving home on 31 March. At RNAS Ford, No. 813 discovered a second Wyvern unit, No. 827 Squadron having been created there on 1 November 1954, with Lt Cdr S.J.A. Richardson as CO, from a nucleus of No. 703W Flight. This flight was commissioned on 4 October 1954 at RNAS Ford with the goal of converting pilots destined for No. 827 Squadron. For its task No. 703W Flight included experienced Wyvern pilots, two of them having served with No. 813 Squadron, and had a complement of four Wyverns, plus permission to use the Station Flight Sea Vampire T.Mk 22. When No. 703W Flight was recommissioned as No. 827 Squadron it comprised 12 pilots but only five Wyverns. In the following months the unit underwent a work-up period – familiarisation flights for the new pilots, general flying,

Above: One of the first production S.Mk 4s is seen during trials of drop tanks. It also sports the revised canopy with metal aft section and bulletproof windscreen that was adopted part way through the production run.

Left: VZ796 overflies Ark Royal, which has a Dragonfly and two Sea Hawks aboard. The 'Ark' had only a fleeting association with the Wyvern, embarking No. 831 Squadron in 1957.

tactical reconnaissance, mine-laying exercises, night flying and weapons training.

In prevision for deck trials aboard HMS *Bulwark*, the CO and Sub-Lt Mercer began mirror ADDLs on 31 January 1955. These two pilots carried out the wire pulling trials on 9 and 10 February. A mirror ADDL programme was organised for the entire squadron. It began on 21 March, all pilots having visited *Bulwark* the previous day to look at the angled deck. From 30 March 1955 all aircraft were stripped down, preparatory to the fitting of recuperators and extra hook dampers. This huge maintenance and modification programme, undertaken by Westland and Armstrong Siddeley work crews, lasted until 29 April, although the squadron technicians had yet to replace propellers, set up engines and carry out the checks. ADDLs then resumed in May 1955. Westland also modified the leading edges of all Wyverns, reducing stall speed to a 89- to 92-kt (165- to 170-km/h) range from over 100 kt (185 km/h).

No. 827 Squadron goes to sea

Embarkation of eight Wyverns from No. 827 Squadron aboard HMS *Eagle* took place on 10 May 1955. After only three days of flying, the unit recorded its first accident when Lt Barstow's aircraft dropped too fast over the round-down during his final approach and smashed the tail: it was Friday the 13th. Three days later a more serious accident occurred. On return from an Air Interception exercise, Lt J.W.F. Jarratt stalled his Wyvern and attempted to go round again but the aircraft crashed into the carrier's island, causing major injuries to the pilot and considerable damage to the ship, which had to retire to Portsmouth for repairs. Meanwhile, No. 813 Squadron, which was practising mirror ADDLs since 12 May and was preparing to join HMS *Eagle* on 18 May, learned that its embarkation was postponed and witnessed the return to Ford of just four aircraft from its sister unit.

At last, the two Wyvern units, along with No. 826 Squadron and its Gannet AS.Mk 1s, embarked on 4 June 1955. They were joined later by two Sea Hawk squadrons and one Skyraider AEW.Mk 1 flight. During five days *Eagle* operated in the Channel, then set course for Gibraltar and entered the Mediterranean on 15 June. The squadrons were deployed at Hal Far for three weeks. During operations from Malta, a No. 827 Squadron Wyvern was written-off on 8 July following the failure of the port

A No. 813 Sqn Wyvern throws clear the bridle as it launches from Eagle *in 1955. Noteworthy are the Youngman flaps which slung down beneath the wing when deployed, and the small flap/airbrake sections.*

No. 813 Squadron was the first operational Wyvern unit. Its aircraft were initially given codes in the 181-192 range (right), and would have worn 'Z' deck letters to signify their assignment to HMS Albion. *When they were reallocated to HMS* Eagle *in 1955, the deck letter changed to 'J' and codes to the 121-129 range (above).*

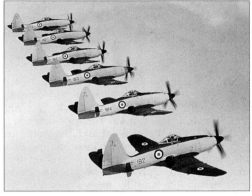

undercarriage lower leg in a landing with Lt Hodgson at the controls.

On 18 July 1955 and on the next day, simulated strikes were carried out by the Wyverns on the US Sixth Fleet off Sicily. USS *Intrepid* and USS *Coral Sea* were successfully attacked with the help of the Skyraider AEW.Mk 1s, the Wyverns reaching their targets without being intercepted by US Navy fighters. The ship's armament work-up programme was then undertaken until the end of the month, including firing of R/Ps with 60-lb (27-kg) concrete heads, dive-bombing with 500-lb (227-kg) MC and 2,000-lb (907-kg) AP bombs.

Serviceability was very good, although some incidents occurred: one Wyvern was accidentally directed into the tail section of another in the forward deck park; one aircraft lost its entire wing tip after take-off, its lock not being properly engaged; and a tailwheel sheared off on yet another as it made a heavy landing. The aircraft went round again, but on the final landing the oleo stub also caught a wire, resulting in structural damage. This type of accident was frequent with the Wyvern. After a short stay in Naples, *Eagle* joined *Albion* on 14 August 1955 in preparation for the forthcoming exercises with the US Sixth Fleet. Two days earlier, No. 813 Squadron lost a Wyvern when Sub-Lt

Steers experienced violent shuddering and RPM swing, and had to eject safely over a vineyard.

'Attacking' the US Navy

During the combined exercise, which began on 16 August, both Wyvern units conducted a strike of six aircraft each, which deployed into six sections to attack USS *Intrepid* some 220 miles (354 km) away. Air opposition was slight, but on a second sortie launched during the afternoon, the Wyverns were bounced by Banshees and appeared to be easy meat for the American jet fighters at a height of 5,000 ft (1524 m), although in real war conditions the Wyverns would normally fly at wave-top height. Next day, the Wyverns demonstrated their superiority over the US Skyraiders and Banshees in rocketing and dive-bombing 'splash targets'. 18 August was the last day of the joint exercises, which included air strikes

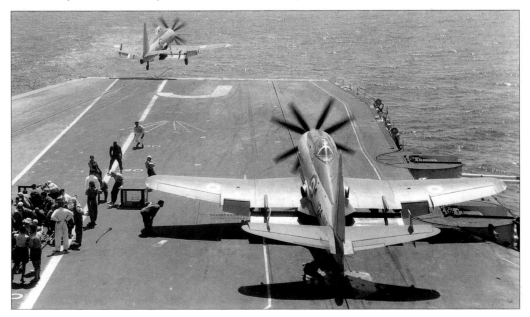

No. 827 Squadron was the second Wyvern unit, allocated codes in the range of 131-139 and Eagle's deck letter 'J'. Squadron aircraft wore a wyvern (a winged dragon of Norse origin) badge on the nose. From May to November 1955 the unit was in Eagle, flying alongside sister squadron No. 813.

for air defence purposes. After an intensive flying programme on 22 August 1955 (No. 813 Squadron performed 26 sorties and No. 827 Squadron attaing 28) and a short stay at Gibraltar, HMS Eagle set course on 8 September for the UK. Before the ship entered Cromarty Firth on 16 September, the Wyverns undertook air defence and mine-laying exercises.

Exercise Sea Enterprise, organised by NATO, took place from 21 September 1955. HMS Eagle sailed in the North Sea and was engaged with three other British carriers. Wyvern units had to wait until 22 September to find targets within range, when a first strike of 12 Wyverns was launched on an airfield near Trondheim. They were intercepted by Royal Norwegian AF F-84G Thunderjets before they struck the target. Continuous strikes were carried out during this

day and the next on ships, bridges, radar installations, gun positions, and other various targets in the Trondheim area.

After a day of rest for replenishment, flying resumed on 25 September with a Wyvern strike on HMS Apollo and HMS Decoy. This strike consisted of six machines from No. 827 Squadron and four from No. 813 Squadron. The Wyverns had to execute a square search before finding the ships. Sub-Lt Teague of No. 827 Squadron realised that he was losing too much fuel through venting and immediately set course to Eagle at economic speed. When he was down to 50 Imp gal (227 litres) he sighted and made a straight approach toward Bulwark, which was five miles nearer than his own carrier, Eagle. Not having enough fuel to go round again, Sub-Lt Teague did not follow the wave-off given by Bulwark and landed on the deck with his hook up. He lost his life as his Wyvern collided with a Skyraider AEW.Mk 1 and dropped over the bow. Notwithstanding this sad loss, a further strike was flown in the afternoon. The exercise continued on

27 September with an attack on RFA Black Ranger by five Wyverns of each unit. On return the CO of No. 827 Squadron badly damaged his aircraft as its tailwheel caught a wire on landing. No. 827 Squadron then had only five serviceable Wyverns, thus bringing the type's participation in the exercise to an end.

Finally, Eagle anchored at Trondheim on 28 September and afterwards at Oslo, departing for further strikes against Norwegian soil on 10 October. A fatal accident again happened to No. 827 Squadron on the next day. Lt Cdr L.P. Watson, who was leading the second strike on Oslo as Senior Pilot, suffered an engine failure two minutes after taking off. At about 300 ft (91 m) he made the decision to eject, but no parachute was seen and the helicopter retrieved a dead body from the sea.

Pending an investigation by Armstrong Siddeley, all Wyverns were prevented from flying from HMS Eagle, and the ship returned to Rosyth on 13 October. Armstrong Siddeley's investigators found corrosion in all the BFCUs (Barometric Fuel Control Units) and it was decided on 20 October that Wyvern flying could resume provided that the BFCUs were regularly inspected. Thus the ship sailed again on 23 October for a participation in a Fighter Defence exercise, for which the two squadrons carried out dummy strikes.

On 4 November 1955 Eagle was moored alongside Devonport Dockyard, marking the end of a five-month cruise. Despite only having five Wyverns left, No. 827 Squadron undertook a short weapons programme, firing 25-lb R/Ps at a towed target on 14 November and dive-attacking with 500-lb MC bombs two days later. At last, Nos 813 and 827 Squadrons disembarked to RNAS Ford on 17 and 18 November, respectively. No. 827 Squadron disbanded on 19 November and No. 813 Squadron followed suit two days later.

A second operational career

As Nos 813 and 827 Squadrons disbanded, two new units were officially commissioned at RNAS Ford on 21 November 1955 to take their places. They were Nos 830 and 831 Squadrons, for which advance parties had actually arrived in October to start the conversion. The COs were Lt Cdr C.V. Howard, coming from No. 764 Squadron, and Lt Cdr S.C. Farquhar, having served with No. 700 Squadron. Several experienced Wyvern pilots from No. 764 Squadron joined both units, including No. 830 Squadron's Senior Pilot, Lt Cdr W.H. Cowling. No. 831's SP, Lt Cdr W.A. Tofts, also knew the Wyvern well as he flew initially with No. 813 Squadron and had worked with the A&AEE.

No. 764 Squadron, also based at Ford, was a Fighter Pilot Holding Unit and had received two Wyvern S.Mk 4s in May 1955 for pilot conversion. In February 1957 the Wyvern Flight of No. 764 Squadron was handed over to the Wyvern Conversion Unit that undertook the task of pilot conversion with six Wyverns until December 1957.

No. 831 Squadron received its first S.Mk 4 from Stretton on 12 November 1955. Nevertheless, as not enough new Wyverns were ready for delivery, both units had to use

Early production Wyverns, like this No. 827 Squadron aircraft, had small inward-folding wingtip sections to keep the folded height within limits. This was deleted from later aircraft.

Above: No. 830 Squadron (370-379 codes) was initially formed with aircraft taken over from No. 813 Squadron, but soon received new aircraft which it took aboard Eagle in April 1956

Right: With two Skyraiders looking on and a destroyer in close attendance, 830's Wyverns prepare to launch on a combat mission to Dekheila on 1 November 1956. Each carries a single 1,000-lb bomb on the centreline.

older aircraft as a stop-gap measure. On 17 November No. 813 Squadron passed on five Wyverns to No. 830. The next day, No. 827 Squadron passed on five to No. 831 as it disembarked from *Eagle*. Now that both units possessed aircraft, even though serviceability was poor and the machines needed a considerable amount of maintenance, the squadrons settled into a normal routine consisting of familiarisation flights, division drills, dummy attacks and ADDLs. New Wyverns were delivered from the RDU (Receipt and Despatch Unit) at Stretton at a slow pace from the end of November. The first accident occurred on 5 December 1955 to Lt J.P. Smith of No. 830 Squadron. During a dummy dive, he discovered a sharp rise in the JPT (Jet Pipe Temperature) and closed the HP. As relighting had no effect, Lt Smith ejected safely while his aircraft, an ex-No. 813 Squadron machine, crashed rather softly.

From mid-December, Westland began modifications to all the new Wyverns delivered to both units, including improvements to the air brakes and fitment of rubberised material to the wing leading edges. During the first months of 1956 the total aircraft in each squadron was raised to nine in number. At the same time routine flying continued, including exercises with the Fleet or army artillery, and live-firing. Several times the Wyverns were grounded following incidents in flight due to flap problems. After the usual ADDL session, pilots from Nos 830 and 831 Squadrons accomplished their first landings aboard *Bulwark* in mid-March 1956. Angled deck and mirror landings appeared to be far easier.

The Wyvern returns to *Eagle*

Both Wyvern squadrons deployed aboard HMS *Eagle* on 16 April 1956, although No. 831 Squadron was there only three days during which Deck Landing Practice was performed. As No. 831 Squadron returned to RNAS Ford, No. 830 Squadron sailed for Gibraltar and then on to the Mediterranean. The unit carried out various exercises, consisting mainly of dive-bombing and photo-recces. From 7 May 1956, half of the squadron disembarked to Hal Far to practise the armament work-up with rocket firing and the dropping of 25-lb (11.3-kg) practice bombs on the Delimara range. The remain-

der of the unit proceeded to the island the next week, but an accident happened to Lt R. King on 17 May. During the landing circuit to Hal Far, one undercarriage leg did not lower. As one leg was locked down, Lt King decided to eject over the sea where the SAR helicopter picked him up. Only five days later another accident claimed the life of Lt J.P. Smith. The pilot was approaching *Eagle* to deck land, having completed an armament exercise on a 'splash' target towed by a ship, when his Wyvern suddenly lost height and crashed into the sea. It was assumed that he ran out of fuel owing to a faulty gauge.

After HMS *Eagle* had successively visited Syracuse, Malta, Istanbul and Beirut, it joined the US Sixth Fleet on 25 June 1956 to carry out Exercise Thunderhead, intended to test the radar system efficiency of southern France and northern Italy. During this exercise, two No. 897 Squadron Sea Hawks ran out of fuel and ditched off Genoa. Immediately, Gannets and Skyraiders made a search, the Wyverns acting as link aircraft to the ship. The two pilots

were finally picked up safely. Exercise Thunderhead was completed on 28 June and *Eagle* reached Malta the next day for a refit. Five Wyverns were disembarked at Hal Far, from where flying resumed with R/P firing and bombing with 25- and 500-lb bombs at the Delimara and Filfla ranges. A photo-recce mission was also organised over Tripoli in Libya. All Wyverns tried to rejoin the ship on 17 July but, due to the absence of wind and the insufficient speed of *Eagle*, they returned to Hal Far. Later in the day the ship was repaired and all No. 830's aircraft were recovered.

Eagle joined the US Sixth Fleet again for another exercise. On 3 August 1956, No. 830 Squadron was back at Hal Far with six Wyverns for a further armament work-up which lasted 11 days. Later, aboard *Eagle*, modifications were undertaken to allow the Wyvern to fire a full load of 16 R/Ps from a double-tier mounting. No. 830 Squadron then returned to Hal Far on 23 August and the pilots tasked to fire or drop the maximum load of 16 25-lb R/Ps or three 500-lb bombs per aircraft. Since the

Columns of smoke rise from bomb blasts as No. 830 Squadron's Wyverns hit the airfield at Dekheila, near Alexandria. This target was struck several times by the unit in the first two days of the campaign.

No. 831 Squadron was formed at the same time as No. 830, and was initially assigned to Eagle, *although was later reallocated to* Ark Royal *(deck letter 'O', codes 380-388). Squadron aircraft are seen on 'Ark' near Rosyth (left), with the Forth Rail Bridge in the background. Above is an 831 machine (the third from last Wyvern built), displaying the carriage of 16 R/Ps. Note also the perforated underwing airbrakes fitted to late-production aircraft.*

beginning of the month, night flying was also carried out, as No. 830 Squadron was intended to become a fully night-capable strike unit with the fitting of Audio ASIs.

On 4 September 1956 No. 830 Squadron re-embarked to carry out strikes from the sea, firing on 'splash' or smoke targets and bombing on the Filfla range. On 13 September, the Wyverns flew ashore to Hal Far for general maintenance, including engine and propeller changes, leaving only two serviceable aircraft until the end of the month. Flying resumed with extensive armament sorties during which the Wyverns were loaded with one 1,000-lb bomb, six 25-lb R/Ps, two inner guns with 120 rpg and two drop tanks. Long-range strikes were also performed. *Eagle* reached Gibraltar on 13 October 1956 for a refit.

Combat at Suez

Eagle's refit at Gibraltar was only half completed when the ship was recalled on 20 October 1956 as a result of the political situation in Egypt. President Nasser had proclaimed on 26 July that the Suez Canal Company was nationalised. As negotiations came to a stalemate, the United Kingdom and France, for whom the canal was of vital importance, were preparing a military operation. At the same time, Israel began to mobilise. The Israelis invaded the Sinai on 29 October 1956 and advanced towards the canal. The next day an Anglo-French ultimatum was sent to Israel and Egypt, requesting each nation to cease fire and withdraw to 10 miles (16 km) from either side of the canal. Of course, Egypt rejected it and Operation Musketeer began on 31 October 1956.

In the meantime, *Eagle* had reached Malta on 25 October where repairs were completed. Four days later, as the ship left harbour, No. 830 Squadron embarked two additional Wyverns and carried out air-to-air gun camera sorties. The carrier was left with only one working catapult when the main driving wire of the starboard catapult broke, sending a No. 897 Squadron Sea Hawk into the sea. On 30 October, all aircraft were painted with 1-ft 0.3-m) wide yellow and black identification stripes. The Carrier Air Group, consisting of

Eagle, *Bulwark* and *Albion* from the Royal Navy, and the Marine Nationale carriers *Arromanches* and *Lafayette*, was on standby on 31 October. Musketeer was planned to begin at 1615 GMT. The first phase of the operation was the neutralisation of the Egyptian Air Force.

On the evening of 31 October only RAF Valiants and Canberras based at Malta and Cyprus carried out high-level bombing. The Carrier Air Group went into action on 1 November. Loaded with a 1,000-lb MC bomb each the Wyverns struck Dekheila airfield, an old Royal Naval Air Station near Alexandria. The attack on the runways was repeated twice during the day, with some strafing and photographing. Again the next day, No 830 Squadron launched three attacks. The first one returned to Dekheila to deliver 1,000-lb bombs on hangars where new aircraft were housed in crates. The two other 2 November strikes were directed against Huckstep Camp near Cairo, where reconnaissance had revealed a large concentration of military armour and transport vehicles. During these first two days, each Wyvern delivered only one 1,000-lb MC bomb on each sortie. The attacks on Huckstep Camp were their first inland incursions, as air opposition was non-existent and danger of air combat for the turboprop aircraft was nil.

Bridge attacks

On 3 November 1956 13 sorties were flown by No. 830 Squadron on the Gamil Bridge, a vital link for Port Said to the west coast and Alexandria. Most Wyverns were loaded with two 1,000-lb MC bombs each, except for four aircraft which carried one 1,000-lb weapon plus two 500-lb MC bombs each. In company with Sea Hawks, the Wyverns dive-bombed their target, but only one hit was recorded on the bridge and that was not sufficient to breach it. British aircraft encountered light flak and Lt D.F. MacCarthy's Wyvern took hits during the bombing dive. The pilot managed to release the bombs and escaped to the sea three miles away before ejecting. He then had to face hostile shore batteries that were attacked by Sea Hawks before his rescue by *Eagle's* SAR helicopter 75 minutes later. The next day, *Eagle* withdrew to refuel and take on ammunition. Lt Barras and Sub-Lt MacKern were ferried to El Adem in Libya by a Skyraider to collect two replacement Wyverns.

Eagle was back again on 5 November to support the airborne assault, in which British paratroopers were dropped on Gamil airfield to progress towards Port Said, while French paratroops had DZs at Port Fouad and in the interior basin. No. 830 Squadron's first two strikes were directed against Coast Guard barracks, on a mortar company and on various gun positions that were menacing the British paras. The third and last strike of the day was again executed on the Coast Guard barracks, where snipers were still active, in a combined low-level and dive-bombing attack using 1,000-lb bombs, 60 lb R/Ps and 20-mm strafing. During the attack, the SP, Lt Cdr W.H. Cowling, felt a distinct thump, probably due to the entry of shrapnel in the air intake and the Python engine began to run rough. Cowling managed to maintain height and ejected within 20 miles (32 km) of the carriers. He was safely picked up by a helicopter that was evacuating army wounded from Gamil.

On 6 November strikes on the beaches targeted for invasion were launched and followed by an intensive ship bombardment in preparation for sea landings. No. 830's Wyverns flew 'cab rank' sorties to support the Royal Marines as they met considerable resistance in Port Said, being directed by HMS *Meon* to various air control teams in the town. Nevertheless, not enough targets were available for all the aircraft and only the second squadron detail was requested for strafing and rocketing on the Police Club. The last mission was flown on El Kantara to provide air cover for the rescue of a No. 897 Squadron pilot who had ejected after being hit by flak during an attack on an army camp.

Although Port Said was taken and the British and French troops began to move down the Suez Canal, the two governments had to accept a ceasefire ordered by the United Nations. As combat had ceased, *Eagle* left for Malta on 7 November to repair the port catapult that had been launching all the aircraft during the operation. A survey of hits on the Wyverns revealed one spinner hit, two nicked propellers, a 0.303-in bullet hole in a tailplane and shrapnel lodged in one air intake. The two aircraft lost during combat were replaced by two new Wyverns when the ship arrived at Malta on 9 November 1956. During five days of operations over Egypt, No. 830 Squadron flew a total of 79 sorties, during which the Wyverns delivered 78 1,000-lb MC bombs plus eight 500-lb weapons, and had fired 338 60-lb R/Ps and 3,870 20-mm rounds.

From 13 November 1956 *Eagle* returned to the Egyptian coast where various exercises were conducted, including firing on 'splash' and smoke targets. On 18 November, one Wyvern was completely destroyed and a second severely damaged as a Sea Venom accidentally fired its guns in the lower hangar and hit a Wyvern drop tank. After short spells at Malta and Cyprus, No. 830 Squadron trained to do CAPs under the direction of Skyraider AEW.Mk 1s – on 14 December an actual interception was accomplished of a Douglas DC-6 of Air Jordan Airlines.

From 19 to 22 December 1956, No. 830 Squadron took part in Operation Harridan, operations consisting of covering British and French forces as they withdrew from Port Said. Wyverns flew low CAPs under the direction of

This No. 813 Squadron S.Mk 4 launches from Eagle *with a single 150-Imp gal (682-litre) tank – complete with shark-mouth markings – on the centreline, and two 100-Imp gal (455-litre) tanks under the inner wing hardpoints. The marking on the fin below the deck letter is the squadron's official badge, which consisted of a diving black eagle over the waves.*

Having been the first Wyvern unit, No. 813 Sqn became the last when it was reformed in November 1956 for service in Eagle. *Assigned codes were in the range 270-278. Initially,* Eagle's *long-standing deck letter 'J' was applied, but this was changed to 'E' in 1958, the last year of operations. Shown above is an aircraft carrying a painting of the popular comic character Dennis the Menace, carrying a bomb inscribed with '813'. The two aircraft at right are flying over the Lofoten islands in northern Norway during the NATO Exercise Strikeback in September 1957.*

Skyraiders. Two armed rocket sorties were launched over the city following calls from teams in Port Said, but the interventions were cancelled and the armament jettisoned into the sea. Following the completion of Operation Harridan, *Eagle* stayed some time in the Mediterranean before returning home in January 1957. No. 830 Squadron returned to RNAS Ford on 5 January and disbanded on the same day.

Phase-out

Since the departure of HMS *Eagle* in April 1956, No. 831 Squadron had been ashore, remaining at RNAS Ford and Lossiemouth. It finally embarked on 9 January 1957, on HMS *Ark Royal*, heading towards the Mediterranean. Until armament training began on the Delimara range on 22 January, the main occupations consisted of DLPs and catapult drills. Three days later the carrier entered Grand Harbour and five Wyverns were disembarked at Hal Far to allow the flying programme to continue. All the Wyverns rejoined *Ark Royal* on 5 February, but as the carrier soon experienced trouble with two of its four turbines, it was ordered to return to Devonport. Before reaching England, where No. 831 Squadron was flown ashore at RNAS Ford on 25 February 1957, the 'Ark' spent some time in the Gibraltar area.

No. 831 Squadron's second trip aboard *Ark Royal* took place from 3 May 1957. This time the carrier stayed in northern waters, crossing the Atlantic and entering Chesapeake Bay on 8 June. The previous day, two of each type of aircraft available aboard the British carrier were flown to the Naval Air Test Center at Patuxent River. Lt Cdr Farquhar, the CO, and Lt Spafford carried out a number of mirror ADDLs on a US starboard-side installed mirror, performed steam catapults and made taxiing arrested trials on the Mk 7 arrester gear. Following these trials, HMS *Ark Royal* and USS *Saratoga* proceeded out to sea and cross-decked on 17 and 18 June. Four No. 831 Squadron pilots, including the two present at the NATC, performed catapult launches, touch-and-go landings and arrested landings with two Wyverns aboard *Saratoga*. At the end of the second day of cross-decking, all aircraft returned to their respective carriers and *Ark Royal* sailed for New York before starting the journey home on 20 June 1957.

The five following months were spent at RNAS Ford, except for two five-day periods; the first aboard *Ark Royal* and the second for Exercise Ash Can, when No. 831 Squadron flew to RNAS Culdrose in the company of two aircraft from the Wyvern Conversion Unit. From 4 November 1957 No. 831 Squadron started a preparatory MADDL programme for its third cruise aboard *Ark Royal*. The Wyverns joined the carrier on 13 November 1957 for an uneventful trip in home waters, although Sub-Lt Edward went over the edge of the flight deck and into the cat walk, damaging a Skyraider AEW.Mk 1 in the process. On 25 November all 40 aircraft of *Ark Royal* were catapulted in about 15 minutes and performed a fly-past before departing to their parent bases. No. 831 Squadron flew direct to Lossiemouth, where the Wyverns were put in storage. The pilots then returned to Ford by train and were engaged in non-flying activities until the official disbandment of their unit on 10 December 1957.

After retirement from a scant five years of service, the Wyvern fleet was gathered at RNAS Lossiemouth. These scenes from 1959 show aircraft from No. 813 Squadron (above) and No. 831 Squadron (below). After a brief period of storage all were scrapped.

This, however, was not the end of the Wyvern's career because there was a third squadron that had been created more than a year prior and was still active. No. 813 Squadron had been reformed on 26 November 1956 under the command of Lt Cdr R.W. Halliday. This officer had earlier experienced troubles with the Wyvern while posted to No. 700 Squadron. He had been flying a Wyvern when the starboard inner flap came off, sending the aircraft into uncontrollable rolls on 6 February 1956. This incident had caused all Wyverns to be grounded for three days. No. 700 Sqdn operated two Wyverns since it had taken over the tasks of Nos 703 and 771

The last of the Wyverns – only this TF.Mk 1 survives today, a cherished exhibit at the Fleet Air Arm Museum. The aircraft was completed but never flew, and was handed to the Department of Design at the College of Aeronautics, arriving at Cranfield in November 1950. The college maintained its ground instructional aircraft well, and in 1965 it was 'rescued' for the museum in excellent condition, although it later endured a spell as an outside exhibit. Today it is safely back inside, apart from the occasional summertime 'photo-call'.

Squadrons on 18 August 1955. The Wyverns remained with this Trials and Requirements Unit until February 1957, when aircraft were dispersed and work handed over to a civilian contractor, the Hurn-based Airwork FRU.

In command of No. 813 Squadron, Lt Cdr Halliday remained at RNAS Ford until the spring when the unit was despatched to RNAS Brawdy on 20 March 1957. Four months elapsed before the nine Wyverns embarked on HMS *Eagle* on 5 August 1957. After having performed operational training aboard the carrier, No. 813 Squadron returned to Ford on the last day of September. A second spell aboard *Eagle* took place from 17 October until 27 November 1957. During the winter at RNAS Ford, a new CO took over command – Lt Cdr R.W.T. Abraham, on 9 December.

Final cruise

The Wyvern's last cruise began on 29 January 1958, when No. 813 Squadron aircraft joined *Eagle* for a trip in the Mediterranean. Before reaching Malta on 14 February for a week of ship self-maintenance, No. 813 Squadron performed only a small amount of flying due to poor weather. The day before entering Grand Harbour, three Wyverns were launched for Hal Far but were diverted to nearby *Ark Royal* owing to a worsening of conditions over the island. Next day, only two reached Hal Far as the third Wyvern failed to start and returned to *Eagle* by lighter.

Flying was still limited from Hal Far due to the shortage of bowsers, aircraft turn around times lasting two hours. Rocketing was carried out at the Delimara range on 20 and 21 February, then *Eagle* left Malta on 25 February. Intensive flying took place the following days, including bombing on Filfla, rocket firing at Delimara, and combined Wyvern and Sea Hawk strikes over Malta. Once more weather

became worse from 28 February. Six Wyverns were surprised by rough weather and diverted to Hal Far and remained there for three days. As they returned to their carrier, Exercise Marjex commenced. No fewer than four carriers were involved: HMS *Eagle* and *Ark Royal*, plus USS *Saratoga* and *Essex*. A programme of strikes was flown on Malta and ships for three days. In March, *Eagle* visited Toulon and conducted trials with French aircraft. Harbour entry and exit manoeuvres were executed, using the running propellers of specially positioned Wyverns on the deck to reduce the area in which the carrier manoeuvred in the Pinwheel operation. This practice was again used on 22 March when *Eagle* entered Gibraltar for a two-day stop.

Passage in the Atlantic took place on 25 March. Although rough weather was encountered, six aircraft flew navigational exercises, these being the last Wyvern sorties. On return to the carrier, the port undercarriage of Lt Cdr Abraham's aircraft collapsed, in turn collapsing the tailwheel and shattering the propellers. The aircraft was ditched over the side three days later. The Wyverns finally left their carrier and flew ashore to their parent base on 29 March 1958. The ground party returned to Ford on 31 March and all personnel went on leave for two weeks. Back from holiday, the first task was to despatch five Wyverns to RNAS Lossiemouth for storage on 18 April. A last maintenance was carried out three days later on the remaining aircraft. No. 813 Squadron disbanded on 22 April 1958.

Among other Wyvern operators were various experimental units: the Air Torpedo Development Unit based at Gosport, the Naval Air Establishment at Thurleigh, and the Handling Squadron at RAF Manby. No. 787 Squadron, the Naval Air Fighting Development Unit at RAF West Raynham, also employed four Wyverns from March to November 1954. All surviving Wyvern S.Mk 4s were collected by the Lossiemouth-based Aircraft Holding Unit and were scrapped.

Ironically, only a piston-engined aircraft survives today – VR137. The last of the TF.Mk 1 pre-production aircraft, it is on display at the Fleet Air Arm Museum at RNAS Yeovilton, having served as a ground demonstrator airframe for vibration testing techniques with the College of Aeronautics at Cranfield.

Geoffrey Bussy

Wyvern operators

The front-line fleet was established as two squadrons, and the type was initially a straight replacement for the Blackburn Firebrand TF.Mk 5. Although four numbered front-line squadrons eventually operated the type, the most to be equipped with the type at any one time was three, and that state of affairs existed only for a very brief period between the reformation of No. 813 Squadron in November 1956 and the disbandment of No. 830 Squadron at the start of January 1957.

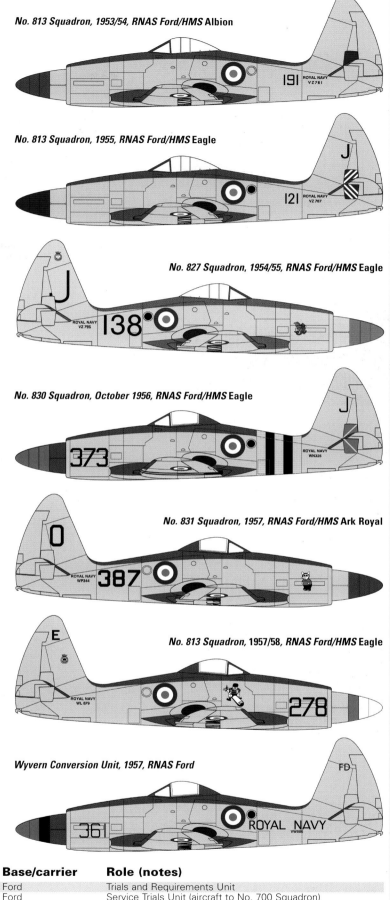

No. 813 Squadron, 1953/54, RNAS Ford/HMS Albion

No. 813 Squadron, 1955, RNAS Ford/HMS Eagle

No. 827 Squadron, 1954/55, RNAS Ford/HMS Eagle

No. 830 Squadron, October 1956, RNAS Ford/HMS Eagle

No. 831 Squadron, 1957, RNAS Ford/HMS Ark Royal

No. 813 Squadron, 1957/58, RNAS Ford/HMS Eagle

Wyvern Conversion Unit, 1957, RNAS Ford

Above: No. 703 Squadron had a number of Wyverns on strength in the service trials role. Here '082' is seen in July 1955 during trials of HMS Ark Royal's new steam catapults. As well as its trials duties, No. 703 formed a dedicated flight (No. 703W) to form the nucleus of a front-line unit (No. 827 Squadron).

No. 764 Squadron (above) added a pair of Wyverns to its Sea Hawk/Sea Vampire fleet in May 1955 in order to provide type conversion training for the turboprop attacker. A few months prior to No. 764's June 1957 move to Lossiemouth, the Wyverns were split off into a dedicated Wyvern Conversion Unit (below), which continued to provide type training at Ford for the remainder of the year, after which no new pilots were converted to the type.

Many Wyverns served with trials units and manufacturer's test fleets at one time or another, including all the prototypes, TF.Mk 1s and TF.Mk 2s. Many passed through the hands of the A&AEE at Boscombe Down and the RAE at Farnborough, while Westland itself performed several test programmes. Other manufacturers to use Wyverns for trials were Rolls-Royce, Napier, Armstrong Siddeley, de Havilland Propellers and Rotol Airscrews. At least five TF.Mk 1s were allocated from 1950 for use as ground instructional airframes – at RNEC Manadon, RNAS Bramcote, RNAS Gosport, RNAS Yeovilton and Cranfield College. Some of the prototypes and TF.Mk 1s ended their days as gunnery targets at Foulness.

The table below is a synopsis of the Wyvern's brief operational career, from initial assignment to No. 813 Squadron in May 1953 to the type's retirement in December 1957.

Unit	from	to	Base/carrier	Role (notes)
No. 700 Squadron	18 August 1955	February 1957	Ford	Trials and Requirements Unit
No. 703 Squadron	June 1954	17 August 1955	Ford	Service Trials Unit (aircraft to No. 700 Squadron)
No. 703W Flight	4 October 1954	1 November 1954	Ford	Intensive Flying Trials Unit (became No. 827 Squadron)
No. 764 Squadron	May 1955	February 1957	Ford	Fighter Pilot Holding Unit (aircraft to Wyvern Conversion Unit)
No. 787 Squadron	March 1954	November 1954	West Raynham	Naval Air Fighting Development Unit
No. 813 Squadron	20 May 1953	21 November 1955	Ford, *Albion*, *Eagle*	Strike Attack Squadron
	26 November 1956	22 April 1958	Ford, Brawdy, *Eagle*	Strike Attack Squadron
No. 827 Squadron	1 November 1954	19 November 1955	Ford, *Eagle*	Strike Attack Squadron
No. 830 Squadron	21 November 1955	5 January 1957	Ford, *Eagle*	Strike Attack Squadron
No. 831 Squadron	21 November 1955	10 December 1957	Ford, *Ark Royal*	Strike Attack Squadron
Wyvern Conversion Unit	February 1957	December 1957	Ford	Operational Conversion Unit

Picture acknowledgments

Front cover: USAF, Dassault, AMI via Luigino Caliaro. **4:** via Tom Kaminski, Lockheed Martin. **5:** Lockheed Martin (two), USAF. **6:** Northrop Grumman, Yaso Niwa, Alexander Mladenov. **7:** via Tom Kaminski (two), Yaso Niwa. **8:** Peter J. Cooper (two). **9:** Peter R. Foster, Alexander Mladenov. **10:** Shlomo Aloni (two), Jon Chuck. **11:** Nate Leong (three). **12:** Marc van Zon (three). **13:** NFA Press Agency, Nate Leong, Timm Ziegenthaler. **14:** Riccardo Niccoli (three). **15:** US Navy, Roberto Yañez. **16-29:** USAF, US Navy, UK MoD, RAAF. **30-31:** Robert Hewson (two). **32:** Marnix Sap, Christiaan Sap. **33:** Emiel Sloot (two). **34:** Emiel Sloot (two), Marnix Sap. **35:** Corné Rodenburg (four). **36:** Corné Rodenburg (two), Marnix Sap, Emiel Sloot. **37:** Lou Drummond via Jim Dunn. **38-39:** François Robineau/Dassault via Henri-Pierre Grolleau (HPG). **40:** Gert Kromhout (three). **41:** Dassault via HPG (two), Peter R. Foster. **42:** Henri-Pierre Grolleau (two), via HPG (two). **43:** David Barrow, Gert Kromhout, via Tom Kaminski (two). **44:** MBDA, Simon Watson/Wingman Aviation. **45:** via Tom Kaminski, Indian Air Force, Simon Watson/Wingman Aviation. **46:** Henri-Pierre Grolleau, MBDA, François Robineau/Dassault, Simon Watson/Wingman Aviation. **47:** Gert Kromhout, Chris Ryan, Indian Air Force (four), Simon Watson/Wingman Aviation. **48-49:** Henri-Pierre Grolleau. **50:** Henri-Pierre Grolleau (two), MBDA, Chris Ryan. **51:** Philippe Roman/Eric Desplaces, Henri-Pierre Grolleau, Gert Kromhout. **52:** Gert Kromhout (three). **53:** EC 3 via Philippe Roman/Eric Desplaces (two), Armée de l'Air via Philippe Roman/Eric Desplaces. **54:** François Robineau/Dassault (two), MBDA. **55:** François Robineau/Dassault, Gert Kromhout (two). **56:** Henri-Pierre Grolleau (three). **57:** François Robineau/Dassault via HPG, Henri-Pierre Grolleau. **58:** François Robineau/Dassault via HPG. **59:** François Robineau/Dassault, François Robineau/Dassault via HPG. **60:** François Robineau/Dassault via HPG. **61:** François Robineau/Dassault via HPG, Henri-Pierre Grolleau. **62:** François Robineau/Dassault via HPG, Henri-Pierre Grolleau (two). **63:** François Robineau/Dassault via HPG, François Robineau/Dassault. **64:** François Robineau/Dassault via HPG (two). **65:** François Robineau/Dassault via HPG, Henri-Pierre Grolleau (six), Gert Kromhout (two), Philippe Roman/Eric Desplaces. **67:** Henri-Pierre Grolleau (two). **68:** Henri-Pierre Grolleau. **71:** Henri-Pierre Grolleau (ten), MBDA, Philippe Roman/Eric Desplaces. **72:** Dassault, Gert Kromhout, Henri-Pierre Grolleau (three). **73:** Peter R. Foster, Henri-Pierre Grolleau, Simon Watson/Wingman Aviation (two). **74:** Simon Watson/Wingman Aviation, Peter R. Foster (two), Henri-Pierre Grolleau. **75:** Henri-Pierre Grolleau, Gert Kromhout, via Tom Kaminski. **76-79:** Chris Knott and Tim Spearman/API. **81-88:** Aleksandar Radic. **89:** Aleksandar Radic (three), Igor Salinger. **90:** Aleksandar Radic (six), Igor Salinger. **91:** Drago Vejnovic via Aleksandar Radic, Aleksandar Radic. **93:** Marko Malec (six). **94:** Marnix Sap, Carlo Brummer. **95:** Carlo Brummer (two), Marnix Sap. **96:** Marnix Sap (three). **97:** Marnix Sap (three), Carlo Brummer. **98:** Carlo Brummer (two), Marnix Sap. **99:** Marnix Sap (three). **100:** USAF, Boeing (two). **101:** USAF (two), Boeing. **102:** Boeing, USAF, General Electric. **103:** USAF, Boeing, USAF. **105:** Boeing, USAF. **106:** USAF, Aerospace. **107:** Boeing, Aerospace. **108:** Boeing, USAF. **109:** Aerospace, USAF (three). **110:** Boeing (three). **112:** USAF (two). **113:** USAF, Aerospace. **114:** James Goodall Collection, David Donald. **115:** USAF (two), NASA, Aerospace. **116-123:** AMI via Luigino Caliaro. **124-125:** Piotr Butowski, Sergey Skrynnikov. **126:** Sergey Skrynnikov, Sergey Skrynnikov via Jon Lake, Swedish Air Force. **127:** Swedish Air Force (two), Aerospace. **128:** US Navy, Swedish Air Force (two). **129:** Hugo Mambour, Anatoliy Artemyev, US Navy, Sergey Skrynnikov. **130:** Sergey Skrynnikov, US Navy, Aerospace. **131:** Swedish Air Force (three), Anatoliy Artemyev. **132:** Serbey Skrynnikov, via Jon Lake. **133:** via Jon Lake, US DoD, Anatoliy Artemyev, Sergey Skrynnikov. **134:** Anatoliy Artemyev (three), Aerospace (three). **135:** Sergey Skrynnikov, Swedish Air Force, Anatoliy Artemyev. **136:** Swedish Air Force (two), Alexey Mikheyev. **137:** Swedish Air Force (three). **138:** Aerospace (two), Kondratenkov/Petrochenko, Sergey Skrynnikov. **143:** Swedish Air Force, Aerospace (two), US Navy, Kondratenkov/Petrochenko (two), Alexey Mikheyev. **144:** US Navy, Swedish Air Force (two). **145:** Swedish Air Force, Anatoliy Artemyev, Sergey Skrynnikov. **146:** Gennady Petrov, TASS, Anatoliy Artemyev (two), Aerospace. **147:** Swedish Air Force, US Navy, Anatoliy Artemyev (two). **148:** Swedish Air Force (two), Anatoliy Artemyev, Peter Batuev Collection. **149:** via Jon Lake, Anatoliy Artemyev, Peter Batuev Collection. **150:** Peter Batuev Collection (two), Anatoliy Artemyev (two). **151:** via Sergey Skrynnikov, Anatoliy Artemyev (two). **152:** US Navy (three), Anatoliy Artemyev. **153:** US Navy (two), Aerospace, Anatoliy Artemyev. **154:** UK MoD, Anatoliy Artemyev, Swedish Air Force (two). **155:** Sergey Skrynnikov via Jon Lake, Peter Batuev Collection, US Navy (two). **156:** Swedish Air Force (two). **157:** US Navy via Jon Lake, via Jon Lake, Swedish Air Force. **158:** via Jon Lake, Hugo Mambour, Anatoliy Artemyev. **159:** LII via Sergey Skrynnikov, Sergey Skrynnikov via Jon Lake, NFA Press Agency (two). **160:** via Jon Lake (two), NFA Press Agency. **161:** via Jon Lake (three), Aerospace. **162:** USAF, Aerospace (two). **163:** Kondratenkov/Petrochenko, Sergey Skrynnikov, Alexey Mikheyev. **164-167:** Wingman Aviation. **169:** Aerospace, Westland. **170:** Aerospace, Geoffrey Bussy, MAP via Geoffrey Bussy (two). **171:** Aerospace (two). **172:** Aerospace (three), via Geoffrey Bussy. **173:** Aerospace (two), Royal Navy. **174:** Aerospace, Royal Navy via Geoffrey Bussy. **175:** Westland, Aerospace, Fleet Air Arm Museum via Geoffrey Bussy. **176:** Fleet Air Arm Museum via Geoffrey Bussy, via Geoffrey Bussy, Aerospace. **177:** Fleet Air Arm Museum via Geoffrey Bussy (two), Imperial War Museum via Geoffrey Bussy. **178:** via Geoffrey Bussy (two). **179:** Fleet Air Arm Museum via Geoffrey Bussy, Imperial War Museum via Geoffrey Bussy. **180:** via Geoffrey Bussy, MAP via Geoffrey Bussy, Peter R. March. **181:** Royal Navy, Fleet Air Arm Museum via Geoffrey Bussy, MAP via Geoffrey Bussy.